ADVANCES in FISHERIES SCIENCE

CULTURE of NONSALMONID FRESHWATER FISHES

Second Edition

ADVANCES in FISHERIES SCIENCE

CULTURE of NONSALMONID FRESHWATER FISHES
Second Edition

Edited by
Robert R. Stickney
Professor
School of Fisheries
University of Washington
Seattle, Washington

CRC Press
Boca Raton   Ann Arbor   London   Tokyo

Library of Congress Cataloging-in-Publication Data

Culture of nonsalmonid freshwater fishes / edited by Robert R.
Stickney. -- 2nd ed.
      p. cm. -- (Advances in fisheries science)
    Includes bibliographical references and index.
    ISBN 0-8493-8633-0
    1. Fish-culture. 2. Freshwater fishes. I. Title: Nonsalmonid
freshwater fishes. II. Series
SH159.C85   1993
639.3′758--dc20   92-25093
                                                                    CIP

© 1993 by CRC Press, Inc.

International Standard Book Number 0-8493-8633-0

Library of Congress Card Number

Printed in the United States 1 2 3 4 5 6 7 8 9 0

Printed on acid-free paper

CRC Series in
ADVANCES in FISHERIES SCIENCE

**PATHOBIOLOGY of
MARINE and ESTUARINE ORGANISMS**
Edited by John A. Couch and John W. Fournie

CULTURE of NONSALMONID FRESHWATER FISHES
Edited by Robert R. Stickney

SERIES PREFACE

ADVANCES IN FISHERIES SCIENCE

Advances in Fisheries Science is a book series published with the objective of providing in-depth treatment of the diverse subject matter that, taken together, forms the scope of fisheries science. Areas of emphasis within the series will include, but not be limited to, aquaculture, fishery management, fishing methods, descriptions of vertebrate and invertebrate fisheries, taxonomy and evolution of commercially important aquatic organisms, policy making with respect to fisheries, and relationships between fisheries and both natural and perturbed environments. Many additional topics of sufficient scope for a book in the series are encompassed within those broad headings. Some of the additional topics are genetics, molecular biology, nutrition, pathology, reproduction, behavior, and the general ecology of aquatic animals and, in some cases, plants.

Books in the series are designed to meet the needs of fisheries professionals and, in most instances, will be appropriate for use in upper division undergraduate and graduate courses in fisheries science.

PREFACE

The art of fish culture goes back millenia. It has its roots in China which continues to lead the world in carp production and, over the past few years, has taken a leadership position in the culture of marine shrimp. In North America, governmental hatcheries have been in operation for something over 100 years, but western hemisphere fish culture is clearly a recent activity when compared with much of the rest of the world. Production of cultured fish for the foodfish market in the U.S. really began to grow in the 1960s when private trout and catfish farming technology was adopted by the private sector and cultured fish began appearing in restaurants and supermarkets.

Fish farming in the United States developed quite differently from traditional agriculture. Many who settled North America came as farmers or soon learned to make a living by farming the vast forestlands and prairies of frontier North America. As agriculture grew, farmers recognized the importance of research in helping them produce more on a given area of land, control pests, develop new crop or livestock varieties, and create new markets for their products. The potential for fish farming was largely developed by researchers in government laboratories and in universities that were interested in learning more about the biology of fishes, including their environmental requirements, nutritional requirements, susceptibility to diseases, responsiveness to selective breeding, and so forth. Commercial fish farming followed the research lead, and many of the techniques employed by commercial fish farmers were developed by researchers and management biologists working on public fish hatcheries.

Fish farming depends on high quality, inexpensive, and readily available water. While significant advances have been made in water reuse systems for use in fish production, most fish farmers in the world today depend on large volumes of water, a commodity that is becoming increasingly scarce in some regions. Water is also being more and more heavily impacted by the activities of man. That impact often translates into pollution, making the water less useful for aquaculture. The competition for existing water by multiple users also impacts fish farmers in many regions. Expansion of fish culture worldwide will be closely associated with the availability of suitable water supplies. Water may well become the major limiting factor in the growth of the industry.

A second limitation on growth will be in the area of animal protein for use in fish diets. Most fishes require a certain amount of animal protein in their diet to satisfy their demand for certain amino acids which cannot be provided from plant proteins. In most cases, that animal protein comes from fish meal, which is also being heavily utilized in the livestock feed industry. The world supply of fish meal is finite, and since much of it comes from the Peruvian anchoveta, populations of which are subject to vast fluctuations in response to El Niño off the west coast of South America, shortages are not uncommon. As food technology advances and techniques are found to process so-called trash fish into surimi (the basic ingredient in artificial crab legs and other analog seafood products), many species that have been used for fish meal in the past could be diverted directly into human food.

We can look for fish farming to expand, but satisfying the need for efficient water treatment and reuse and for developing diets containing alternate proteins will play a major role in how much the industry will be able to grow. Fish are viewed favorably by the public. They are seen as being healthy for man, they are flavorful, the quality control on most products is excellent, and they are now available throughout the year. Per capita seafood consumption in the United States has risen over the past few years and can be expected to continue its upward trend so long as there is enough high quality product to satisfy the demand.

This book concentrates on the culture of nonsalmonid freshwater fishes. The fishes which fit within those limiting words are similar in many respects, though the details of culturing them are distinctly different in most instances. In this book, what is currently known about such species as channel catfish, carp, tilapia, largemouth bass, walleye, striped bass, and northern pike is summarized. The material is not meant to provide a novice with everything needed in order to become a successful fish farmer. The book is written under the assumption that the reader has at least a modest foundation in biology and chemistry. Those with technical backgrounds in fish culture should be able to apply what is presented without too much difficulty. References are provided throughout to take the reader back to the primary literature, which will provide many additional details. This volume is a revision of a book by the same title that was published in 1986. The authors felt that sufficient progress has been made with respect to some of the species considered that a revision was in order. Those familiar with the earlier edition will find that for some species the bulk of the information presented in 1986 pertains today, though there has been new research developed in association with each species discussed and we have attempted to incorporate the results of that research into the present volume.

Robert R. Stickney
Bothell, Washington

EDITOR

 Robert R. Stickney, Ph.D., is Professor in the School of Fisheries, University of Washington, Seattle, Washington. Dr. Stickney received his B.S. degree in Zoology at the University of Nebraska in 1967, his M.A. in Zoology (Limnology) at the University of Missouri in 1968, and his Ph.D. in Biological Oceanography at Florida State University in 1971. His dissertation research, conducted at the Skidaway Institute of Oceanography, was on the lipid requirements of the channel catfish.

Dr. Stickney served as a Research Associate and Assistant Professor at the Skidaway Institute after his graduation until 1975, when he assumed the position of Assistant Professor in the Department of Wildlife and Fisheries Sciences at Texas A&M University. He directed the Aquaculture Research Center at Texas A&M and advanced through the ranks to Professor in 1983. He accepted directorship of the Fisheries Research Laboratory at the Southern Illinois University (SIU) in January 1984 and served in that capacity and as Professor in the Department of Zoology at SIU until September 1985, when he assumed the position of Director of the School of Fisheries at the University of Washington. He stepped down from the directorship in July 1991 to assume a more active role in teaching and research.

Dr. Stickney has been involved in a breadth of research throughout his career with publications on subjects varying from phytoplankton and periphyton to marine mammals. His primary interests have centered around the environmental and nutritional requirements of fishes. He has been involved in aquaculture research on channel catfish, tilapia, largemouth bass, freshwater shrimp, flounders, trout, salmon, and Pacific halibut.

As a result of his aquaculture expertise and interest, Dr. Stickney has been involved in developmental projects in many nations. His professional activities have taken him to the Philippines, China, Japan, Nepal, New Zealand, Jamaica, Haiti, Brazil, Mexico, Canada, Norway, Egypt, and Israel.

Dr. Stickney is author of some 100 scientific papers and is author, co-author, or editor of a number of books. He has been co-editor of the CRC journal, *Reviews in Aquatic Sciences,* and has been named editor of the new CRC journal, *Reviews in Fisheries Sciences.*

Involvement with professional societies has been important to Dr. Stickney. He has served as President of the Fish Culture Section and Education Section of the American Fisheries Society and President of the World Aquaculture Society. He currently serves on the Board of Directors of the Western Regional Aquaculture Consortium, an entity established by the U.S. Department of Agriculture.

CONTRIBUTORS

Gary J. Carmichael
U.S. Fish and Wildlife Service
Region 2
Division of Fisheries
P.O. Box 1306
Albuquerque, New Mexico 87103

James T. Davis
Extension Fisheries Specialist
Texas Agricultural Extension
    Service
Texas A&M University
College Station, Texas 77843

Kerry G. Graves
U.S. Fish and Wildlife Service
Region 4
Private John Allen National Fish
    Hatchery
111 Elizabeth Street
Tupelo, Mississippi 38801

Roy C. Heidinger
Fisheries Research Laboratory
Department of Zoology
Southern Illinois University
Carbondale, Illinois 62901

Terrence B. Kayes
Department of Forestry, Fisheries
    and Wildlife
12 Plant Industry, East Campus
University of Nebraska-Lincoln
Lincoln, Nebraska 68583

J. Howard Kerby
Fish Culture and Ecology
    Laboratory
National Fisheries Research Center
U.S. Fish and Wildlife Service
Kearneysville, West Virginia 25430

Robert B. McGeachin
1208 Glade St.
College Station, Texas 77840

John G. Nickum
Aquaculture Coordinator
U.S. Fish And Wildlife Service
Washington, D.C. 20240

Bill A. Simco
Department of Biology
Memphis State University
Memphis, Tennessee 38152

Robert R. Stickney
School of Fisheries
University of Washington
Seattle, Washington 98195

Joseph R. Tomasso
Department of Aquaculture, Fisher-
    ies and Wildlife
308 Long Hall
Clemson University
Clemson, South Carolina 29634

Harry Westers
P.O. Box 8
Rives Junction, Michigan 49277

J. Holt Williamson
U.S. Fish and Wildlife Service
Region 6
Division of Fisheries
P.O. Box 25486
Denver, Colorado 90225

TABLE OF CONTENTS

DEDICATION

If people feed fish, then fish will feed people. Aquaculture is currently being viewed as both a source of relief for a human population facing a protein shortfall and a threat to the environment and livelihood of commercial fishermen. We live at a crossroads. The reality is that a balance must be struck to assure that mankind can produce the required protein at an optimal level while preserving environmental diversity and integrity. This book is dedicated to those students, past, present, and future, who will help meet that goal.

Chapter 1

INTRODUCTION

Robert R. Stickney

TABLE OF CONTENTS

8633-9/93/$0.00 + $.50

I. SCOPE

This book is a revision and update of a volume of the same title edited by Stickney (1986). The purpose of the book is to provide students and practitioners of fish culture with information on the environmental and nutritional requirements of various nonsalmonid freshwater fish species, along with information on culture technology, disease susceptibility and control, and reproductive strategies. Primary literature citations are used liberally and can serve as an entrée into the extensive literature base that is available on each of the species covered.

Aquaculture is the rearing of aquatic organisms under controlled or semicontrolled conditions (Stickney 1979). Thus, aquatic plants and animals, freshwater and marine species, organisms reared for commercial sale, as well as those raised by government agencies for stocking are all included within the broad definition. In this text, only nonsalmonid freshwater finfish are discussed. A companion volume (Stickney 1991) addresses the culture of the salmonids (salmon and trout). Some of the species discussed herein are produced for direct human consumption, while others are sold as bait. Still others are stocked to provide recreational fishing opportunities. Aquarium species are not included, and most of the species discussed are of interest and importance in the U.S. Various additional species of freshwater fishes that are cultured around the world could have been included but were considered to be beyond the scope of this volume.

Beginning in the 1960s, commercial culture of channel catfish has become a reality in the U.S. Tilapia were introduced to the U.S. in the 1960s but did not enjoy significant popularity among commercial fish farmers until the following decade. Interest in tilapia culture grew even more in the 1980s and continues strong in some regions of the country. Many species, such as largemouth bass, walleye, and northern pike, are cultured by state and federal agencies for stocking into public waters. Each of the 50 states has an agency (Department of Fisheries, Department of Wildlife, Department of Fish and Game, Conservation Department, Department of Parks and Wildlife, etc.) which stocks fish in public, and in many states, private waters. Additionally, the federal government has a number of hatcheries which provide fish for stocking into public waters such as large impoundments. Striped bass are of interest to both commercial aquaculturists and to those involved with governmental fish-stocking programs. Similar to the salmonids in that they spawn in freshwater but commonly spend the majority of their lives in the marine environment, striped bass are included in this book since they are not strictly marine fish.

Included among the species discussed in the chapters that follow are fishes which can be placed into one of two categories relative to their temperature optima. Warmwater species are those which grow most rapidly when the water is in the range of about 26 to 30°C. Included are channel catfish, tilapia, carp, and largemouth bass. Coolwater species are those which seem to perform best at temperatures between about 20 and 25°C, which would include the striped bass, walleye, yellow perch, northern pike, and muskellunge. The salmonids are classified as coldwater fish and tend to have a performance optimum below 20°C.

Many species that enjoy popularity as culture species around the world have been brought into the U.S., but are not being reared except in some instances by the aquarium fish industry. Some of those species, such as the Oscar (*Cichla ocellaris*) and Nile perch (*Lates* spp.), have been evaluated and even stocked as sport fish by state agencies; other species, like the notorious walking catfish (*Clarias* spp.), are considered to be a nuisance or to potentially outcompete native species and are banned. None of those fishes is of much significance in the U.S., and they are not discussed in the chapters that follow.

The controversial grass carp (*Ctenopharyngodon idella*) and its hybrids are cultured in a number of states both by commercial fish farmers and fishery management agencies. Outlawed in a number of states, there are increasing exceptions being made for the introduction of sterile hybrids or triploid grass carp (the latter have a third set of chromosomes and are functionally sterile). The most widely cultured group of fishes in the world are the carps, so a chapter on that group (Chapter 4) has been included. The techniques employed by carp culturists can be adapted to other species of fish having small eggs. Buffalo are included in the carp chapter since cultural practices applied to buffalo are similar to those used by carp culturists. Buffalo was once thought to have significant culture potential in the U.S., and there has been some renewed interest in the fish in recent years.

The taxonomy used throughout the text follows that of the American Fisheries Society (Robins 1991). Most notably, with respect to the species discussed in this book, is the retention of the genus *Tilapia*. Trewavas (1973) advocated placement of many species of tilapia within the genus *Sarotherodon*. Later (Trewavas 1982), that proposal was modified with respect to most species of culture interest. The recommendation was to place those species in the genus *Oreochromis*. Since the American Fisheries Society has not accepted the change in generic status for the species of interest, this text follows Robins (1991) as well.

In each chapter, one or more related or highly similar species or groups of species are discussed. Aspects of the culture requirements of each type of fish include the types of culture systems that are employed, water quality requirements, reproduction and breeding techniques, nutritional requirements when known, feeding strategies employed, common diseases and their treatment, and specialized information of various kinds. A general introduction to fish culture is presented in this chapter, allowing the authors of the chapters that follow to discuss specialized circumstances that are employed by those who rear the various species of fishes that are described.

II. AQUACULTURE PHILOSOPHY

Aquaculture can be conducted for a number of reasons. The two most common in the U.S. have either a profit motive or are geared toward recreational fishery enhancement. Globally, the extremes, perhaps, are subsistence aquaculture and aquaculture as a hobby. Subsistence culturists are typically persons living in developing nations. They may be individual families, groups of families, or entire small villages, the members of which work together to provide supplemental animal protein in their diets and perhaps a small amount of money in their pockets by growing fish. Typically, subsistence culturists, being poor, employ little in the way of modern technology and depend upon species such as tilapia and carp that can be easily cultured under a variety of less than desirable conditions.

Hobbyists, and the so-called backyard aquaculturists, commonly employ the ultimate in available technology, and since they tend to be in no way dependent on the animals that they culture, they engage in the activity as a means of entertainment. Commercial ventures which produce aquarium fish for the hobbyist trade, of course, fall in the category of those who have a profit motive for engaging in aquaculture.

Commercial fish farmers, even those living in developing nations, engage in fish culture as a means of making a living. While it has often been said that aquaculture can provide a means of feeding the hungry in the world, in reality, aquatic animals cannot generally be raised inexpensively, and in nearly every case, they demand a high price in the marketplace. Private aquaculturists are not in business to provide inexpensive food to the poor, even when they number themselves as members of that portion of the human population. While governments have often become involved in helping people develop commercial aquaculture operations, virtually no government is in the aquaculture business with the objective of providing inexpensive animal protein to undernourished humans.

Government aquaculture operations produce fish that are used for stocking private farm ponds, public lakes, streams, and reservoirs, and (in the case of salmon, striped bass, and a few other species) marine waters. While there is little in the way of commercial fishing in most of the inland waters of the U.S., some of the coastal stocking programs augment natural stocks of fish that may be the targets of commercial as well as sport fishermen.

Public hatchery philosophy in the U.S. is beginning to change. It was once possible for a private farmer in virtually any state to obtain free fish from the appropriate state agency to stock farm ponds on the farmer's private land. In many states, access to farm ponds for fishing is available to the public with the farmer's permission, but most people who do not own farms have probably never walked up to a farmhouse and asked permission to go fishing. Many that have were turned away. The practice is being increasingly called into question by taxpayers who feel that they should not be subsidizing the stocking of fish in private waters. Thus, the private sector is being asked to meet the demand for fish to stock private waters. That is not the case with striped bass, red drum, spotted seatrout, and salmon. For those species, public hatcheries continue to be developed. It can be argued that the U.S. taxpayer is subsidizing fisheries that are only used by a few, but with the exception of some objections to Pacific salmon hatcheries by those who feel that cultured salmon are inferior to those spawned and reared in the wild, there is little resistance on the part of taxpayers to having their money spent to enhance marine fisheries. Inland fisheries that are enhanced with federal dollars are exclusively recreational in nature.

Not all profitable fish culture ventures provide food for the table or ornamental fish for the hobbyist. One of the largest parts of the aquaculture enterprise in the U.S. involves the production of baitfish. Thousands of hectares of water are being used to produce the minnows that are used as bait by hundreds of thousands of recreational anglers. Goldfish are also produced for the bait industry, but since goldfish are also found in pet stores, they fill more than one niche in the fish market.

III. GETTING STARTED

For those who are interested in becoming commercial fish farmers, the decision is not one which should be made until the pros and cons have been carefully studied. Raising fish profitably is a lot more than merely digging a pond, throwing in some young fish, and standing back while the money pours in. Fish farming is hard work. It is truly a form of agriculture — one in which the farmer often does not see his livestock and never knows how many fish are in a pond until they are harvested. Decisions about which species to raise and where to raise it or them are of primary importance. If a person already has a piece of land, he or she may look for an appropriate species to raise on that land. If a person is particularly attracted to a certain species, a suitable site should be found on which the species of choice can be profitably reared. Some of the factors that should be considered by anyone interested in becoming a commercial fish farmer are outlined in Table 1.

One of the first things that a prospective fish farmer will have to deal with is the permitting process. Fish farming may be subject to various federal regulations. Some of the agencies that may be involved include the U.S. Army Corps of Engineers, the Environmental Protection Agency, and the U.S. Fish and Wildlife Service. The process of permitting varies significantly from state to state. In some states it is straightforward and simple, while in others there may be a plethora of agencies that must be contacted before the proper permits can be obtained. Depending upon the state and location of the proposed aquaculture facility, the process might be as simple as paying a small fee for a fish farming license to as difficult as requiring several years and the expenditure of hundreds of thousands of dollars on lawyers and permit fees. In many states, the state department of agriculture can assist prospective fish farmers through the sometimes serpentine process. Table 2 describes the permitting process in the state of Washington.

IV. CULTURE SYSTEMS

Culture systems can be thought of as lying along a continuum of intensity with farm ponds (which require little management) lying at one extreme and closed recirculating water systems (requiring continuous monitoring) at the other. Figure 1 shows how intensity increases with respect to culture system type. Production can range from a few hundred kilograms per hectare per year in extensive systems to on the order of 10^6 kg/ha per year in highly intensive systems. The intensity of a culture system increases as a function both of increasing fish production and increasing need for technology to keep the system operating properly.

A. PONDS

The majority of freshwater fishes produced by aquaculturists are raised in ponds (Figure 2). Fish culture ponds range in size from fractions of a hectare to tens of

TABLE 1
Some Considerations for the Prospective Fish Farmer

Financing	• A prospectus should be developed which details the costs of land and capital expenditures for fish stock, buildings, pond construction, operating funds, labor, financing, harvesting, marketing, and insurance; the prospectus should contain depreciation schedules and a profit and loss estimate.
	• The prospective fish farmer should be capitalized to the extent of at least $4500/ha of water that will be placed into production.
	• An overhead figure of $250/ha monthly, exclusive of fingerling costs, can be anticipated.
Site	• The land should provide the type of relief required to allow gravity drainage of water from whatever type of culture system is employed.
	• Water of the proper type (freshwater, brackishwater, marine) and quantity should be available at a reasonable cost.
	• When ponds are to be constructed, soils should be analyzed to determine that their physical properties are suitable for water retention.
	• Sediment cores should be taken to determine that the desired type of soil is available throughout the selected site of the fish farm.
	• Lands adjacent to the fish farm should not be subjected to aerial spraying for insect and weed control, and the soil on the fish farm should contain no toxic residues.
	• The site should not be subject to flooding.
	• State and federal permits should be obtained in advance of construction.
	• Potential losses from poaching and predation should be assessed and plans made to deal with them.
Fish source	• The prospective farmer will have to decide whether to produce fingerlings or purchase them.
	• If breeding is to be conducted on site, sufficient space for maintenance of broodstock and for fingerling production must be provided.
	• If fingerlings are to be purchased, the availability of reliable sources of good quality (disease-free) fingerlings must be determined.
Feeding	• Feed of the proper quality and quantity must be available upon demand throughout the growing season.
	• Decisions must be made as to how to adjust feeding rates during the growing season unless demand feeders are going to be used.
	• The culturist should determine whether sinking or floating feed will be used.
Harvesting	• The most economical harvesting method for the target species should be used.
	• Special equipment needs for handling fish from harvesting to processing should be assessed.
	• Holding facilities for harvested fish may be required and should be factored into the cost analysis as appropriate.
	• The need for and size of transportation tanks and trucks should be assessed.

Marketing
- The proximity of the fish farm to a suitable market should be a factor in site selection.
- Alternative processing and marketing strategies for the species to be cultured should be evaluated.
- Quality control should be exercised to avoid loss of customers as a result of off-flavors or improperly sized fish for the existing market.

Management
- A trained biologist competent to diagnose and treat diseases should be readily available or on the staff of the fish farm.
- Suitable laboratory equipment for water quality analyses, disease diagnosis, and other routine activities should be available.
- Emergency backup power should be available in case of electrical failure.
- All-weather access to all facilities should be provided.
- Areas for expansion should be considered in the initial development phase

Table adapted from an unpublished checklist for prospective farmers prepared by James T. Davis.

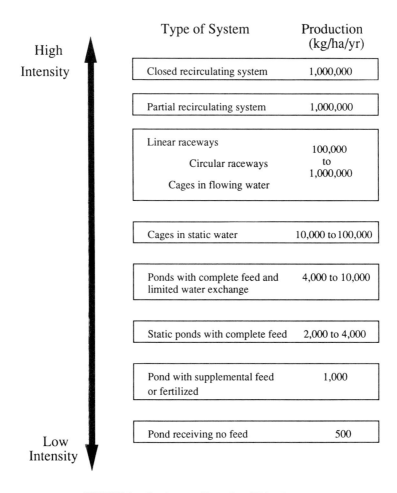

FIGURE 1.   Continuum of intensity of fish culture systems.

TABLE 2
The Aquaculture Permitting Process in Washington State

State Permits

Hydraulic project approval	• Issued by Department of Fisheries or the Department of Wildlife • Required of any project that will use, divert, obstruct, or change the natural flow or bed of any of the salt or freshwaters of the state
Water quality certification	• Required by the Department of Ecology • Covers in-water construction that involves discharge of pollutants and needs a permit from the U.S. Army Corps of Engineers
Water quality standards modification	• Issued by the the Department of Ecology • Covers the discharge of pollutants causing water quality to fall below standards
National pollutant discharge elimination system (NPDES) permit	• Issued by the Department of Ecology • Needed by any person who conducts a commercial or industrial operation that results in the disposal of solid or liquid waste materials into the waters of the state
Aquacultural identification of private sector products	• Department of Agriculture • Designed to identify privately produced products that would otherwise be regulated by the Departments of Fisheries and Wildlife
Registration of aquatic farmers	• Department of Fisheries • Purpose is to collect statistical data on the cultivation of aquatic products by the private sector
Fish disease control	• Department of Fisheries (developed jointly with the Department of Agriculture) • Purpose is to protect the aquaculture industry and wildstock fisheries from loss of productivity due to aquatic diseases

Federal Permits

Dredge and fill	• U.S. Army Corps of Engineers • Covers placement of structures, excavation, and deposition of dredge or fill material into navigable waters to prevent private parties from obstructing those waters

Note: Additional permits (local, state, and federal) may be required for aquaculture operations established in or adjacent to the marine waters of the state.

hectares. In general, relatively small ponds (less than a hectare) are typically used by researchers because of the relative ease and economy with which replicated experiments can be conducted in small ponds of uniform size. Small ponds are also commonly used for the holding and spawning of broodfish, production of fry, and production of fingerlings by commercial fish farmers. Most production ponds do not exceed 5 to 10 ha because extremely large ponds are difficult to manage. Also, if there is a problem in a large pond the economic loss can be devastating.

Unlike the typical farm pond that is commonly formed by constructing a dam across a low spot in a watershed (Figure 3) and which is, therefore, of irregular shape, fish culture ponds are often square or rectangular. Culture ponds are typically located in areas that feature relatively little relief, have levees with specific side slopes, and have bottoms that slope toward a drain. While farm ponds generally depend on rain runoff to keep them filled with water, culture ponds typically receive well water or surface water that is pumped into them. In most instances, the levees of culture ponds extend above the surrounding ground elevation so runoff water is not a significant input.

Sufficient quantities of water should be available to fill all of the ponds on the farm within a reasonable period of time, maintain the water level throughout the growing season, and provide supplemental water during emergencies. While the optimum amount of water required for maintenance of optimum pond conditions varies depending upon the source of information that is consulted, many culturists insist upon having between 100 and 200 l/min of water available for each hectare of surface area devoted to ponds. Water may be put into a pond at virtually any location, but during harvesting operations it is particularly handy to have water available at the drain end to help keep fish alive during the last stages of capture.

A well-designed culture pond should have a drain design that allows it to be emptied within a period of no more than 2 or 3 d. Drains may be as simple as a standpipe connected to an elbow, so it can be tilted up and down to control water level, or as complex as a kettle (also called a monk) constructed of concrete. Some such structures feature catchment basins with stairways to allow easy access and egress (Figure 4) and may use gate valves on the drain line. Standpipes may also be included in kettles to carry off overflow water. Pond bottoms should slope toward the drain, with a 1% bottom slope being sufficient.

Ponds may be dug into the ground, placed on top of the existing ground level, or partially excavated to lie partially above the original ground level. A core trench should be dug under the base of each pond levee or under the perimeter levee around the facility if there is such a structure. A core trench is a ditch dug to a depth of about a meter below the elevation of the pond bottom. Once dug, it is filled and compacted with the same type of soil that will be used to complete the pond levee. This results in a plug of well-compacted soil that goes from beneath the lowest elevation in the pond to the top of the levee. Water will sometimes seep under levees that have not been core trenched if there is differential compaction that the water can follow.

FIGURE 2.   A typical fish rearing pond.

FIGURE 3.   A farm pond that was constructed by placing a dam (background) across a low area where rainwater naturally runs off the land. Such ponds can be used for aquaculture, but are most commonly employed for livestock watering and recreational fishing.

Pipes that penetrate levees (such as drain or inflow lines) should have an antiseep collar placed around them (Figure 5). The antiseep collar performs much like a core trench in that it keeps water from following the outside of a pipe and breaching the levee by creating an obstacle around which the seeping water will not pass.

FIGURE 4. A drain structure that features a catchment basin (foreground) a pipe for introduction of new water (upper right), two means of draining, and a stairway to aid workers entering and leaving the pond.

Levees can be seen as triangles which should have base to height ratios of 2:1 or 3:1. That is, for every meter in elevation, the levee base should be 2 or 3 m wide. As the levee is constructed, it should be compacted as new layers of material are applied. Improperly compacted levees can seep water and may fail. Developing the

FIGURE 5.   A drain line being placed in a pond that is under construction. The steel plate is an antiseep collar. The pipe is PVC.

proper side slopes is important during construction because a bulldozer operator can drive lengthwise along such a levee. The completed levee will also not be too steep for easy access by the personnel who will be working therein, will not promote the growth of submerged vegetation, and will not be overly prone to erosion, particularly once it has been planted with grass.

All pond levees should be sufficiently wide that they can be mowed, and at least one side of each pond should have an all weather road on it. The road will be used for access during stocking, feeding, visual examination, collection of water samples, and harvesting. General characteristics of ponds are summarized in Table 3.

The production capabilities of ponds vary widely depending upon the species under culture and the management strategy employed (see Figure 1). Farm ponds typically produce a few hundred kilograms per hectare per year with little or no management and no feed provided to the resident fish. By feeding complete diets, carefully monitoring water quality, and providing additional water and aeration as required, production levels of more than 4,000 kg/ha per year are often seen, and levels in excess of 10,000 kg/ha per year have sometimes been obtained.

B. CLOSED RECIRCULATING WATER SYSTEMS

Closed recirculating water systems, or reuse systems, represent the highest level of technology that is currently being applied to aquaculture. Following passage

TABLE 3
Characteristics of a Fish Culture Pond

Location	Select land with a gentle slope and lay out ponds to take advantage of existing terrain.
Construction	Ponds may be dug into the ground, they may be partially above and partially within the ground, or they may be below original ground elevation; side slopes and bottom should be well-packed during construction to retard erosion and seepage; soil should contain a minimum of 25% clay; rocks, grass, branches, and other foreign objects should be eliminated from levees.
Core trenches	A core trench should be dug beneath the perimeter levee of the facility, and it is desirable to provide a core trench under each pond levee to prevent lateral water seepage.
Pond depth	Depth should be 0.5 to 1.0 m at the shallow end, sloping to 1.5 to 2.0 m at the drain end; deeper ponds may be required in northern regions where the threat of winterkill below deep ice-cover exists.
Pond configuration	Ponds should be rectangular or square.
Drain	A means of rapidly draining the pond should be provided; draining should require no more than 3 d.
Inflow lines	Inflow pipes should be of sufficient capacity to fill the pond within 3 d; if surface water is used, the incoming water should be filtered to remove undesirable organisms and debris.
Total water volume	Sufficient water should be available to fill all ponds on the facility within a few weeks and to maintain them full throughout the growing season; sufficient water should be available to provide 100 to 250 l/min/ha of total pond area.
Levees	Construct ponds with 2:1 or 3:1 slopes on all sides; levees should be sufficiently wide to mow; at least one side should be wide enough (2 to 3 m) to provide vehicular access; road levees should be gravel; grass should be planted on all levees.
Orientation	Orient ponds to take advantage of wind mixing, or in areas where wind causes extensive wave erosion, place the long axis of each pond at right angles to the prevailing wind; use hedge or tree wind-breaks as necessary.

through a chamber containing the fish, the water is treated to improve its quality and is then placed back in the fish-holding chamber. The water treatment is similar to what takes place in a typical sewage treatment plant. Such systems have been successfully used for many years by researchers but have found limited use in commercial aquaculture to date, largely because of the initial expense of setting them up and their high demand for energy required by pumps and accessory equipment. Reuse systems also require backup units on all the mechanical apparatus in case of equipment failure. Emergency power is also required during electrical failures.

A typical recirculating water system features one or more culture units, compartments in which suspended material can be settled for removal, a biological filter, at least one pump to move the water through the system, and such auxiliary components as aeration and pathogenic bacterial control through such means as ultraviolet (UV) light sterilization or ozonation (Figure 6). The biofilter is the heart of the recirculating system. It provides a large amount of surface area for the colonization of the genera of bacteria that detoxify metabolites produced by the fish in the culture chambers. The primary function of the biofilter is the conversion of ammonia to nitrate through a nitrite intermediate. That function is accomplished by the bacteria *Nitrosomonas* and *Nitrobacter*. The reactions are as follows:

$$NH_3 \xrightarrow{\text{Nitrosomonas}} NO_2^- \tag{1}$$

$$NO_2^- \xrightarrow{\text{Nitrobacter}} NO_3^- \tag{2}$$

Both ammonia, particularly in the unionized form (NH_3), and nitrite (NO_2^-) are toxic to fish, while nitrate (NO_3^-) can be present at high concentrations without affecting performance or survival. The efficiency of biofilter performance is sensitive to changes in pH, as are the fish in the system. Organic acids which accumulate in closed systems will force the pH down if the system is not properly buffered. Limestone or crushed oyster shell is often incorporated to provide a carbonate source which will combine with hydrogen ions and maintain pH within normal limits.

Sizing of biofilters to provide sufficient treatment capacity for the biomass of fish that will be produced in the system has received a considerable amount of study. Biofilter size will be a function of the species present, their biomass (which is continuously changing as the fish grow, so the system needs to be designed for ultimate carrying capacity), water temperature, water flow rate, and other factors. While sand and gravel biofilters have long been used by fish culturists, they tend to clog with particulate matter, causing the water to pass through restricted channels. The clogged portions of such filters can quickly become anaerobic, leading to ammonia production rather than removal. Filter media with a high percentage of void space are more reliable and will not clog. Any material that will support colonization by the desired bacteria can be used. Fiberglass and various types of plastics are typical types of effective media that allow the free flow of water through biofilters. Hess (1981) discussed the performance of various types of biofilter media in nitrifying ammonia to nitrate. Mechanical filtration (e.g., with a sand filter) can be incorporated into a recirculating system (Figure 7), though effective removal of particulate matter, including bacterial mats that slough off the biofilter, can usually be achieved by settling chambers. Design criteria and formulas for sizing biofilters have been discussed in detail by Wheaton (1977). Various other aspects of closed

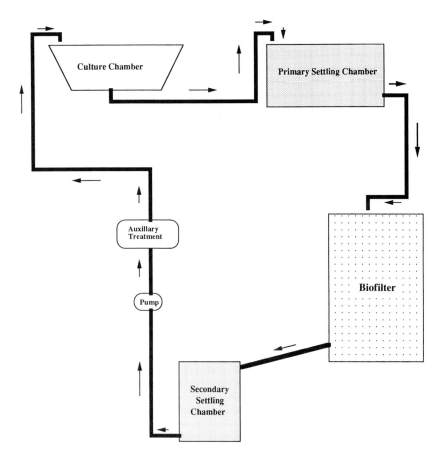

FIGURE 6. Diagram showing the components of a closed recirculating water system.

water systems can be found in Allen and Kinney (1981) and Lucchetti and Gray (1988).

Biofilters of various types have been employed by fish culturists. Submerged biofilters are units wherein the medium on which the bacteria grow is continuously under water. A downdraft biofilter is much like the trickling filters seen in sewage treatment plants. Water is distributed onto the filter media from above, trickles over the medium, and exits from the bottom of the filter. The filtration medium is not under water but is continuously wetted by inflowing water. Atmospheric oxygen can circulate through the medium to maintained aeration. A rotating biodisc filter is comprised of a series of circular plates made of fiberglass or some other suitable material on which bacteria will colonize. The discs are spaced along an axle that is turned by a gear reduction motor. Most rotating biodisc filters are set in tanks of water that cover about one half of each disc. As the unit rotates, the medium is exposed to the atmosphere half of the time and submerged in water the other half.

FIGURE 7.   Mechanical filters, such as the sand filters shown here, are sometimes used to remove particulate matter in recirculating and flowthrough water systems. In this case, pond water is being filtered before being run through indoor circular culture tanks (not shown).

This provides intermittent submersion of the bacteria to prevent desiccation while also exposing them to the atmosphere thereby keeping the system aerobic. Such units are commonly seen in conjunction with sewage treatment plants and can be purchased commercially (Figure 8).

Before fish are introduced into a recirculating water system, the biofilter should be activated. This can be done by adding ammonium salts to the system, placing a few fish in the culture chamber, or putting fish feed into the system. The idea is to provide an ammonia source to stimulate the colonization of the appropriate bacteria. Commercial products which are supposed to stimulate the colonization of biofilters are also available. Typically, *Nitrosomonas* becomes well established before *Nitrobacter*, so after a few days of operation there may be a reduction in ammonia with a concomitant increase in nitrite, but little nitrate production. Since nitrite is highly toxic, the culturist should not stock the culture system until the level of that chemical is sufficiently reduced.

Ammonia removal can also be affected by the use of ion exchange media, such as clinoptilolite, a natural zeolite (reviewed by Lucchetti and Gray 1988). The material eventually becomes saturated with ammonia and can be regenerated by flushing it with brine solution (NaCl at 40 ppt or above) at 20 to 30 times the volume of the zeolite material (Smith et al. 1981). Regeneration may be required every 3 to 4 weeks.

Continuous aeration is an important feature of any recirculating water system. Aeration is generally provided within the culture chambers to provide dissolved oxygen for maintenance of the fish and in the biofilter to maintain aerobic conditions for the bacteria. Both air compressors and regenerative blowers work well, though blowers tend to be more reliable and require less routine maintenance.

Diseases are difficult to treat in recirculating water systems, since chemicals used to kill disease organisms also destroy the biofilter bacteria. If the system is shut down so that the treatment is confined to the culture chamber, water quality may quickly deteriorate. To reduce the threat from circulating pathogenic bacteria in closed systems, treatment of the water with UV light or ozone is often incorporated into the design. In most cases, sterilization would occur just before the water enters the culture chamber.

UV sterilization requires passage of the water before an UV light source in a relatively thin film since the effectiveness of UV light diminishes rapidly as water depth increases. Commercial UV systems are available for handling various quantities of water. The lethal dose for the bacterium *Aeromonas salmonicida* is $3620 \, \mu W \, s/cm^2$. For the parasitic protozoan *Ichthyophthirius multifiliis*, the lethal concentration is $1.7 \times 10^6 \, \mu W \, s/cm^2$ (Vlasenko 1969). Most culture systems do not have enough UV energy available to accomplish the level of control required for parasites. Typically, commercial systems provide at least $30,000 \, \mu W \, s/cm^2$ at specified rates of flow.

The efficiency of UV systems is decreased markedly in turbid water. Sand filtration in advance of UV exposure may be recommended in situations where gravity settling of suspended solids is ineffective.

UV lamp efficiency drops about 10% during the first month of use (Lucchetti and Gray 1988), after which the efficiency rate decline slows. After about 6 months, the lamps are usually at 70% of their original energy level. The lamps should be changed at intervals of between 6 months and 1 year. Quartz tubes which keep the lamp surface from coming in contact with the water or clear plastic pipes which

FIGURE 8. A commercial biodisc filter.

carry water past UV lamps should be cleaned or replaced periodically, as organic matter will plate out on them and reduce the ability of the UV light to penetrate into the water.

Ozonation is becoming a method of choice by many aquaculturists. Commercial ozonation systems are available that operate efficiently and effectively.

At concentrations of 0.56 to 1.0 mg/l, ozone will kill most pathogenic organisms within 1 to 5 min (Dupree 1981). In addition to killing pathogens, ozone is effective in reducing biochemical oxygen demand, ammonia, and nitrite (Colberg and Lingg 1978, Sutterlin et al. 1984). Care must be taken to ensure that fish in the culture system are not exposed to ozone. This can be accomplished by providing continuous aeration to water that has been ozonated. While there has been little in the way of definitive tolerance levels established for most nonsalmonid fish species, fathead minnows are killed when exposed to 0.2 to 0.3 mg/l of ozone (Arthur and Mount 1975).

C. CAGES

Under certain circumstances, it may be desirable to rear fish in cages (Figure 9). Cages can be constructed of welded wire, plastic mesh, or nylon netting placed over a framework and equipped with floats. Cages have been used to hold and rear fish in streams, lakes, reservoirs, power plant intake and discharge canals, and even in large ponds. Cages used in conjunction with freshwater fish may be quite small (as depicted in Figure 9), or they may have volumes of several cubic meters. Net pens, which are currently being utilized around the world in the marine environment for the culture of salmon and various other fish species are, in reality, very large cages.

FIGURE 9.   A small cage designed for use in research.

The major use of cages in freshwater is for research and to hold fish in situations where their containment and recapture would otherwise be difficult or virtually impossible. Free circulation of water through each cage is required for the maintenance of water quality. Given the same volume within a cage, a much higher density of fish can be held in flowing than in static water because rapid exchange will maintain the quality of water to which the fish are exposed. In a static system (such as a pond or lake), local problems with increased ammonia or reduced dissolved oxygen can occur, sometimes with catastrophic results.

Disadvantages of cage culture include difficulty in treating fish for diseases, the need to provide feed on a daily basis to each individual cage (usually by boat), initial relatively high costs for constructing the cages, the need for continuous maintenance, the need to replace the units every few years, and the ease with which fish can be subjected to poaching. Cages have found considerable use by researchers, but few commercial cage-culture operations have been developed in the U.S. Cages are more widely used, particularly for holding small fish, in various other nations. In the Philippines, for example, young fish are often held in fine-meshed cages called happas until they are large enough for stocking into ponds.

Concentrations of several hundred fish per cubic meter have routinely been maintained in cages by fish culturists. The actual density utilized will depend upon the species under culture, the size to which the fish are to be reared, and the flow characteristics through the cage. A reasonable level of production might be 100 kg/m^3 or more.

D. RACEWAYS

Raceways are culture chambers through which water constantly flows. In most cases, the flow rate is relatively rapid, leading to a turnover rate of a few minutes to several hours. As a result of the rapid turnover rate in raceways, the fish are constantly exposed to new water, which should be of high quality. Thus, it is possible to maintain high densities of fish in relatively small volumes of water. Raceways may be linear or square, with water entering one end and exiting the other (Figure 10), or circular. Circular (and sometimes square) raceways often have center drains (Figure 11). Introducing water tangentially to the surface of a circular raceway causes the entire water volume to move in either a clockwise or counterclockwise manner, thereby imitating a linear raceway by providing a flow.

Raceways may be used in single pass systems (the water flows through one linear raceway or circular tank and is then discharged from the culture system), or they may be set up in series wherein water exiting one raceway is aerated and then enters another raceway. The process can be continued through a number of raceways, but since there is no biofiltration involved, there is continuous degradation of water quality from one raceway to the next. Generally, fewer than ten raceways can be accommodated in such a series, and even then aeration should be provided between raceways. Still, productivity within the culture system will decrease in the downstream direction because of the accumulation of metabolites and resulting degraded water quality.

FIGURE 10.   Linear raceways running down a hillside. Water drops over spillways located between each raceway in the series. The inverted bottles over raceways are demand feeders.

Raceways are used as the culture chamber component of recirculating water systems and in partial recirculating systems. In the latter type of water system, a percentage of the flow passes through a biofilter and is recirculated; the remainder is discarded and replaced with new water. The proportion of recirculated water may be adjusted as the biomass in the system increases. By increasing the proportion of new water added, the efficiency of the biofilter can be maintained.

Relatively small circular raceways are commonly used by researchers, with larger units having been adapted for use by some commercial and many public aquaculturists. Raceways, both linear and circular, are particularly useful in hatcheries. Linear production raceways are often fabricated from concrete, while smaller linear and circular units commonly found in hatcheries and research laboratories may be of concrete, aluminum, or fiberglass. In most cases, water depth in raceways is maintained at 1 m or less. Very tall circular raceways, called silos, have been developed for use in certain circumstances, but their popularity is not widespread.

The carrying capacity of a raceway depends upon water flow rate, the species under culture, and the quality of the incoming water. It is possible to maintain much higher densities of fish in raceways than in ponds or other static water systems, but fish in a raceway may quickly succumb if there is even a brief interruption in water flow.

FIGURE 11.   A circular fish culture tank about 2 m in diameter. The tank shown in static (does not receive a continuous flow of water). Excess water exits such tanks through the central standpipe.

V. WATER QUALITY

For optimum growth and the maintenance of good health, fish require water of a certain minimum quality. The specific minimum requirements vary from species to species, but there are some general characteristics that apply to all of the species discussed in this book. For example, the optimum temperature for growth of the species discussed in the later chapters generally falls within the 20 to 30°C range. Most can tolerate temperatures outside of that range. Some, such as striped bass and many species of tilapia, can also tolerate wide ranges in salinity, while others have relatively limited salinity tolerance. In freshwater environments the measurement of salinity is not necessary, but careful monitoring of temperature is demanded by most hatchery managers because metabolic rate of the fish is so dependent upon the temperature of water within the culture system.

Brief descriptions of some of the most important water quality variables which should be measured by freshwater fish culturists are provided in the following sections. Table 4 presents information on the preservation of water samples prior to analysis. Complete details about the chemistry involved with each determination discussed can be found in Boyd (1979). Many water quality variables need to be determined only once or at intervals of several weeks or months. Examples are hardness and alkalinity when their initially measured levels are sufficiently high to

TABLE 4
Methods of Preservation of Water Samples

Determination	Preservative	Holding time
Alkalinity	Cool to 4°C	24 h
Ammonia	Cool to 4°C; add 1 ml/l	24 h
	concentrated H_2SO_4, 40 mg/l	7 d
	$HgCl_2$	7 d
Biochemical oxygen demand	Cool to 4°C	6 h
Carbon dioxide	Cool to 4°C	2 h
Chemical oxygen demand	Add 1 ml/l concentrated H_2SO_4	7 d
Chlorophyll *a*	Cool to 4°C	12 h
Dissolved oxygen	Fix immediately for titration	6 h
Hardness	Cool to 4°C or add 1 ml/l HNO_3	7 d
Nitrate	Same as ammonia	
Total phosphorus	Cool to 4°C	7 d
Soluble orthophosphate	Cool to 4°C	2 h
Settleable solids	None required	24 h

Adapted from Boyd, C. E., Water Quality in Warmwater Fish Ponds, Alabama Agricultural Experiment Station, Auburn University, Auburn, 1979, 1.

be of no special concern. Fish culturists often do not routinely measure pH or nutrient levels in their water, but those measurements may be necessary in instances where those variables can be expected to fluctuate greatly. For example, pH should be monitored in closed recirculating water systems, and nutrient levels should be monitored when a specific fertility level is being maintained. Routine measurements generally include temperature, dissolved oxygen, and in some systems, ammonia.

A. TEMPERATURE

Culture system water temperature should be routinely monitored since management decisions with respect to such activities as spawning and amount of feed that should be provided are often made primarily on the basis of that variable. Daily records from representative culture units (ponds, raceways, tanks, cages) should be maintained and if some units within the system are known to vary significantly from the mean, those units should be individually monitored on a routine basis. In systems where water temperature is controlled by heating or chilling, continuous temperature monitoring may be required to maintain the system within design limits.

Temperature can be measured with a glass thermometer containing either mercury or alcohol, an in-line temperature gauge, or with a thermistor used in conjunction with a temperature meter which may also automatically record the readings. Computerized systems have been developed which will automatically adjust the flow rates of warm and cold water through mixing valves and maintain the desired temperature within the system.

Outdoor culture systems, such as those which incorporate large culture units, are usually operated at ambient water temperature. If geothermal water is available in combination with a cold water source, the two can be mixed to provide the optimum temperature on a year-round basis; such situations are uncommon, though highly desirable. Recirculating water systems are typically placed within the confines of a building where temperature control can be affected by controlling the air temperature, the water temperature, or both. Temperature control in cages is not possible in many instances, but there have been attempts to employ cage culture in conjunction with the heated effluent from fossil fuel and nuclear power plants. The technique has received a considerable amount of attention by researchers interested in trying to extend the growing season by taking advantage of the extra heat from the power plant during spring and fall when the ambient water temperature in the source water body (usually a lake or stream) is too cold for optimum fish growth. Problems with gas bubble disease (caused by supersaturated gases in the heated water, particularly during winter when the heat is most valuable to the culturist) have been severe in many instances. That problem, coupled with the fact that few power plants have been designed with a fish culture application in mind, have led to general abandonment of them as aquaculture sites.

B. ALKALINITY

Alkalinity is defined as the total amount of acid required to titrate the bases in a water sample. Included in the total alkalinity of a water sample are carbonate, bicarbonate, hydroxide, silicate, phosphate, ammonia, and various organic compounds; the majority of the bases are attributable to carbonate and bicarbonate, so an alternative definition is that alkalinity is the total bicarbonate and carbonate in a water sample. It is also a measure of the buffering capacity of the water.

Measurement of alkalinity is by two-step titration. The first step takes the water sample to its phenophthalein endpoint and measures carbonate alkalinity (reported as milligrams per liter of calcium carbonate [$CaCO_3$]). The amount of titrant used in the second step (indicated with the indicator methyl orange), when added to the first, provides a measurement of total alkalinity (also reported as milligrams per liter of $CaCO_3$). The difference between total and carbonate alkalinity is a measure of the bicarbonate alkalinity.

Most water sources remain constant or nearly constant with respect to alkalinity over long time periods. Occasionally, culture systems are developed with sources of water that are of extremely low alkalinity. When total alkalinity is less than about 20 mg/l, it may be necessary to add a source of carbonate to the ponds. $CaCO_3$ can be added to ponds to increase alkalinity (and hardness). Methods of calculating the amount of $CaCO_3$ required under various conditions have been presented by Boyd (1979).

As previously discussed, closed recirculating water systems are subject to reduced pH as organic acids accumulate and as microflora affect the levels of bicarbonate and carbonate in the water. Pond pH is affected by photosynthesis, which removes carbon dioxide (CO_2) from the water, and respiration, which adds carbon dioxide. When CO_2 is dissolved in water, it forms a weak acid (carbonic acid) that tends to drive the pH down. When CO_2 is removed by plants during photosynthesis, there is less acidity and the pH tends to rise. The dissociation of CO_2 in water is shown by the following formula:

$$CO_2 + H_2O \rightleftharpoons H^+ + HCO_3^- \tag{3}$$

Bicarbonate ion (HCO_3^-) further dissociates in a reversible reaction into hydrogen ion (H^+) and carbonate ion ($CO_3^=$).

$$HCO_3^- \rightleftharpoons H^+ + CO_3^- \tag{4}$$

As long as there is a supply of carbonate and bicarbonate available in the water, hydrogen ions will be released or removed from solution and the pH will not change. This process is known as buffering and can be extremely important. If the pH is not maintained within the range of approximately 5.0 to 9.0, some species of fish may be killed (Randall 1991). The addition of calcium carbonate to a pond as well as somewhere within closed recirculating water systems provides a source of carbonate, and then bicarbonate as the calcium carbonate goes into solution:

$$CaCO_3 \rightleftharpoons Ca^{++} + CO_3^- \tag{5}$$

$$CO_3^- + H^+ \rightleftharpoons HCO_3^- \tag{6}$$

The carbonate-bicarbonate buffer system can normally be depended upon to maintain the pH in fish culture ponds within the desirable pH range of approximately 6.0 to 9.0.

C. HARDNESS

The concentration of divalent cations, in particular calcium and magnesium, expressed as equivalent $CaCO_3$, is a measure of total water hardness. Hardness does not normally vary temporally, but in instances where the water is extremely soft (contains less than 20 mg/l total hardness), it may be necessary to add gypsum or crushed limestone to provide the cations needed for maintenance of the buffer system. Limestone (calcium carbonate) will increase both hardness and alkalinity, since upon dissociation it provides both divalent calcium and carbonate ion.

In extremely hard water (hardness of a few hundred milligrams per liter), calcium carbonate may precipitate on the surfaces of materials that come in contact with the water. This can cause the eventual constriction of water pipes and make glass aquaria opaque, sometimes after only a few days. Glass can be cleaned with weak hydrochloric acid, but the problem will recur.

D. AMMONIA

Ionized ammonia accumulates in water as a result of the elimination of ammonium ion (NH_4^+) through the gills by fishes within the culture system

(Hochachaka 1969). Once ammonium ion enters the water, much of it is converted to un-ionized ammonia (NH_3) which can then be nitrified to nitrate by bacteria. Ammonia is also lost from water into the atmosphere. Un-ionized ammonia is the more toxic form, so the relative percentages of the two forms is an important factor in determining whether the fish are being subjected to undue stress (Chipman 1934, Wuhrmann and Woker 1948). The percentage of total ammonia contributed by the un-ionized form is affected primarily by temperature and pH, though there are other less important factors. As either temperature or pH increase, so does the relative amount of un-ionized ammonia in a given sample of water. Measurements of ammonia (colorimetric or with a specific ion electrode) provide only a value for total ammonia, but tables are available which allow the fish culturist to determine what percentage of the observed value is in the un-ionized form (Emerson et al. 1975, Boyd 1979, Thurston et al. 1979).

In general, if the water remains within the normal temperature range for the nonsalmonid fish under culture and the pH is no more than slightly alkaline, problems with ammonia toxicity can be avoided if the total ammonia level does not exceed 1.0 mg/l. Research with blue tilapia has shown that at least that species can develop some degree of resistance to elevated ammonia if exposed to sublethal levels for a period of time (Redner and Stickney 1979). A summary of the effects of ammonia on a variety of fishes was recently provided by Russo and Thurston (1991).

E. NITRITE

Nitrite toxicity has not historically been a problem with respect to pond culture, though it has been of considerable concern in recirculating water systems, particularly during the early stages of biofilter colonization (Stickney 1979). Fishes suffering from nitrite toxicity have frequently been observed in recirculating systems in which the bacterial flora has not been adequately established on the biofilter medium prior to fish stocking. Fishes suffering from nitrite toxicity exhibit chocolate-colored blood, hence the common name for nitrite toxicity syndrome, brown blood disease. When nitrite levels become sufficiently high, hemoglobin reacts with nitrite to form methemoglobin which is unable to combine with oxygen. Thus, the tissues are deprived of oxygen and the affected fish are eventually asphyxiated. Fish suffering from methemoglobinemia may begin to swim erratically just prior to dying (Konikoff 1975).

Advancing technology in aquaculture has led to ever-increasing production levels. With respect to the catfish industry, improved cultural practices have resulted in the production of 4000 kg/ha or more in some pond facilities. As catfish farmers began to push the limits of their pond systems, methemoglobinemia began to appear in the fish, particularly during the latter part of the growing season when biomass per unit volume of pond space was peaking.

The level at which nitrite toxicity occurs varies widely from one fish species to another. As reviewed by Russo and Thurston (1991), studies conducted to date show that salmonids are much more susceptible to nitrite toxicity than other tested species. Bioassays run for 48 or 96 h have shown toxicity at nitrite levels from below 0.2 mg/l to nearly 200 mg/l, depending upon the species being studied. Median tolerance limits (the level of a toxicant that will kill 50% of the animals exposed over a given period of time) for channel catfish exposed to nitrite have been reported by Konikoff (1975). The median tolerance limits after 24, 48, 72, and 96 h were 33.8, 28.8, 27.3, and 24.8 mg/l, respectively. All of the exposure trials were conducted with 40 g fish at 21°C.

Nitrite toxicity is pH and chloride dependent (Russo and Thurston 1991). Adding calcium chloride or sodium chloride to the water has been an effective means of reducing nitrite toxicity in aquaculture ponds (Tomasso et al. 1979, 1980; Huey et al. 1980; Schwedler and Tucker 1983). Levels of either chemical applied at 60 mg/l have been efficacious.

F. DISSOLVED OXYGEN

Perhaps the most critical water quality variable over which the fish culturist has at least some measure of control is dissolved oxygen (DO). Unlike temperature, DO is largely under biological control. Oxygen can be dissolved in water through diffusion from the atmosphere (sometimes aided by mechanical forces provided by nature in the form of wind or the culturist who might use aeration devices) and through photosynthetic activity.

In pond systems, DO levels are typically lowest near dawn as a result of continuous respiratory demand by both plants and animals during the night in the absence of photosynthetic oxygen production. As photosynthesis begins to introduce oxygen into the water after dawn, the DO level will begin to rise. The DO will often increase throughout the daylight hours and then begin to decline toward dusk. A typical diurnal pattern of dissolved oxygen fluctuation is presented in Figure 12. The example in Figure 12 shows a range of between slightly less than 2 mg/l and about 6 mg/l. Actual patterns in a given pond will vary from day to day and may show a wider or narrower range than in the example. Net 24-h gains or losses in DO level may be influenced by any activity that affects the rate of photosynthesis,

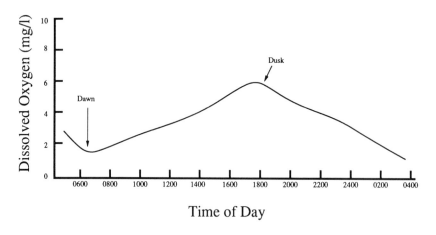

FIGURE 12.   Graph showing a hypothetical diel oxygen curve for a fish culture pond.

respiration, or mechanical mixing of the water. Some of the factors which can lead to a net reduction in DO, such as shown in Figure 12, are a period of overcast weather, increased turbidity, and the presence of herbicides in the water. All have a negative effect on phytoplankton productivity, thereby reducing oxygen production. Net increases in DO can occur when phytoplankton productivity is increasing, when there is wind mixing, or when the culturist takes some action to enhance DO.

For virtually all warmwater fish species, it is desirable to maintain a DO level of at least 5 mg/l, though many species such as channel catfish and tilapia can tolerate and even thrive at lower oxygen levels. Stress in warmwater fishes often begins to occur when the DO falls below about 3 mg/l, and mortality will begin at some lower DO level. The conscientious fish farmer routinely checks each pond for signs of oxygen depletion in the morning at about dawn, particularly during the height of the growing season when the water is warm, fish densities are high, and the respiratory demand on oxygen is great. Many farmers now monitor DO throughout the night, either utilizing personnel to observe the ponds and collect DO measurements as appropriate, or by employing automated systems to monitor DO concentration. Predicting nighttime oxygen depletions is possible, but the technique is not completely reliable, so careful monitoring continues to be important (Boyd et al. 1978).

Fish that are experiencing hypoxia will often appear to be gasping at the surface of the water. By measuring the DO level in the water, action to prevent an oxygen depletion can be taken before the fish begin to exhibit that behavior, which indicates that they are already severely stressed.

DO can be measured chemically utilizing a method known as the Winkler titration which is inexpensive but requires a few minutes, some glassware, and a series of chemicals that are added to the water sequentially. The presence of oxygen is indicated when the water turns yellow (straw colored) after the chemicals have been added. A precise measurement of the DO level is obtained through a titration procedure. Highly reliable and suitably accurate DO meters are currently available and widely employed by fish farmers. They are normally battery operated and often have associated with them a thermistor for temperature measurement in conjunction with an oxygen electrode. The combination of a meter which reads both temperature and DO is handy since the meter can be calibrated to compensate for the influence of temperature on DO. In salt water, a salinity adjustment is also required, and at high elevations, an altitude adjustment is also employed. DO meters which can be adjusted for each variable are available.

As temperature, salinity, and altitude increase, the capacity of the water to hold oxygen is reduced. The saturation level may, under some conditions, be no more than about 5 mg/l (e.g., warm seawater). Oxygen probes in each pond on a culture facility can be connected to a computer which will monitor diurnal fluctuations. The computer can be programmed to sound a local alarm to alert on-station personnel if the DO falls below a set level in any pond, or it can even dial a telephone and alert a sleeping farmer located remotely from the fish farm that a problem is imminent.

Emergency aeration procedures to overcome oxygen depletions include the use of paddlewheel aerators (Figure 13) or some other type of aeration device, running

FIGURE 13. Paddlewheel aerators are commonly employed to maintain adequate levels of dissolved oxygen in fish ponds or for emergency aeration.

oxygen-rich new water into the pond, recirculating the pond water through a pump which sprays the water back in forcefully, injecting compressed air or air provided by a regenerative blower, employing agitators of various kinds, or injecting compressed or liquid oxygen. Aeration devices may be used routinely at night (Hollerman and Boyd 1980) or as needed when an emergency situation is imminent. In most instances only a small percentage of the ponds on a farm will experience an oxygen depletion on the same day, so it is not necessary to have aeration equipment constantly present in each pond. The equipment used should be portable so it can be moved to the needed location quickly and efficiently.

VI. NUTRITION

Various techniques are utilized to feed fish. Natural food can be supplied in the form of algae, rotifers, brine shrimp, minnows, crayfish, and a variety of other items. Today, live food is generally used only for species which will not accept prepared feeds. That situation may exist only during a portion of the life cycle of a given species. For many species, particularly those with very small eggs and larvae, live food is often required during the early stages of culture. In any case, the goal of the fish culturist is to provide sufficient quantities of food of the proper quality to promote rapid growth. In natural waters, the carrying capacity of the environment may be quite low (typically no more than a few hundred kilograms per hectare). If both fertilization and supplemental feeding are employed, standing crops of a few thousand kilogram per hectar are commonly achieved.

The early feeding stages of most of the species discussed in this text can be reared at relatively high densities if the water is properly fertilized. The quality and quantity of fertilizer required will be dependent on the type of food organisms being promoted. For example, inorganic fertilizer is often used to support blooms of unicellular algae, while zooplankton blooms may be best supported by a fertilization regimen that employs organic compounds. Methods and frequency of fertilization for various culture strategies are discussed as appropriate in the chapters that follow.

Training fish to accept prepared feeds is required for some species. In some cases, that has not been successfully accomplished, perhaps because the texture, color, or method of presentation of the prepared feed does not stimulate the fish to initiate feeding behavior. Some research has been conducted to find attractants that can be added to prepared feeds.

The specific nutrient requirements of the various fishes discussed in this book are well documented in some cases and very poorly known in others. For fish nutritionists to define the nutritional requirements of a given species with any degree of confidence, they must evaluate such things as the needs of the fish for proteins and the amino acids that comprise them, energy, lipids, carbohydrates, vitamins, and minerals. In addition, the feed may have to be presented in a specialized form or contain some type of attractant to make it palatable to the fish.

The general nutritional requirements of many of the fishes discussed in this book are quite similar, but sufficient individual differences exist that broad generalizations can only be made with caution. Information on the known nutritional requirements of each species covered is discussed in the chapters that follow.

VII. DISEASE

The types of infectious diseases that impact fish are the same as those found in various other organisms, including man. The causes of disease in fish generally can be attributed to viruses, fungi, bacteria, and parasites. Fish have a fairly well-developed immune system and tend to be resistant to disease unless they are stressed. Any type of stress can trigger a disease epizootic. For example, if a pond of fish experiences an oxygen depletion, a disease epizootic may appear after a period ranging from 24 h to 2 weeks. Other commonly encountered stresses are rapid changes in temperature, exposure to high levels of ammonia, exposure to excessive nitrite, and the presence of high levels of suspended solids. Handling is always a cause of stress in fish.

There are a number of diseases found in a broad variety of fishes, but many disease organisms are species-specific or impact a narrow range of species. Thus, information on diseases is also detailed in the chapters that follow.

Only a few treatments are available for use on fish in the U.S. Registration of chemicals for use in disease control is an extremely expensive process and can take a number of years. There have been a number of fish vaccines developed in addition to treatment chemicals, but the arsenal of treatments legally available to fish farmers is very limited. Thus, it is extremely important to minimize stress in cultured fish. That is difficult since, virtually by definition, fish culture involves the rearing of fish under high densities, which means that the chance of exposing the animals to stress is always present.

VIII. GENETICS AND SELECTIVE BREEDING

There can be virtually no improvements in culture stocks if the life cycle of the fish species under culture is not completely within the control of the producer. If the culturist is required to go out into nature for broodfish, there is little, if any, opportunity to practice selective breeding. For most of the species discussed in this book, the production of numerous generations in captivity has become a reality. Yet, for others such as striped bass and a few coolwater species, it continues to be common practice to collect broodfish from the wild. For sportfishes, that approach may not only be the most practical but also the most desirable, since development of a domestic strain is not the goal of the culturist. On the other hand, selective breeding of sportfishes may lead to a fish that exhibits better fighting characteristics, or is more or less easy to catch.

With respect to foodfishes, a major goal is often to achieve more rapid growth. Higher dressout percentage and increased fecundity may also be goals of fish breeders. To date, little progress has been made in these areas with respect to most of the species under culture.

Presently, research and development in the areas of selective breeding and genetics of cultured fishes must be considered as infant disciplines. Some active programs are in existence, but in general, facilities are limited and the time required for success is long. Also, funding is somewhat difficult to obtain. With the development of genetic engineering over the past several years, and the more recent application of the technique to fishes, some of the seemingly unattainable goals of fish breeders may soon become reality. Genetic engineering holds the promise of an entirely new dimension in which fish culturists might operate in the future, though there are a number of hurdles that must be jumped since resistance to the development of transgenic species and their use in aquaculture has developed. Fears of mixing transgenic cultured fish with wild populations through escapement are real and undoubtedly will prove well-founded in some, or even most, instances. The issues can be expected to grow as the application of molecular genetics to fishes and their culture expands.

IX. RECORD KEEPING

One of the facets of fish farming that is often overlooked is the need to keep detailed records of not only business transactions, but also of the cultural practices employed. Record keeping should include the amount of feed provided to each pond, raceway, or cage each day; each water quality determination made in each culture chamber; incidences of diseases, along with the dosage and type of treatment chemical that was applied, if any; dates and amounts of fertilizer added; and so forth. The date or dates that fish are stocked in a particular culture chamber, numbers and species stocked, biomass, and any special information (e.g., the fish may have been sex-reversed, hybrids, or display some other unique characteristic), dates that spawning was initiated and completed, and the dates and amount of biomass harvested should be recorded. Such records help the fish culturist keep track of what is in the culture system on a day to day basis and can be extremely important in helping the farmer identify trends over periods of months and years.

Microcomputer programs have been developed or are under development in many regions of the U.S. to assist the fish farmer with such record-keeping chores. County Extension Service Offices, the Cooperative Extension Service at Land-Grant Universities, and the U.S. Department of Agriculture can assist farmers in finding sources of those programs.

X. QUALITY OF THE FINAL PRODUCT

The responsibility of the fish culturist does not end when the fish are handed over to a live-hauler, processor, or distributor. Until the consumer passes judgement — whether that judgement is made by an individual sitting down to dinner in the home or in a restaurant, a fisherman placing a minnow on a hook, or a sportsman who ultimately catches a cultured fish on a rod and reel — the fish farmer should continue to be vitally interested in the product even after it leaves the culture facility. The production of off-flavor fish may cost the producer a large number of customers. Similarly, the farmer may be responsible for selling sickly minnows that do not survive long enough to make it from the bait dealer to the old fishing hole. Finally, hatchery fish stocked for recreational fishing should not only survive to catchable size (sometimes they are stocked at that size), they should also behave, particularly when hooked, like a native fish of the same species. As with any business, only the satisfied customer will return to the same product time after time.

REFERENCES

Allen, L.J. and Kinney, E.C., Eds., *Proc. Bio-engineering Symp. Fish Culture*, Fish Culture Section, American Fisheries Society, Bethesda, MD, 1981, 1.

Arthur, J.W. and Mount, D.I., Toxicity of a disinfected effluent to aquatic life, in *1st Int. Symp. on Ozone in Water and Wastewater Treatment*, International Ozone Institute, Waterbury, CT, 1975, 778.

Boyd, C.E., *Water Quality in Warmwater Fish Ponds*, Alabama Agricultural Experiment Station, Auburn University, Auburn, AL, 1979, 1.

Boyd, C.E., Romaire, R.P., and Johnston, E., Predicting early morning dissolved oxygen concentrations in channel catfish ponds, *Trans. Am. Fish. Soc.*, 107, 484, 1978.

Chipman, W.A., Jr., The role of pH in determining the toxicity of ammonium compounds, Ph.D. dissertation, University of Missiouri, Columbia, 1934, 1.

Colberg, P.J. and Lingg, A.J., Effect of ozonation on microbial fish pathogens, ammonia, nitrate, nitrite, and BOD in simulated reuse hatchery water, *J. Fish. Res. Board Can.*, 34, 2033, 1978.

Dupree, H.K., An overview of the various techniques to control infectious diseases in water supplies in water reuse aquacultural systems, in *Proc. Bioengineering Symp. Fish Culture*, Allen, L.J. and Kinney, E.C., Eds., Fish Culture Section, American Fisheries Society, Bethesda, MD, 1981, 83.

I. INTRODUCTION

The production of channel catfish (*Ictalurus punctatus*) surpassed that of any other commercially cultured foodfish species in the U.S. a few years ago and continues to grow. Channel catfish (Figure 1) are native from Montana eastward to the Ohio Valley and southward through the Mississippi Valley to the Gulf of Mexico and into Florida (Eddy and Underhill 1974). The channel catfish is also native to Mexico. In addition to its use in commercial aquaculture, the species is widely produced in state and federal hatcheries and stocked for recreational fishing.

The natural habitat of channel catfish is moderate- to swift-flowing streams, and the fish originally preferred relative clear water (Jordan, 1969), though it is now frequently found in sluggish, more turbid streams and in lakes. Its present distribution throughout most of the U.S. is attributable, in part, to widespread stocking programs.

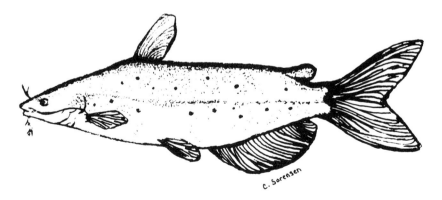

FIGURE 1.   The channel catfish can be distinguished from other species in the genus by the presence of spots on its sides. (Drawing by Cheryll Sorensen. From Stickney, R. R., *World Aquaculture*, 22, 2, 1992. With permission.)

Other species within the family Ictaluridae that have received the attention of fish culturists include the blue catfish (*I. furcatus*) and the white catfish (*I. catus*). Both of those species, along with their hybrids and hybrids between them and channel catfish, have been produced. Some of the hybrids have been reared commercially, but the bulk of the industry is currently centered on channel catfish with less emphasis being placed on blue catfish. Various bullhead species are also contained in the genus *Ictalurus,* and there has been interest expressed from time to time in their culture. No active significant production of bullheads seems to be occurring either in the commercial or private sector. Another related species, the flathead catfish (*Pylodictis olivaris*), has been the focus of some research, but its cannibalistic behavior has kept it from becoming an active candidate for production by fish culturists.

II. DEVELOPMENT OF THE COMMERCIAL CATFISH INDUSTRY

Channel catfish have long enjoyed popularity as a foodfish in the southern part of their range, and the development of commercial production occurred in Alabama during the late 1950s. Dr. H.S. Swingle, a professor at Auburn University, was the first to demonstrate that channel catfish could be reared in ponds and sold for a profit (Swingle 1957, 1958). According to Swingle's estimates, a producer would have to receive $1.10/kg for the fish at harvest.

Initial development of the commercial catfish industry was relatively slow. By 1960, there were about 160 ha of commercial catfish production ponds in the U.S. (Meyer et al. 1973). There was impressive growth during the 1960s. There were 950 ha in production by 1963 and 16,000 ha by 1969. The industry became mature in the 1980s. By 1990 some 63,000 ha of catfish ponds were in production (Anonymous 1991). Annual foodfish production figures (in tons) for catfish are presented in Figure 2. Production figures for 1976 through 1988 are from Rhodes (1989), with the 1988 value being a projection based on processing through August of that year. The values for 1989 and 1990 came from the review in *Aquaculture Magazine* of the status of aquaculture around the world (Anonymous 1990, 1991a). Sales of catfish reached $323 million during 1990 (Anonymous 1991b). Interestingly, while the costs of all goods and services have gone up significantly since the 1950s, catfish today only bring something less than $1.80/kg to the producer. The fact that catfish culturists continue to make a profit relates to improvements that have been made in management strategies and in the feed that is provided.

The catfish industry first became established in central Arkansas in farming areas that had been primarily known for rice production. The water table was relatively shallow, the water plentiful, and the soil had good water-holding characteristics. The industry grew to the point that by 1970 the water table was falling alarmingly, and the farmers began looking for alternate locations. They found the Delta region of Mississippi.

The Mississippi Delta was also a farming region; however, the main crop was not rice, but cotton. Large expanses of flat land, again with a shallow and seemingly

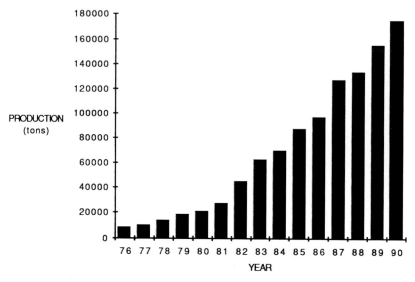

FIGURE 2.   Annual U.S. channel catfish production figures from 1976 through 1990.

limitless volume of water, was present. While Arkansas continued to be a major catfish producing state, many farmers moved to Mississippi or opened subsidiary farms, and production in the Mississippi Delta soon surpassed that of Arkansas and then surpassed total production within the other catfish producing states combined. Several major processing plants and large feed mills were established within easy driving distance of the Delta fish farms. The industry blossomed, particularly during the 1980s (see Figure 2) when aggressive marketing campaigns and painstaking quality control measures were implemented.

Mississippi is, and can be expected to remain, the state having the largest production. A number of other states produce significant numbers of catfish: notably Arkansas, Louisiana, and Alabama. Significant numbers of catfish farms are found in various other states as well. The status of the industry throughout the nation was reviewed by Busch (1985). His analysis showed that the potential for culture of catfish was good to fair in the southeastern portion of the U.S., generally fair to poor in the midwest and west, and poor in the northeast.

The growing season is much shorter as one moves out of the southern tier of states, so the time required to grow catfish from egg to market can be considerably longer than 18 months unless warm water is readily available and inexpensive. Geothermal or effluent from a power plant are two possible sources of heated water. Some producers in the north have been successful in developing local markets, selling fingerlings for stocking or running fee-fishing operations.

III. CULTURE SYSTEMS

While most catfish production occurs in earthen ponds, commercial producers and particularly researchers have also used cages, raceways, and tanks. The basic characteristics of these water systems have been discussed in Chapter 1. Examples of culture systems that have been adopted by the channel catfish-farming industry are presented in the subsections below.

A. PONDS

Mississippi farms tend to be large, though there is considerable variation as is true of most other states. Farms sometimes involve several hundred hectares and are comprised of numerous ponds of various sizes (Figure 3). The large farmers usually sell their product exclusively to live-haulers or take the fish directly to processing plants themselves. At the other extreme are so-called mom-and-pop operations which may be comprised of one or a few small ponds that produce fish for a local market, fee-fishing operation, or for sale directly to neighbors. In most instances, production ponds are no more than about 5 ha in area, though some in excess of 10 ha are being utilized. Large ponds are difficult to manage efficiently. Severe disease or water quality problems in a large pond can be economically disastrous because of the large inventory of fish that might be involved. Small ponds (less than 1.0 ha) are often used for holding broodstock, spawning, and fry rearing.

On the flatlands of such places as the Mississippi Delta and central Arkansas, ponds receive little or no runoff and are generally rectangular in shape or nearly so with the levee tops extending above the natural ground elevation. Watershed ponds, those which are formed by damming a low area, are common in some parts of the

FIGURE 3. Aerial; view of part of a commercial catfish farm in the Delta region of Mississippi. (Photo courtesy of Thomas Wellborn.)

country. Watershed ponds may dam a continuous watercourse such as a small stream or an intermittent watercourse that flows only after precipitation. The catfish industry in Alabama, for instance, has been developed to a large extent in watershed ponds. As might be expected, the size of the watershed, soil characteristics, and slopes of the sides of watershed ponds vary significantly from one site to another (Boyd 1985).

B. RECIRCULATING AND FLOW-THROUGH TANK SYSTEMS

Prior to the first of several oil crises that the U.S. has been exposed to since the early 1970s, it appeared as though a catfish producer might be able to heat water a few degrees, then pass it through tanks or raceways of fish, then down the drain without recirculation, and still make a profit. With the current cost of energy and the prospect that the cost will only increase in the future, flow-through systems that require supplemental heating are no longer attractive. There is still considerable interest in recirculating systems, both from the standpoint of conserving water and heat.

The key to a properly functioning recirculating water system is in the design and operation of the biofilter. All types of biofilters (described in Chapter 1) have been used by catfish producers and the research community. For laboratory use, an effective recirculating water system can be constructed in a matter of hours utilizing two or three fiberglass tanks, scrap pieces of PVC pipe for the biofilter medium, and a little plumbing expertise. Such a system is depicted in Figure 4. Temperature control can be achieved by adjusting air temperature in a building or by heating or cooling the water. A system that can provide both chilled and heated water by using a heat pump can be much more economical than one which employs electrical coils or other sources of energy (Stickney and Person 1985).

Small systems can be used for studies conducted in circular tanks or flow-through aquaria (Figure 5). Many research systems are of the single-pass type where the water that exits in the culture chamber is discarded, though recirculating systems have sometimes been employed. Circular tanks have found only limited utility in catfish production systems except as temporary holding facilities. Tanks have been widely employed for research. The pioneering work with large fiberglass circular tanks for catfish production was conducted in Georgia by Andrews et al. (1971), and the technology continues to be widely employed not only for catfish but for a wide variety of aquatic organisms. Circular tanks are commonly seen on aquaculture facilities virtually around the world.

C. RACEWAYS

Raceways have long been used by trout culturists, but are relatively uncommon in conjunction with catfish culture operations, though some highly successful commercial enterprises have employed them. Those producers that have employed raceways usually have operations that are associated with abundant supplies of flowing water of suitable quality that is available at little or not cost. For example, if geothermal water is available at or near the ground surface and can be obtained through artesian flow, its use in raceways would be ideal. In one such commercial

FIGURE 4. Portion of a small recirculating water system in a research laboratory. Water enters the circular biofilter tank on right from another similar tank that serves as a primary settling basin. The biofilter tank, which is filled with pieces of PVC pipe to support bacteria, is well aerated as indiated by the turbulence on the surface. Water leaving the biofilter tank enters a reservoir (circular tank on left) from which it is pumped through a sand filter (lower left) and then to the aquaria in the background.

FIGURE 5. Replicated aquaria (in this case, each has a capacity of approximately 40 l) are widely employed by catfish researchers. Exchange rates are typically 1 to 2 h and supplemental aeration is usually provided. Such systems may be of the single pass or recirculating type.

operation, concrete raceways are used that each receive 26.5 m³/min of water and support channel catfish densities of 200 to 400 kg/m³ (Ray 1981).

Another possibility is the use of raceways which utilize a large impoundment as a water source and receive water that is subsequently pumped back up into and recirculated through the reservoir. One such system was established for research purposes in Tennessee (Hill et al. 1974). Two series of 33-m long raceways were each stocked with 2000 channel catfish. Gradual deterioration of water quality occurred down the progression from the upper to the lower raceway in each series, but average production was over 1000 kg per section of raceway.

There is no set size for catfish raceways, but Beleau (1985) indicated that most have length to width to depth ratios of 30:3:1. Most are constructed from concrete.

D. CAGES

The use of cages by catfish farmers is limited, but commercial producers have employed the technique in instances where fish are reared under special circumstances. For example, the cage culture technique might be employed in a body of water that cannot be easily seined, such as in lakes, reservoirs, or reaches of streams where space is leased from a state (such as has been allowed in Arkansas), or in the intake or discharge canals of power plants (Tilton and Kelly 1970; Pennington 1977). Cages have been utilized by researchers for replicated experiments in a number of instances. Stocking density is often several hundred fish per cubic meter, with 500/m³ apparently being within the optimum range (Schmittou 1969).

Problems associated with cage culture include difficulty in treating diseases, loss of fish in the event of cage damage, and the need to provide continuous security to prevent poaching. In most cases, cages are equipped with covers to keep predatory birds from taking fish as well as keep the fish from jumping out (Beleau 1985). A major advantage of cages is ease of harvest.

The density of a given body of water or area of water will vary considerably based upon ambient water quality and the rate at which water within each cage is exchanged. Cage density and the number of fish that can be contained within a cage can be increased in flowing water because of the rapid exchange rate. There will be degradation of water quality as the water flows through a cage; thus, placing a number of cages in close proximity to one another may lead to stress in the downstream fish.

IV. POLYCULTURE

The majority of channel catfish farming involves monoculture, i.e., the rearing of catfish alone. However, the concept of polyculture, where two or more compatible species are raised together, has been applied by some catfish farmers. Among the animals that have been cultured with channel catfish are buffalo, paddlefish, common carp, grass carp, silver carp, sunfish, striped bass, golden shiners, tilapia, blue catfish, white catfish, mullet, crayfish, and freshwater shrimp (Perry and Avault 1972a, 1972b, 1975; Green 1973; Williamson and Smitherman 1975; Merkowsky and Avault 1976; Kilgen 1978; Lewis 1978; Clady 1981; Gabel et al. 1981; Tuten and Avault 1981; Wilson and Hilton 1982; Huner et al. 1983). The general conclusion that has been drawn from such stockings is that catfish production is not significantly altered, and total production is augmented by the increased biomass attributable to the additional species in the system. Overall income to the farmer is also increased (Dunseth and Smitherman 1977; Henderson 1979). In some cases, the presence of other species may actually stimulate feeding by the catfish (Clady 1981).

Not all of the polyculture studies have led to the conclusion that polyculture of other species with channel catfish is beneficial. Research has shown that stocking blue tilapia (*Tilapia aurea*) at densities equal to or greater than 25% of the weight of channel catfish in a flow-through laboratory system is impractical (Wilson and Hilton 1982). Similarly, in ponds, combinations of catfish and tilapia do not appear to be particularly desirable (Perry and Avault 1972a). Combinations that included blue tilapia, carp, and hybrid buffalo have led to a reduction in catfish yields (Williamson and Smitherman 1975). However, when channel catfish were stocked in ponds with hybrid tilapia, catfish growth was not affected (Gabel et al. 1981), though such water quality variables as pH, turbidity, and chlorophyll *a* content were increased in the presence of tilapia hybrids. Thus, commercial producers should evaluate any proposed polyculture combination before implementing a large-scale program involving the technique.

Polyculture has the major disadvantage that the species cultured must be separated for marketing. The process can be labor intensive, and the excessive handling that is sometimes required can lead to losses of submarketable fish and mortality of marketable fish prior to delivery at the processing plant. The farmer should determine in advance that each of the species proposed for polyculture will be marketable. Processing plants that specialize in catfish may not accept other species, so alternative processing and marketing channels may have to be developed.

The most commonly cited advantage for polyculture, in addition to increased overall production and potential profit, involves the fact that properly selected, the polyculture species can utilize niches within the pond that are not being used by the catfish. Various types of fishes and invertebrates will feed on virtually every part of the food chain, which is the basis of carp polyculture that goes back for millenia in China. However, in heavily stocked catfish ponds, natural food organisms are often unavailable, so prepared feed may be required in any case.

Polyculture may also be employed as a management tool. Perhaps the best example is the introduction of grass carp into catfish ponds to control rooted aquatic vegetation and filamentous algae (Kilgen 1978; Lewis 1978). Grass carp are strictly herbivorous after the fry stage and will not prey upon channel catfish or desirable invertebrate species (Huner et al. 1983).

V. STOCKING DENSITY

A typical strategy employed by channel catfish farmers is to stock fry at high densities and rear them to selected fingerling sizes during the first growing season. Thereafter, the fish are restocked at lower densities for growout to market size during the second growing season. Alternatively, an initial high stocking density may be used with subsequent partial harvest and restocking as the fingerlings grow (Busch 1985). The density at which fry are stocked significantly affects the size of fingerlings that can be produced over the same amount of time (Snow 1981). Fingerlings ranging from 5 to 10 cm can be produced in about 120 d when fry are stocked in ponds at 250,000 to 375,000/ha . If fingerlings of 20 cm are desired over the same growing period, fry should be stocked at from 35,000 to 50,000/ha (Brown 1977). Fry catfish can also be reared in tanks or troughs (Stickney 1972; Murai 1979) for later stocking into ponds or more intensive growout systems.

Pond growout stocking densities vary widely depending on the type of management employed. Final harvest biomass generally ranges from 1500 to 4000 kg/ha per year, with stocking rates to achieve those figures typically ranging from 2000 to in excess of 20,000 fish per hectare. Great improvements in pond management, the quality of catfish feed, and disease control have contributed to the fact that production levels per unit area of pond have improved greatly over the years. For example, Swingle (1958) reported that 1702 kg/ha could be produced when catfish were stocked at 4940 fish per hectare and that 2019 kg/ha resulted from a stocking density of 9880 fish per hectare . Two decades later, Tucker et al. (1979) produced 2990 kg/ha at a stocking rate of 4942 fish per hectare and 4100 kg/ha when they stocked the fish at 10,007 kg/ha .

Most producers rear fish to 0.4 to 0.5 kg, and fish in that size range are preferred by many processing plants. Fish of larger and smaller sizes are desired in some markets, so the farmer should determine in advance the best size to produce for the market that the fish will enter.

In recent years, the technique of intermittent harvesting has been introduced. Instead of harvesting the entire cohort of fish stocked in a growout pond during the spring, the pond is seined periodically and marketable fish are removed. Since catfish of the same initial size at stocking will not reach market size as a cohort, the faster growing individuals are marketed first. New fingerlings may be introduced to replace harvested fish, which will lead to an even greater range of sizes within the pond. Selective harvesting is often conducted every few weeks. Seines with mesh sizes large enough to retain marketable fish while allowing submarketable animals to escape are employed. Pond productivity will eventually decrease because of the accumulation of organic matter associated with fish wastes and uneaten feed. Also, some fish will become stunted and, while they will continue to consume feed, will not grow to become marketable. Thus, at intervals of approximately three years, ponds managed for intermittent harvesting should be totally harvested, drained, thoroughly dried, and disced before being put back into production.

Stocking densities employed in intensive culture systems are highly variable. Stocking will depend on initial size of the fish and the amount of water flowing through the system. For cages and tanks, it has been shown that neither cage size nor fish density seem to affect catfish growth or food conversion efficiency if water quality is maintained (Allen 1974; Kilambi et al. 1977; Lewis 1969; Schmittou 1969).

VI. WATER QUALITY

Virtually every substance that is soluble in water could conceivably impact the performance of fish under culture, but aquaculturists routinely measure only a few variables. Each variable continues to be evaluated independently of others, though synergisms undoubtedly occur. The water quality variables discussed are among the most important one affecting channel catfish growth and survival. They are also important for the other species considered in this book.

A. TEMPERATURE

Channel catfish grow most rapidly at temperatures between 26 and 30°C (Kilambi et al. 1970; Andrews and Stickney 1972; Andrews et al. 1972), though relatively rapid weight gains can be attained at temperatures from 24 to about 32°C. Depending upon time of year and fish size, the maximum temperature that can be tolerated may range from 33.4 to 37.5°C (Schwedler et al. 1985). While the temperature range for optimum growth is somewhat narrow, the range in temperature over which channel catfish can survive is broad. The species can be found in naturally occurring populations throughout the latitudes covered by the continental U.S., excluding Alaska. Feeding activity slows when channel catfish are exposed to temperatures above or below the optimum range. Rapid changes in ambient temperature, such as those that occur in ponds during spring and fall, are stressful to the fish and may precede a disease epizootic. The immune system of the fish functions best when they are within their optimum temperature range (Clem et al. 1984).

B. DISSOLVED OXYGEN

It is generally accepted that a concentration of dissolved oxygen (DO) of 5.0 mg/l or above is sufficient to support optimum growth of fishes. Channel catfish are somewhat more tolerant of lower levels of DO than some other species, but producers and researchers alike commonly agree that levels below 5.0 mg/l are undesirable. Farmers often institute emergency aeration procedures when the DO level falls below 3.0 mg/l. Fluctuations in culture ponds between 15 mg/l in the afternoon and 3 mg/l or less at dawn are not uncommon (Boyd et al. 1979). At least one study has shown that food consumption and growth are reduced in pond-reared channel catfish exposed to a mean constant DO level of 3.5 mg/l or less (Carlson et al. 1980). In tanks, a minimum DO level of 3.0 mg/l has been recommended (Allen 1976).

The saturation DO level in water varies considerably in relation to temperature, salinity, and altitude. Saturation DO is reduced as any of the three variables increases. Andrews et al. (1973b) evaluated the effect of various levels of oxygen saturation (36%, 60%, and 100% saturation) on growth of channel catfish and found that food consumption and food conversion efficiency were considerably reduced at 36% DO saturation.

Different species of fishes often exhibit differences with respect to their tolerance to low DO. With respect to catfishes, Dunham et al. (1983) found that hybrids may have somewhat different ability to tolerate low DO than the parental stocks (Table 1). The apparently increased ability of blue catfish × channel catfish hybrids to better tolerate low DO than were channel catfish was hypothesized as being due to differences in the ability of hemoglobin to transport oxygen in the blood of the two types of fish. The authors recommended that hybrid catfish be used in culture to avoid oxygen depletion-reduced mortality.

Standard practice in the industry is to obtain DO data from ponds at or before sunrise during periods when problems are anticipated. Periods of cloudy, calm, and warm weather are often precursors to DO problems since photosynthesis is limited in such times. The late summer is a period when oxygen depletions are common since the majority of the fish have attained or are approaching their maximum size for the year, fish densities are high, feeding rates are at or near maximum, and the water temperature is at its annual peak. During the summer many producers have crews available to obtain DO readings throughout the night or data are collected with oxygen probes and processors that transmit the data to a central location on the fish farm. When the DO falls below a preselected minimum value, emergency aeration procedures are put into effect. Paddlewheel aerators that run off the power takeoffs of tractors or which are electrically operated are commonly utilized by catfish farmers and are effective in overcoming oxygen depletions (Boyd and Tucker 1979). Aeration of catfish ponds has been shown to increase production (Loyacano 1974), particularly when done on a nightly basis in heavily stocked ponds (Hollerman and Boyd 1980).

Certain management techniques can influence subsequent levels of DO in ponds. For example, it has been shown that application of the herbicide simazine at 13.4 kg/ha to the bottom of channel catfish ponds before flooding can result in extended periods of low DO and reduction in fish yields along with reduced food conversion ratios as compared with untreated ponds (Tucker and Boyd 1978).

Measurement of DO throughout the night or only near dawn is somewhat labor intensive unless expensive automated equipment is used for that purpose. Investi-

TABLE 1
Comparison of Mortality Percentages for Channel Catfish and Channel Catfish × Blue Catfish Hybrids Exposed to Dissolved Oxygen Levels of <1.0 mg/l

Environment	Channel catfish mortality (%)	Hybrid catfish mortality (%)
Ponds	50.5	7.5
Cages	87.5	51.0
Tanks	100.0	33.0

Adapted from Dunham, R. A., Smitherman, R. O., and Webber, C., *Prog. Fish Cult.*, 45, 55, 1983.

gators have attempted to relate DO changes to various limnological variables. Variations in DO have been shown to relate to solar radiation, the concentration of chlorophyll *a*, and percent oxygen saturation at dawn, as well as to Secchi disc visibility (Romaire and Boyd 1979). Based on findings of that nature, Boyd et al. (1978b) developed a computer programming technique to predict early morning DO levels in catfish ponds. Predictions can be obtained from input data on planktonic oxygen consumption, chemical oxygen demand, and temperature. In instances where plankton represent the major source of turbidity, Secchi disc visibility may be used to estimate chemical oxygen demand or oxygen consumption. Boyd et al. (1978b) presented a simple graphical technique for estimating dawn DO by obtaining a DO value at dusk and another 2 or 3 h later. Plotting the two points on graph paper and extrapolating appeared to provide a good approximation of what the DO level would be early the following morning.

C. AMMONIA

Chronic growth depression and mortality associated with elevated ammonia concentrations are more commonly reported in intensive culture systems such as tanks and raceways than in ponds. Yet, with increased stocking rates in ponds over the past few years, decreased growth rates associated with elevated levels of ammonia may be on the increase. Most studies of the effects of ammonia concentration on channel catfish growth and survival have been conducted under controlled laboratory conditions, but the results of those studies should be applicable to commercial pond facilities.

The toxicity of un-ionized and total ammonia on channel catfish has been evaluated in only a few instances (Knepp and Arkin 1973, Colt and Tchobanoglous 1976, Robinette 1976, Tomosso et al. 1980a). According to Robinette (1976), the level of un-ionized ammonia that will lead to 50% mortality in channel catfish over a 24-h period is 2.36 mg/l, though growth was impaired at un-ionized ammonia concentrations of 0.12 mg/l or higher. The concentration of un-ionized ammonia that causes growth reduction in channel catfish was challenged by Mitchell and Cech (1983) who determined that the residual chloramine level in charcoal-dechlorinated municipal water (which was used in the Robinette study) has a synergistic effect with ammonia. In any case, relatively low levels of un-ionized ammonia in culture water can lead to growth depression in catfish.

As the level of ammonia in water increases, growth decreases in a linear fashion (Colt and Tchobanoglous 1978). Such other water quality factors as pH, hardness, and temperature also affect ammonia toxicity (Tomosso et al. 1980a). Research has led to the recommendation that elevated pH should be avoided in culture systems to reduce the percentage of un-ionized ammonia (Tomosso et al. 1979b). In most pond and raceway situations, ammonia does not lead to significant problems for catfish culturists (Worsham 1975).

D. NITRITE

The condition known as brown blood disease in channel catfish has been found in association with exposure of the fish to elevated levels of environmental nitrite.

When exposed to nitrite, hemoglobin in the blood of the fish is converted to methemoglobin, which imparts a chocolate brown color and interferes with the oxygen-carrying capacity of the erythrocytes. The condition is known as methemoglobinemia. When that condition becomes sufficiently severe, death from asphyxiation may result. Early reports of methemoglobinemia came only from intensive culture systems wherein flushing rates were inadequate or biofilters were not functioning properly (inadequate levels of the bacterial genus *Nitrobacter*, which converts nitrite to nitrate). However, in recent years the increased stocking densities adopted by pond culturists, particularly in Mississippi, have led to the development of nitrite problems in ponds.

According to Tucker and Schwedler (1983), nitrite concentrations range from 0 to 4 mg/l or more in commercial catfish ponds. The concentration is usually lowest during the summer and is greatest during the other seasons of the year. While a major source of nitrite is from the conversion of ammonia to nitrite by bacteria of the genus *Nitrosomonas*, Hollerman and Boyd (1980) showed that denitrification of nitrate in the sediments of ponds can be a significant source of nitrite.

The concentration of methemoglobin in catfish blood can be as high as 89% of total hemoglobin according to Schwedler and Tucker (1983). Huey and Beitinger (1982) found that channel catfish have the ability to convert methemoglobin back to hemoglobin since the fish contain the enzyme methemoglobin reductase. Thus, if catfish which have developed a sublethal level of methemoglobinemia are transferred to water containing little or no nitrite, the condition can be reversed (Huey et al. 1980).

Half of the 40 g channel catfish fingerlings exposed to a nitrite concentration of 33.8 mg/l at 21 °C will die within 24 h, with levels of 28.8, 27.3, and 24.8 mg/l nitrite leading to death of 50% of a population at 48, 72, and 96 h (Konikoff 1975). While mortalities may not be as dramatic, much lower nitrite levels will eventually cause severe problems for catfish farmers over extended periods of time. For example, exposure to 5 mg/l nitrite for 5 h will result in a 42.5% level of methemoglobin in water that has a low level of chloride ion (Tomasso et al. 1979b). After being exposed for 24 h to the same nitrite level, the methemoglobin level may reach 90% (Huey et al. 1980). Even a nitrite level as low as 1 mg/l will lead to 35% methemoglobin formation within 24 h (Huey et al. 1980). While low levels of nitrite are not considered to be directly lethal, they are certainly stressful to the fish, and increased mortality has been reported during a 31-d study with catfish exposed to 3.7 mg/l nitrite. In the same study, growth was retarded when the fish were exposed to 1.6 mg/l (Colt et al. 1981).

In order to effectively treat methemoglobinemia, the percentage of methemoglobin in the blood should be known. A sample of about ten fish from a given pond will generally be adequate for making an estimate of the average methemoglobin level in the population (Tucker 1983). Chloride is an effective chemical that can be used in the prevention and treatment of methemoglobinemia (Tomasso et al. 1979a, 1979b, 1980b; Huey et al. 1980). The chloride ion may be added as calcium chloride, potassium chloride, or sodium chloride, though there is some difference in the degree of protection with respect to the various compounds. A ratio of 16 chloride ions to 1 nitrite ion has been shown to completely repress nitrite-induced methemoglobin formation (Tomasso et al 1979a). Sodium bicarbonate also appears to repress nitrite-induced methemoglobin formation (Huey et al. 1980), but sodium sulfate has no apparent protective effect.

E. SALINITY

The abundance of brackish water in many coastal, and even some inland areas, has led aquaculturists to experiment with the utilization of those waters for the culture of both marine and freshwater fish species. Inland brackish water is common in portions of the southwestern U.S. (e.g., parts of Texas and New Mexico), and some interest in developing those brackish water resources for fish culture, including the culture of catfish, has been demonstrated within recent years.

In pond studies, Perry and Avault (1970) found that blue, channel, and white catfish could all be successfully reared in salinities of 8 parts per thousand (ppt). The fish could tolerate salinities as high as 11 ppt for extended periods, but growth appeared to be reduced at the higher salinity. Both blue and channel catfish have been collected from natural waters containing salinity slightly in excess of 11 ppt (Perry 1967).

Channel catfish eggs can tolerate salinity up to 16 ppt (Allen and Avault 1970b), but the tolerance falls to 8 ppt at the time of hatching. Once yolk-sac absorption has occurred, salinity tolerance increases to 9 or 10 ppt and will further increase to 11 or 12 ppt by the time the fish are 5 to 6 months of age. Blue catfish appear to be slightly more tolerant to salinity than channel catfish (Allen and Avault 1971), but both species can survive for several days at 14 ppt, as can hybrids among blue, channel, and white catfish (Stickney and Simco 1971). Currently, there is not any emphasis being placed on the rearing of any catfish species in brackish water within the U.S.

Some pronounced benefit can be gained by exposing catfish to low salt levels. The use of sodium chloride concentrations of 2 ppt in catfish culture water has been shown to protect the fish against the protozoan parasite *Ichthyophthirius multifiliis* (Allen and Avault 1970a; Johnson 1976). A small amount of salt in the water may even enhance catfish growth (Lewis 1972).

F. pH CONTROL

Pond water which has high alkalinity and low hardness can develop high pH following fertilization (Boyd et al. 1978a). When the photosynthesis rate in ponds is high, pH readings as high as 10 have been observed (Boyd 1979). As a means of reducing pH in such situations, hardness may be increased or alkalinity reduced. To increase hardness, agricultural gypsum (calcium sulfate) may be added at the rate of 2000 mg/l without apparent harm to fish (McKee and Wolf 1963). Mandal and Boyd (1980) recommended an application of agricultural gypsum of four times the difference between total alkalinity and total hardness (in milligrams per liter). Alum (aluminum sulfate) has been used effectively to reduce pond alkalinity (Boyd et al. 1978a). About 1 mg/l of alum will reduce calcium carbonate alkalinity by 1 mg/l.

VII. REPRODUCTION AND GENETICS

Many commercial catfish farmers maintain their own brood stock and produce fingerlings. Attempts to improve the performance of channel catfish through genetic manipulation are a relatively recent activity, and major advances have yet to be achieved. In the commercial sector, selective breeding generally involves retaining the fish which demonstrate the best performance characteristics, usually based on which of the fish produced during a given year grow most rapidly, and maintaining them for eventual use as broodstock. While breeding fish selected on the basis of rapid growth appears to make good sense, little improvement appears to have been made over a number of generations of such selection. Selecting the largest fish from a population that has produced small, medium, and large fish has typically produced offspring that grow to small, medium, and large sizes. Rather than having been selected for growth potential, there is some reason to believe that the broodstock were selected for aggressiveness. If that is the case, the larger fish grew more rapidly because they were the ones that fed to satiation and deprived the less aggressive individuals an opportunity to obtain sufficient feed for the same growth rate. Catfish demonstrate compensatory growth; that is, if the more slowly growing fish are separated from the more rapidly growing ones, the smaller fish will quickly catch up since they will not be restricted from food by the previously dominant members of the population.

Genetic engineering has only recently been applied to fish. Transgenic catfish are being developed, and the potential for such fish to make an impact is great, assuming that resistance to the application of molecular biological techniques to cultured fishes can be overcome.

A. REPRODUCTION

Reproduction activities with channel catfish include selection of broodstock, establishment of proper conditions for spawning, hatching of the eggs, preparation of ponds, and stocking of the fry. General requirements for spawning of channel catfish include water temperature in the range of 21 to 29°C (Clemens and Sneed 1957; Brown 1977; Piper et al. 1982), with the optimum being about 27°C (Brown 1977). Spawning occurs in the spring, beginning during April in the southern extremes of the range (Busch 1985) and in June or July in more northern latitudes. Spawning activity is generally complete by August, though some spawning may occur during that month. It is possible to obtain off-season spawning by exposing channel catfish to an artificial winter by cooling the water to 17°C or below for a period of time (actual minimum exposure time has not been determined) and then gradually warming the water to within the normal spawning temperature range (Brauhn 1971).

1. Broodstock Maintenance, Selection, and Stocking

Channel catfish broodstock (Figure 6) are often maintained in one or more communal ponds at densities as low as 12 fish per hectare throughout most of the year (Toole 1951; Snow et al. 1964) and are stocked prior to spawning at densities

FIGURE 6. Broodfish are usually not more than a few kilograms in weight, making them relatively easy to handle.

FIGURE 7. Catfish spawning pens. The pond at left has yet to be filled and the nests (wooden boxes with concrete blocks to keep them submerged) are visible.

of about 375 fish per hectare (Nelson 1960). Typically, fish from 0.9 to 4.5 kg are utilized as broodstock (Martin 1967) since larger fish are difficult to handle. Under some conditions, catfish may become sexually mature during their second year of life, but most are in their third year before first spawning.

Maintenance of broodstock between spawning seasons and preparing the fish for spawning involves providing them with ample food of the proper type. Currently, complete catfish feeds appear to meet the requirements of broodstock, but in the past, when complete feeds were unavailable, a variety of live feeds along with fresh and frozen meat or fish were provided (reviewed by Busch 1985).

Channel catfish are most commonly spawned in open ponds, though pen spawning has also been widely applied, particularly when the culturist desires to become involved in a relatively refined level of selective breeding. Aquarium spawning presents a third alternative, but that has largely been used by researchers and is not utilized to any extent by commercial or enhancement hatcheries. Raceways and tanks have also been used for spawning (Carter and Thomas 1977).

In the pond spawning technique, adult fish are allowed to select their own mates from among the population that has been stocked within the pond. Of course, there is some selective breeding involved, since the fish that were retained for broodstock will have been selected on some basis from among fish produced on the facility or purchased from some other source. When the pen technique is employed (Figure 7), the culturist selects each male and female that will be mated during the spawning season and places them in an enclosure (pen) within a pond, thereby imposing an increased level of selection on the breeding process. Similarly, in the aquarium spawning technique, the culturist places a pair of adult fish in a suitably sized aquarium and injects hormones to induce ovulation and spawning (Clemens and Sneed 1962).

The augmentation or replacement of broodstock implies that a selection process is occurring. In most instances, fish selected for retention and use as broodstock are chosen because they exhibit one or more characteristics viewed as desirable by the culturist. It has been demonstrated with respect to channel catfish that if only the largest fish are chosen from among a group of fish spawned during the same season, the group will be dominated by, or even composed completely of, males (Beaver et al. 1966; Brooks et al. 1982). The disparity will occur because male catfish grow more rapidly than females. Selection of broodstock should involve individuals sufficiently large to be sexed with a reasonable degree of accuracy. The sexes can be distinguished fairly easily when the fish approach spawning condition (Clemens and Sneed 1957), but sexing subadults is more difficult. A reliable technique which involves microscopic examination of the vent region was developed by Norton et al. (1976). Once the fish reach maturity the males develop a genital papilla behind the anus. As the spawning season approaches, males develop a heavy muscular pad on the back of the head, and their heads are wider than their bodies. Females develop distended abdomens as the ovaries become enlarged. The vent of females becomes reddened and swollen when the females are ready to spawn (Busch 1985).

Studies have been conducted to determine the best stocking ratio of male to females in spawning populations (El-Ibiary et al. 1977; Bondari 1983b). There appears to be no difference in the efficiency of reproduction when males to female ratios of 1:1, 1:2, 1:3, or 1:4 are employed (Bondari 1983b). That conclusion was based on spawn weight, egg weight, and various hatching traits. Ratios at which brood catfish are stocked in ponds vary from one culturist to another. Commonly used ratios of males to females in ponds are 1:2, 2:3, and 3:4. When the pen or aquarium spawning method is used, broodfish are stocked in a 1:1 ratio with a pair assigned to each spawning chamber.

2. Spawning and Egg Hatching

In pond and pen spawning, some type of spawning container must be provided. The spawning container should be large enough for a pair of broodfish to enter and lay their eggs. Various items and materials have been used for catfish spawning, with milk cans being among the most popular, though now becoming difficult to obtain. Other types of containers have included nail kegs, drain tiles, wooden boxes, earthen crocks, and grease cans (Clapp 1929; Lenz 1947; Toole 1951; Crawford 1957; Stickney 1979). For open pond spawning, containers are spaced at intervals around the pond, usually with the open end facing the middle of the pond. The containers are normally placed in water from 0.30 to 0.75 m deep, since placing them in deeper water makes inspection more difficult (Busch 1985).

Females ranging from 0.45 to 1.8 kg can be expected to produce eggs numbering about 8800/kg of body weight, while in larger fish the number of eggs drops to approximately 6600/kg (Clemens and Sneed 1957). Smitherman et al. (1984) reported fecundities ranging from less than 3000 to an excess of 12,000 eggs per kilogram of female in fish from two distinct strains.

The male catfish will clean out the spawning container and attempt to entice a female into the container. Once a female has completed spawning and the male has fertilized the eggs, he will chase the female from the nest and begin the process of guarding the eggs. The eggs are laid in an adhesive mass (Figure 8). The male catfish will position himself over the egg mass and fan it with his fins to maintain a flow of oxygen-rich water.

Most culturists inspect the spawning containers at intervals of from 24 to 96 h and move any egg masses found to indoor hatcheries. This allows the culturist to separate the two generations of fish before they become intermingled within the spawning pond and provide the culturist with an easy way of determining the number of fry that are produced. If the fish are allowed to spawn in the pond and no tally of spawning success is maintained, the culturist will not know how many fry were produced until the fish are harvested for restocking into nursery or growout ponds.

Inspection of spawning containers typically involves lifting them to the water surface and slowly pouring out the water. This procedure reduces the chances that fish which are in the act of spawning will abandon the nest, perhaps never to return. The culturist who sticks a hand into the nest to feel for an egg mass stands the chance of being bitten by an aggressive and protective male catfish. Such bites, while not dangerous, can be painful and startling.

In the hatchery, egg masses may be broken into two or more pieces, depending upon their size, and are placed in hatching troughs that receive well-oxygenated

FIGURE 8.   A channel catfish egg mass.

flowing water. Paddlewheels are typically used in conjunction with hatching troughs. The paddles turn slowly and oscillate the water, simulating the action of the male catfish, and assisting in aeration of the egg mass. The first paddlewheel hatching trough was powered with a water wheel (Clapp 1929). Modern paddles are operated with electric motors and gear reducers which allow the axle on which the paddles are mounted to turn at approximately 30 rpm (Figure 9). Flow-through troughs without paddlewheels can be used successfully, and oscillating lawn-sprinkler heads mounted upside down over hatching troughs have even been employed. Jar hatching has been used successfully (Suppes 1972), but has not been adopted by fish culturists to any extent.

The time required for hatching is heavily dependent on ambient water temperature. Within the normal spawning temperature range of the species, channel catfish eggs can be expected to hatch in from 5 to 10 d (Toole 1951).

Aquarium spawning is normally accomplished by injecting the females with hormones. Hormone injections are also often used in conjunction with pen spawning, though noninjected catfish will frequently spawn in pens. The injections will not work unless the fish are fully mature and egg development is well advanced. Hormones will sometime stimulate ovulation, but they do not induce egg development.

Popular hormones for spawning catfish include fish pituitaries (usually carp, but also various other species) and human chorionic gonadotropin (HCG). Sneed and Clemens (1960) tried pituitary injections from various sources to induce ovulation in channel catfish. Their recommendation was the provision of 4.4 mg of pituitary per kilogram of female catfish by intraperitoneal injection every 24 h until spawning

FIGURE 9. Hatching troughs containing egg masses in hardware cloth baskets. Paddlewheels keep aerated water flowing through the egg masses.

occurred (usually within 10 d). Spawning typically occurred within 16 to 24 h after the last injection. HGC provided at dosages of 1760 IU/kg of female in single or multiple injections were recommended by Sneed and Clemens (1959). Other authors have had success with HCG at dosages as low as 300 IU/kg of female (Nelson 1960).

Catfish egg development has been discussed by various authors, e.g., Clemens and Sneed (1957), and Saksena et al. (1961). Egg diameter is positively correlated with fry growth during the first 30 d following hatching, but not thereafter (Reagan and Conley 1977). It appears as though other factors (genetic and/or environmental) exert more influence on fish growth after 30 d.

Egg masses are commonly held in hardware cloth baskets within spawning troughs. When the fry hatch they fall through the hardware cloth and form aggregates on the bottom. First feeding follows yolk-sac absorption, which generally occurs about a week after hatching (Figure 10). The fish will swim up to the

FIGURE 10.   A group of channel catfish sac fry. Within another 1 to 3 d these fish will begin to accept prepared feed.

surface at first feeding, a clear signal that feed should be provided. While there may be adequate supplies of natural feed in fry ponds, it is always a good idea to have the fish adapted to prepared feed before moving them out of the hatchery. Several days to 2 weeks is sufficient to ensure that the fish are eating and growing well. Trout starter rations are usually provided, though catfish fry formulations have been developed (Winfree and Stickney 1984a, 1984b).

Hormones have been fed to channel catfish fry in an attempt to produce sex-reversed populations. The results have been distinctly different than for other species, such as tilapia. For example, if 17-α-ethynyltestosterone, a hormone that will produce all-male populations of *Tilapia aurea* (Guerrero 1975), is fed to sexually undifferentiated channel catfish over a 21-d period following yolk-sac absorption, all-female populations can be produced (Goudie et al. 1983).

3. Fry Stocking

Catfish ponds are prepared in the spring to receive fry. Preparation usually includes fertilization to promote plankton blooms and may involve an application of diesel fuel or a mixture of cottonseed oil and diesel fuel (1:4) to the pond surface to control aquatic insects (Bryan and Allen 1969). The oil floats on the surface of the pond and impairs respiration of insects that obtain oxygen at the air-water interface. Insecticides have also been used to eliminate insects (Busch 1985).

Fertilization not only promotes the growth of phytoplankton and zooplankton, which provide the fry catfish with a natural food base, the presence of plankton blooms limits the penetration of light into ponds and helps reduce the development of undesirable rooted aquatic plants. Ponds often respond well to fertilization with

50 kg/ha of 16-20-4 or 16-20-0 (percentages of nitrogen, phosphorus, and potassium or N-P-K) fertilizer applied every 10 to 14 d until a Secchi disc reading of 30 cm is obtained (Stickney 1979). Other recommendations for catfish fry-pond fertilization include 1.1 to 2.2 kg/ha of 10-34-0 or 13-38-0 liquid fertilizer applied at intervals of 8 to 14 d (Tucker 1984) and a combination of 112 to 222 kg/ha of 20-20-0 inorganic fertilizer, 224 to 561 kg/ha of organic compounds such as plant meals and manure, and several bales of hay per hectare (Dupree and Huner 1984).

Once the ponds have been properly prepared and the fry have reached the stage at which the fish culturist desires to move them, they are typically stocked at densities of from 125,000 to 625,000/ha (Martin 1967). Catfish are usually held at high densities until they reach fingerling size after which they are transferred to production ponds, usually during the growing season subsequent to that in which they are hatched. Dupree and Huner (1984) recommended an initial feeding rate of 11 to 28 kg/ha for fry ponds. While that level provides feed in excess of the needs of the fish, it also ensures that the fish will have feed available if needed. Feed that is not ingested by the fish will provide additional nutrients for pond fertilization.

B. GENETICS

The genetics of channel catfish has received less attention than other aspects of its culture, though the amount of information available has increased significantly in the past decade. A comprehensive review of catfish genetics was produced by Bondari (1990).

Many catfish producers select from among their most rapidly growing fish for broodstock replacement since they assume that growth rate is genetically controlled. Yet, in virtually every sibling group of channel catfish, there are individuals that grow rapidly, those that are intermediate in growth, and those that grow slowly. The relationship does not appear to change significantly even after several generations of selection. There is some reason to believe that selection for rapid growth may, in fact, be selection for aggressiveness, since fish that are able to compete effectively for food are generally those that grow most rapidly. The theory is supported by the fact that when the larger fish within a population are removed, some of the smaller ones will exhibit compensatory growth by becoming the most aggressive feeders within the new population. However, Bondari and co-workers (1985) reported that one generation of divergent selection for rapid growth based on fingerling selection created genetic differences among selected lines of catfish. Thus, broodfish might best be selected early in the life history of the fish rather than waiting until the fish are of harvestable size.

As with other organisms, phenotypic traits exhibited by catfish may be under genetic control, environmental control, or in many instances, some combination of the two. The genetics and selective breeding of channel catfish has been reviewed in detail by Smitherman and Dunham (1985).

Various traits have been evaluated in channel catfish with the intent being to apply selective breeding to obtain improvement (El-Ibiary et al. 1976). While size at a given age is most often utilized as the characteristic upon which selection is based, it has been demonstrated that body size is not strictly under genetic control

(Bondari 1983c). Bondari (1983d) indicated that a combination of within and between family selections should be effective in improving the genetic character-istics with respect to body weight and total length. He also indicated that female fish which show early growth inferiority should not be employed as broodstock. In later studies, Bondari (1986) applied divergent selection and was able to demonstrate that heavier, more food-efficient fish could be produced after three generations of selection.

It is likely that most of the catfish broodstock presently being utilized in the U.S. represent a relatively small gene pool which can be traced back to governmental hatcheries. For that reason, comparison of groups of progeny obtained from different hatcheries may not reveal distinct genetic differences since all the fish may have come from a common original source. To overcome this problem, investiga-tors have attempted to obtain broodstock from isolated natural populations or have attempted to improve performance through crossbreeding various catfish species native to the U.S. Such attempts at genetic improvement have not always been successful. For example, Broussard and Stickney (1981) compared two domestic and two wild strains of channel catfish and found that higher growth rates occurred in the domestic strains, though no significant differences in survival were obtained among the strains.

Geographic region of origin, as well as species, may play a role in catfish performance. In at least one study, channel catfish from northern regions outper-formed their southern conspecifics during winter, though white catfish outper-formed not only channel, but also blue × catfish (Dunham and Smitherman 1981). At the fingerling stage, hybrids from blue × channel catfish crosses raised in ponds have been shown to grow more rapidly than either parent (Giudice 1966; Yant et al. 1975; Dunham et al. 1987a; Dunham et al. 1990). Giudice (1966) and Yant et al. (1975) found that the hybrids were easier to seine, more uniform in size, and yielded a better dressout percentage. During the early part of the catfish life history, stocking density seems to play an important role. Dunham et al. (1990) found that channel catfish fry grew more rapidly at low densities than channel catfish × blue catfish hybrids, but that the hybrids grew faster at high densities. Density did not have an influence in comparative studies with fingerlings. While fingerling hybrids grow faster in ponds, channel catfish growth is more rapid in cages (Dunham et al. 1990).

Spawning success of hybrid blue × channel catfish has been demonstrated to be much poorer than that of either parental group (Tave and Smitherman 1982). Further, blue catfish female × channel catfish male offspring cannot be produced as easily as the reciprocal cross. The blue male × channel catfish female hybrid has been shown to grow more rapidly than the more difficult cross and has other desirable characteristics from a culture standpoint (Brooks et al. 1982). The heritability of various traits in channel catfish-blue catfish hybrids appears to be most heavily influenced by the male of the species (Dunham et al. 1982). Among the male-dominated traits are swimbladder shape, anal fin-ray number, external appearance, growth rate, morphometric uniformity, and susceptibility to capture by seining.

Induction of polyploidy in catfish has been utilized in attempts to improve performance. Wolters et al. (1981) induced triploidy by cold-shocking fertilized eggs at 5°C for 1 h, beginning 5 min after fertilization. Triploid fish were significantly heavier than diploids at 8 months of age and older (Wolters et al. 1982). Triploids also had better food conversion efficiencies than diploids. There were some abnormalities noted in gonad histology among both sexes of triploids as compared with diploid fish.

Albinism has been common in some commercial catfish hatcheries and is rare in others. The trait is generally thought to be inherited in the same manner that has been determined for other species; that is, albinism is the result of a homozygous recessive gene. Environment may also be an important factor since the incidence of albinism in channel catfish has been shown to increase with exposure of either eggs or adult fish to various heavy metals (Westerman and Birge 1978). Metals that have led to increased albinism include arsenic, cadmium, copper, mercury, selenium, and zinc, all at concentrations ranging from 0.5 to 250 mg/l.

Various skeletal deformities have been reported from channel catfish (Smitherman et al. 1978; Broussard 1979; Green et al. 1979; Bondari 1983a). They include fish which are tailless, partially tailed, and triple tailed and those which have crooked backs, are eyeless, have malformed mouths, or are pugheaded. Siamese twins (Smitherman et al. 1978) and fish with rayed adipose fins (Tave et al. 1990) have also been reported .

Fish seem to be a good model for the creation of transgenic organisms — those in which novel genes are introduced into the recipient organism's genome (Maclean and Penman 1990). Transgenic catfish containing the metallothionein-human growth hormone gene were developed by Dunham et al. (1987b). Special facilities have been constructed at Auburn University to provide pond space for the growout of such transgenic species. Significant and rapid changes in the characteristics of catfish may be produced through the application of genetic engineering. Whether transgenic fish will be approved for use in commercial fish culture remains to be seen.

VIII. FEEDING AND NUTRITION

Feed represents the highest variable cost associated with commercial catfish farming, accounting for as much as 55% of annual operating expenses (Joint Subcommittee on Aquaculture 1983). Large farms may require feed inputs of several tons per day, particularly late in the growing season. Modern catfish feeds are comprised of relatively few ingredients, though price can vary considerably on a temporal basis because of fluctuations in the price of the feedstuffs used in the feeds and in the costs of energy associated with producing those feeds.

Feeding practices for channel catfish have been fairly well-standardized, and the nutritional requirements of catfish have been established in some detail (Stickney 1979, National Research Council 1973, 1977, 1983; Stickney and Lovell 1977;

Robinson and Wilson 1985; Robinson 1989; Lovell 1988, 1989b). Research continues in both areas, however, since even small improvements in feed quality or in the manner in which feed is presented can have considerable impact on the economics of production.

A. TYPES OF CATFISH FEED

Prepared feeds for catfish are of two basic types. A typical feed may be prepared by pressure pelleting or by passage of the mixed ingredients through an extruder. The major difference between the two techniques involves the amount of pressure and heat to which the feed mixture is exposed. Pressure pellets are exposed to far less heat and pressure than extruded products; thus, the former type tends to retain heat-labile vitamins and other ingredients much better than the latter.

Pressure-pelleted feeds sink when introduced into water, while extruded pellets may either sink or float depending upon the type of process that is used in their manufacture. When certain formulations exit the extruder, the carbohydrates in the diet become almost instantaneously gelatinized, resulting in the entrapment of air within the pellets causing them to expand. The result is a pellet with a density less than that of water, so the feed floats. In some cases, pellets will float virtually intact for up to 24 h.

Because the amount of energy involved in the manufacture of extruded pellets is significantly greater than that employed in the pressure-pelleting process, extruded diets tend to be the more expensive of the two. Another difference in cost relates to the fact that extruded diets must be overfortified with heat-labile ingredients if the product is to meet the minimum requirements of the fish.

The major advantage of utilizing expanded diets is the flotation which enables the fish farmer to observe feeding activity. If the fish go off feed it may be a sign of water-quality depletion or disease. Most commercial catfish farmers in Mississippi and elsewhere utilize floating feed.

Both pressure-pelleted and expanded feeds can be ground to small sizes for use in feeding fry and the early fingerling stages. It is generally acknowledged that the formulation utilized for young fish should not be the same as that utilized for older ones. In general, young, rapidly growing fish require higher levels of nutrients and energy as compared with older fish. Diets specifically formulated for fry and early fingerling channel catfish have been formulated (Winfree and Stickney 1984a, 1984b), though many producers utilize trout starter-feeds or finely ground catfish feed for their young fish.

B. FEEDING REGIMES

Within the optimum temperature range of the species, channel catfish <1.5 g should be fed at 3-h intervals throughout the day when maintained in the hatchery under conditions where natural food is not readily available. Once the fish are larger than 1.5 g, feeding frequency can be reduced to four times per day (Murai and Andrews 1975). Robinson and Wilson (1985) recommended feeding fry with finely ground feed that is almost a powder (400 to 600 µm). When the fish are 1.5 cm, they can be offered #1 crumbles (600 to 900 µm). Crumble size is increased as the fish grow; the maximum size pellet being used today is about 6.4 mm in diameter. A feeding rate as high as 50% of body weight daily is typically used for first-feeding fry. Wasted feed should be siphoned from the troughs daily to avoid water-quality degradation and the development of bacteria and molds. Murai and Andrews (1976) recommended reducing the feeding rate to 10% of body weight daily when the fish reach 0.25 g and further reducing it to 4% when the fish reach 4 g in intensive culture systems.

Most commercial producers feed pond-reared catfish *ad libitum* with floating pellets. By monitoring food consumption over periods of about 30 min, the amount the fish are prepared to consume before reaching satiation can be determined. That typically turns out to be about 3 to 4% of body weight daily, though studies with demand feeders have shown that there is a good deal of daily variation in feeding rate, at least of fish of about 400 g (Tackett et al. 1988). Feeding to satiation twice daily has been shown to produce the best growth and food conversion as compared with higher and lower feeding frequencies (Andrews and Page 1975). *Ad libitum* feeding will produce better growth of channel catfish than rates equivalent to 50 and 75% of the *ad libitum* rate (Andrews, 1979).

Swingle (1958) recommended that no more than 34 kg/ha of feed should be offered to catfish in ponds daily, regardless of how much the fish might eat. That recommendation, which was followed by many catfish farmers for a number of years, was based on concern about water-quality degradation when more than the recommended amount of feed was used. With the development of feeds that are more water-stable and better meet the nutritional requirements of the fish, along with improved management strategies, daily feeding rates began to increase and may exceed 100 kg/ha late in the growing season (Busch 1985). In a study comparing water quality at feeding rates that reached 34, 56, and 78 kg/ha/d, no DO problems occurred at the lowest feeding rate, but dawn DO was typically below 2.0 mg/l in ponds fed at the highest rate (Tucker et al. 1979). Mortality was higher in the ponds fed at the highest rate, and net profit was considerably reduced as compared with the low and intermediate rates of feeding. Best profit was obtained from ponds fed a maximum of 56 kg/ha/d.

As environmental temperature increases, so does the metabolic rate of catfish, at least up to a point. Widely used recommendations for feeding rate as a function of temperature were presented by Bardach et al. (1972). They recommended no feeding when the temperature is less than $7°C$, feeding 1% at 7 to $16°C$, 2% at 16 to $21°C$, and 3% at 21 to $32°C$, when the temperature is in the best range for catfish growth. Feeding rate should be reduced to 1% when the temperature exceeds $32°C$. In climates where there are no extended periods of ice cover during winter and at least some warm days occur, maintenance of prewinter body weight, and even some growth, can be obtained by feeding catfish on warm days or every other day (Lovell and Sirikul 1974). A winter feeding program based on maximum daily air temperature that can be applied to any geographical location was developed by Tackett et al. (1987).

Studies with demand feeders have shown that spring onset of feeding did not occur until the temperature reached $12°C$ and that fall feeding activity stopped when the temperature fell below about 22°C (Randolph and Clemens 1976b). When the fish do not have to trigger a demand feeder, they will ingest feed at considerably lower temperatures.

Some reduction in winter feed-costs may be realized by employing a lower protein feed than the formulation used during the growing season. In some cases, a 25% protein diet may perform as well as one containing 35% protein during winter (Robinette et al. 1982).

Distinct feeding behavior patterns have been observed in catfish ponds. During much of the time, the fish occupy home areas, but they make daily trips to the area where feed is made available, taking specific routes from the home area to the feeding areas. Home areas tend to be in shallow water during the spring and fall and in deeper water during the summer and winter (Randolph and Clemens 1976a). When demand feeders are used, the larger fish feed first (Robinette et al. 1982). A similar pattern may occur when feed is provided once or twice daily by the farmer, particularly if the fish are fed in a restricted portion of the pond.

Many fish farmers feed only 6 d a week, and for some, a 5-d-a-week feeding regimen is employed. Studies have demonstrated that even deprivation of feed for a single day will cause the fish to alter their feed intake (Randolph and Clemens 1978). For each day off feed, it will take an additional day for a fish to return to its original feed intake level. In addition, each day of feed deprivation requires 2 d for recovery of the original growth rate.

C. NUTRITIONAL REQUIREMENTS

Animal feeds are formulated to contain three energy-bearing components: protein, carbohydrate, and lipid. In addition, most feeds contain added vitamins and minerals. Early catfish feeds were supplemental in nature. When catfish were stocked at low densities, there were sufficient quantities of natural food organisms in ponds to meet some of the nutritional requirements of the fish. In those instances, prepared feeds provided additional protein and energy, but little else. As culture intensity increased, natural food items became increasingly rare, and nutritional deficiency problems began to develop. Vitamin deficiencies, such as lordosis and scoliosis in fish deficient in vitamin C (Figure 11), became increasingly common. Complete feeds, which meet all the known nutritional requirements of the fish, were the solution to such problems.

1. Protein

The most expensive component of a catfish diet is protein. It must be provided not only in the proper quantity, but must contain the ten essential amino acids required by catfish (arginine, histidine, isoleucine, leucine, lysine, methionine, phenylalanine, threonine, tryptophan, and valine) in the proper amounts (National Research Council 1973). Catfish have a particularly high requirement for lysine that can most effectively be met with the incorporation of animal protein into the diet (National Research Council 1983). In addition, protein must be properly balanced with energy if it is to be effectively utilized. A major goal of the fish nutritionist is to maximize the use of dietary protein in the development of new animal tissue

FIGURE 11. Channel catfish fingerlings exhibiting signs of vitamin C deficiency. Externally, the vertebral column of the fish may be deformed (lower fish), while internally actual fractures and hemorrhaging into the tissues may occur (upper fish).

(growth) and minimize its use as an energy source by providing the proper levels of carbohydrate and lipid.

Modern complete channel catfish rations for fingerlings in growout facilities generally contain about 32% crude protein. Fry and early fingerlings require higher levels of protein, perhaps 50% at first feeding. The protein level can be reduced over the first several weeks of feeding until the 32% level is reached (Winfree and Stickney 1984a). Dietary protein should be supplied, at least in part, by some type of animal by-product. Fish meal, meat and bone meal, and poultry by-product meal are examples of ingredients that have been effectively utilized in catfish feeds. Fish meal is the most widely used animal protein. More exotic animal proteins, such as shrimp by-product meal (Robinette and Dearing 1978), could also be used if available in sufficient quantity at a competitive price. In general, the contribution of animal protein is kept as low as possible (usually under 10%), but is sufficiently high to ensure that the level of the first limiting amino acid (lysine) is met (Wilson and Robinson 1982). From a practical standpoint, meeting the lysine requirement will bring the other essential amino acids into the formulation at or above their minimum required levels.

Most of the protein in catfish diets come from various types of plants. The most widely utilized plant proteins are soybean meal and corn meal. Wheat, at a level of 2% or so, is sometimes utilized as a binder and provides some additional crude protein. Other plant proteins which have received some use, particularly during periods when their prices are sufficiently low to bring them into a feed formulation on a least-cost basis, are peanut meal (commonly used in Mississippi) and cottonseed meal.

Cottonseed meal typically contains a toxic chemical known as gossypol. Growth of channel catfish may be inhibited by dietary levels of cottonseed meal above 17.4%, or by diets containing a free gossypol level above 0.09% (Dorsa et al. 1982). Glandless cottonseed meal (reviewed by Hendricks and Bailey 1989) can often be used in fish diets since the gossypol problem and other disadvantages are reduced.

The ratio of protein to energy in fish diets is an important one since the farmer wants to provide the proper amount of energy for efficient growth while sparing protein for growth. Metabolic energy should come, to the extent possible, from nonprotein sources which are less expensive (carbohydrate and fat). At the same time, the total protein in the diet must meet the amino acid requirements of the fish, so a balancing act takes place with respect to feed formulation. Today, computers make the job easier once the requirements have been determined.

Optimum protein to energy ratio can be expected to vary with protein source and the percentage of protein in the diet. Reis et al. (1989) evaluated diets with protein levels of 26, 31, 35, and 39% protein, or 91, 107, 120, and 127 mg of protein per Kilocalorie of digestible energy. Food conversion efficiency was best in fish fed the 35% diet. The study concluded that the optimum protein to energy ratio in a diet containing soybean meal, fish meal, and corn meal as protein sources was 120 g of protein per Kilocalorie of digestible energy.

2. Carbohydrate

Energy is obtained from carbohydrates, but sugars and starches are stored only in small concentrations within the bodies of fishes. Relatively high levels of carbohydrate (up to 40%), often primarily in the form of starch, can be found in channel catfish feeds. The presence of starch does not seem to affect protein digestibility, but dietary cellulose does reduce starch digestion (Smith and Lovell 1972). Feeds containing high levels of fiber do not appear to provide added nutritional benefit to catfish (Leary and Lovell 1975), and nutritionists generally consider fiber to have no food value for fish (Stickney and Lovell 1977, National Research Council 1983). Cellulase enzyme activity (apparently acquired from intestinal microflora) has been found in channel catfish (Stickney and Shumway 1974), but the amount of cellulose digested during passage of food through the gut can be considered negligible.

Carbohydrates can be utilized in catfish diets to spare protein. That is, carbohydrate can be utilized for various metabolic processes as an energy source, allowing dietary protein to be deposited into new tissues (growth). Lipid can also be used in protein sparing.

Many of the dietary ingredients utilized as plant protein sources also contain high levels of carbohydrate. Examples are soybeans, corn, and wheat. Thus, when plant meals are added to fish diets, a significant proportion of the energy they provide is in the form of carbohydrate, largely starch. Nutritionists generally formulate to balance protein and amino acid levels, so the carbohydrate present in dietary ingredients is not given much consideration.

3. Lipid

Dietary fats fall into several biochemical categories that can be collectively referred to as lipids. On a per unit weight basis, there is approximately twice the number of calories in a lipid than are found in either a protein or carbohydrate. Thus, lipid is an excellent source of dietary energy. In addition, many dietary lipid sources are relatively inexpensive, readily digestible, and may contain fatty acids essential for proper growth and metabolism.

When practical dietary lipids such as beef tallow, oilseed oils, or fish oil are fed to channel catfish at environmental temperatures below about 25°C, there does not appear to be much effect of lipid quality or quantity on growth (Stickney and Andrews 1971, Gatlin and Stickney 1982). However, at temperatures within the optimum growth range of the species, diets containing beef tallow and fish oil outperformed such oilseed products as safflower and soybean oils (Stickney and Andrews 1972) and dietary lipid quantity became more important (Yingst and Stickney 1980).

Stickney and Hardy (1989) reviewed the literature on fish lipid nutrition and concluded that channel catfish may require both linoleic acid family (n-6) and linolenic acid family (n-3) fatty acids. Satoh et al. (1989) fed purified fatty acids to channel catfish and were unable to determine whether linoleic or linolenic family fatty acids were essential. They did conclude that growth was enhanced when the fish received n-3 fatty acids. High dietary levels of linolenic acid may actually lead to poor growth (Stickney and Andrews 1972; Stickney et al. 1983).

4. Vitamins

Included among the vitamins required by channel catfish are both water-soluble (B complex and vitamin C) and fat-soluble ones (vitamins A, D, E, and K). Vitamin premixes are routinely added to practical catfish feeds, particularly when the feed is formulated for fish stocked into situations where natural food is unavailable or present in low quantity. The levels of each vitamin required for proper growth and health have been outlined and widely reported (National Research Council 1973, 1977, 1983; Stickney and Lovell 1977; Robinson and Wilson 1985).

The method of diet preparation can have an influence on the amount of certain vitamins present in the finished feed. Vitamin C, for example, is heat-labile (though heat-resistant forms of that vitamin are now commercially available). Overfortification with that vitamin has been common, with greater overfortification being necessary when extruded pellets are produced than in the pressure-pelleting method because extrusion exposes the pellets to heat for a longer period of time. Robinson (1984) provided recommendations for vitamin additions to extrusion-processed feeds for channel catfish. Doses of vitamin C well above those normally present in finished feeds have been shown to impart some disease resistance to channel catfish at 23°C (Durve and Lovell 1982).

Some vitamins, such as vitamin D, K, B_{12}, and choline are apparently required at low levels or their requirements are impacted by other chemicals or organisms. While it has been difficult to demonstrate a requirement for vitamin D, deficiency signs including reduced weight gain, poor food-conversion efficiency, and an effect on bone mineralization have been observed when extremely low levels of the vitamin are fed (Lovell and Li 1978, Andrews et al. 1980). Deficiency signs related to vitamin K have not been observed (Murai and Andrews 1977).

Microorganisms in the intestinal tract of channel catfish are able to synthesize vitamin B_{12}, which is then absorbed (Limsuwan and Lovell 1981), so a dietary requirement has not been established. Dietary methionine appears to spare choline to some extent (Wilson and Poe 1988), but a level of 400 mg/kg of diet is still recommended for choline.

5. Minerals

Most of the minerals required by channel catfish can be obtained from the water and are present in the natural feed ingredients used in practical diet formulation. Supplementation of certain minerals is common, however. In research diets of a semipurified or purified nature, mineral supplements are added that contain the major minerals (calcium, phosphorus, sulfur, sodium, potassium, and magnesium) and various trace minerals (iron, copper, iodine, manganese, cobalt, zinc, molybdenum, selenium, and fluorine).

Calcium and phosphorus are widely used as supplements to practical rations. They are routinely added to ensure that the feed meets the recommended daily minimum intake levels for those minerals (Andrews et al. 1973a; Lovell 1978).

The availability of minerals in feed can vary considerably depending on the form in which the mineral is present. For example, phosphorus is much more readily available from monosodium or monocalcium phosphate than from dicalcium phosphate, fish meal, soybean meal, or cereal products (Murai and Andrews 1979). Certain other substances in feeds can affect the availability of minerals. Phytic acid (which occurs in a number of feed ingredients), for example, reduces the availability of zinc. Supplementation of catfish diets with zinc at 150 to 200 mg/kg of diet have been recommended (Gatlin and Wilson 1984; Gatlin and Phillips 1989).

Sodium chloride does not seem to affect catfish growth when fed at levels of up to 2% of the diet (Murai and Andrews 1979). The elaboration of deficiency signs for other minerals may also be difficult to elicit. For example, no deficiency signs were discerned from channel catfish fingerlings reared on diets containing various levels of copper from 0 to 32 mg/kg of feed when the ration contained a background level of 1.5 mg/kg which could not be eliminated (Murai et al. 1981). For some minerals, background levels in virtually all natural waters appear to be sufficient to meet the requirements of the fish. For example, calcium deficiency signs have only been induced in catfish reared in well water that contained virtually no calcium (Robinson et al. 1986). Robinson (1989) summarized the mineral deficiency signs that have been observed in channel catfish.

6. Meeting the Requirements

Catfish nutritional requirements can be met with relatively simple feeds. Most contain no more than four major ingredients along with supplemented vitamins, minerals, and sometimes lipid. Fat is often sprayed on the finished feed. Examples

TABLE 2
Channel Catfish Feed Nutrient Level
Recommendations

Nutrient	Recommended level
Protein (%)	32
Lysine (% of protein)	5.1
Methionine + cystine (% of protein)	2.3
Energy (kcal/g protein)	8–10
Carbohydrate (%)	25–35
Lipid (%)	6 or less
Vitamins	
Thiamin (mg/kg diet)	11
Riboflavin (mg/kg diet)	13.2
Pantothenic acid (mg/kg diet)	35
Nicotinic acid (mg/kg diet)	88
Folic acid (mg/kg diet)	2.2
B-12 (mg/kg diet)	0.01
Choline (mg/kg diet)	275
Vitamin C (mg/kg diet)	375.6
A (IU)	4400
D_3 (IU)	2200
E (IU)	66
K (mg/kg diet)	4.4
Minerals	
Phosphorus, available (%)	0.5
Magnesium (%)	0.05
Zinc (mg/kg diet)	200
Selenium (mg/kg diet)	0.1
Manganese (mg/kg diet)	25
Iodine (mg/kg diet)	2.4
Iron (mg/kg diet)	30
Copper (mg/kg diet)	5
Cobalt (mg/kg diet)	0.05

Adapted from Robinson, E. H., *Rev. Aquat. Sci.,* 1, 365, 1989.

of practical ration formulations were provided by Robinson (1989). Fish meal or meat and bone meal are used as animal protein sources in catfish formulations. The remaining protein typically comes from soybean meal and corn augmented by rice bran or wheat. A summary of the recommended nutrient levels for channel catfish was provided by Robinson (1989). That information is summarized in Table 2.

IX. DISEASE

For purposes of this discussion, the term disease includes maladies caused by such diverse organisms as viruses, fungi, bacteria, protozoan parasites, and metazoan parasites. Also included are nutritional deficiency diseases such as broken back syndrome associated with insufficient dietary vitamin C (reviewed by Robinson and Wilson 1985) and gill disease in fry possibly associated with pantothenic acid deficiency (Brunson et al. 1983). In some instances, the presence of a pathogenic organism can act synergistically with a water quality problem such as nitrite-induced methemoglobinemia (Tucker et al. 1984; Hanson and Grizzle 1985). Reviews which do go into detail on specific pathogenic organisms and treatment protocols are available (Plumb 1979; Macmillan 1985; Schwedler et al. 1985).

Epizootics frequently occur after fish have been stressed. Some epizootics take as little as 24 h to develop, while others may take from 3 d to approximately 2 weeks. The aquaculturist should always be watchful for signs of a disease outbreak following any type of stress to the fish.

During the mid-1960s, when the commercial fish farming industry in the U.S. was developing, only about 5% of the farmers reported serious problems with diseases (Meyer 1967). As fish farming has grown, disease incidence has increased both in terms of frequency and diversity of organisms found to cause epizootics. In Mississippi, for example, a 1980 study found that about one third of the catfish farmers reported severe epizootics (Reagan 1982). Increases in disease incidence have been seen in various other aquaculture species. New diseases are found which had not been reported from wild populations or those maintained only at low densities. High intensity aquaculture is a form of stress in itself. Coupled with incidents of degraded water quality, handling, inefficient feeds in some instances, and other stressors, disease incidence normally increases. As a general rule, diseases in cultured catfish can be avoided or kept within acceptable limits if good culture practices are employed.

When a disease epizootic develops, treatment with effective therapeutic agents may be the only means of resolving the problem. Fish culturists in the U.S. have a paucity of treatment chemicals available to them for use on food fish (Table 3). Bacterial resistance to antibiotics has reduced effectiveness of the available products. The need for additional therapeutic agents for the treatment of fish diseases remains critical, though progress has been slow because of the large costs involved in obtaining clearance by the U.S. Food and Drug Administration. In many

TABLE 3
Compounds Approved by the U.S. Food
and Drug Administration for Use to Treat
Channel Catfish Diseases

Compound	Diseases subject to control
Acetic acid	Parasites
Formalin	Parasites
Sodium chloride	Osmoregulatory enhancement
Oxytetracycline	Antibacterial
Sulfadimeoxine and ormetoprim	Antibacterial

Data from Meyer, F. P. and Schnick, R. A., *Rev. Aquat. Sci.*, 1, 693, 1989.

cases the costs of demonstrating the efficacy and safety of a therapeutic compound exceed what can be expected from profits over anything approaching a reasonable rate on return for the investment.

With the arsenal of therapeutic agents being small, and with increasing pressure on the fish culture community to reduce the amounts of such chemicals being released into the environment, the pressure has grown to handle disease problems in other ways. Disease can often be avoided by good management, but even the best managers will occasionally experience a system failure, unanticipated or uncontrollable stressful change in water quality, or introduction of a disease from outside the fish farm.

The answer to avoiding at least some disease problems may lie in the development of substances that will provide immunity to the fish. Vaccines may be delivered through injection, in the feed, or by immersion. Each technique provides a different level of immunity, with injectable vaccines being the most effective, but also the most expensive because each fish has to be handled individually. Methods of immunizing large numbers of fish, even by injection, have been developed, and the cost per fish makes employment of vaccines economically attractive in some instances.

The development of fish vaccines has been rapid and recent. Recently, Plumb (1988) described vaccines against a common catfish bacterial disease, *Edwardsiella ictaluri*, which were of an experimental nature. Such vaccines are now on the market and available for use by fish farmers (Robert L. Busch, personal communication, 1991). Catfish can also be immunized against another common bacterium, *Aeromonas hydrophila* (Thune and Plumb 1982).

Bacteria are not the only organisms against which immunity can be developed in channel catfish. An antigen has been developed against the protozoan bacterium *Ichthyophthirius multifiliis*, so immunization against Ich or white-spot disease is also possible (Goven et al. 1981). Injections of serum containing anti-channel catfish virus (CCV) neutralizing activity has been shown effective in providing immunity to juvenile channel catfish (Hedrick and McDowell 1987), and a vaccine has been developed from the soluble envelope of the CCV virus itself (Awad et al. 1989).

X. HARVESTING AND PROCESSING

Once the fish have reached market size, they must be captured, transported to the processing area on the farm or to a processing plant, dressed and perhaps further processed, and then made available for sale. The responsibility of the catfish farmer does not end when the fish leave the farm. Instead, the culturist should recognize future sales and the ultimate success of the business are jeopardized unless the product meets the standards of the processor and, as importantly, the consumer. Many catfish farmers have not been too concerned about quality control in the past, but with the occurrence of significant off-flavor problems throughout the industry in recent years, both producers and processors have become acutely aware that if the product does not meet exacting quality standards, sales will be lost. This is particularly important as farm-raised channel catfish begin to appear in some of the best-known and largest fast-food restaurant chains in the nation.

There has been no mandatory inspection of seafood products in the U.S. in the past, but mandated inspection programs that have been discussed by the U.S. Congress seem to be widely acceptable to the industry. Health inspections of seafood processing plants have been occurring for a number of years, but product quality inspection is new to at least parts of the industry. In some instances, processors have been paying to have their products inspected, even though that was not mandated under law.

A. HARVESTING

The harvest of channel catfish from tanks, raceways, and cages is a relatively simple matter. Water volume can be sufficiently reduced in those types of intensive culture chambers to allow for capture by dipnetting. Large raceways are also commonly harvested by dipnetting after the fish have been crowded to one end. In ponds, the most widely used culture chamber for channel catfish, harvesting with seines is a standard technique.

A typical seine is at least a few tens of centimeters deeper than the maximum depth of the pond that is to be harvested, has a float line at the surface which is buoyed by plastic or cork floats, a mud line at the bottom comprised of several strands of rope that become saturated with water to help hold the net on the bottom, and a bag in the middle into which the fish become concentrated during the seining operation. A seine should be about 1.5 times longer than the width of the pond.

In the early years of the industry, and in some areas at present, common practice has been to harvest all of the fish in a given pond simultaneously. While growth is not uniform, once the average size of fish stocked at about the same size reaches 0.5 kg, most of the individuals will be marketable. In Mississippi and other places where large volumes of catfish are produced, partial harvesting has become the standard. The technique involves seining each pond at intervals of a month or so once the largest fish become marketable. Seines with mesh size sufficiently large to allow submarketable fish to escape are employed. As the population in a pond is reduced, new fingerlings are introduced, thereby leading to any even broader range of fish sizes. The partial harvesting process provides the processors with fish throughout the year rather than in the late fall or early winter which was when virtually all the fish reached the market until the new system was introduced.

Partial harvesting can be continued for a period of up to a few years before the pond must be drained and totally harvested. Over time, fecal wastes and other organic matter will accumulate causing water-quality deterioration and leading to gradually reduced rates of production. Some fish become stunted in each population, and the number of runts in a pond will increase as new fish are added over a period of time. Those fish continue to consume feed, so they need to be eliminated periodically. When the ponds are drained and totally harvested they are thoroughly dried to oxidize organic matter and may be disced and limed. Any required levee work should also be done at that time.

Total pond harvest typically involves an initial seine haul through the entire pond, followed by reduction of the water volume by about 50%, after which the pond is seined once again. Depending upon the size of the pond and number of fish recovered, the water level may be reduced one or more times and then seined. Eventually, the remaining water is drained and the last of the fish are removed.

Small ponds can be seined by hand, but large ponds require seines that can only be towed mechanically. Trucks and tractors are used to pull large seines.

Large seines may capture tons of fish, so special net enclosures called live cars may be employed to hold the fish during loading operations. The bag end of the seine is fitted to the live car into which the fish pass (Greenland 1974). The live car forms a net-pen which serves as a corral for the fish and allows the seining crew to move to another pond, if necessary. Dip nets and baskets or fish pumps can be used to transfer fish from either a seine bag or a live car to the hauling truck.

Catfish are hauled to market alive in trucks of various sizes that are specifically equipped for that purpose. Fish hauling trucks are equipped with large holding tanks (which are commonly insulated) and provided with some form of supplemental aeration equipment. Agitators, air blowers, compressed gas cylinders (air or oxygen), liquid oxygen tanks, or some combination of those sources of aeration may be available on the truck. It is important that the fish reach the processing plant alive and in good condition. Farmers may haul their own fish or arrange to have the fish hauled by a custom live hauler who will charge a price based on the weight of the fish to be transported.

B. PROCESSING

The large commercial catfish processing plants in the U.S. are, as would be expected, located within close proximity to the major catfish-producing areas, thus they are predominantly located in Mississippi. In some states, even some with significant levels of production, there are no large processing plants. Farmers may process their own fish or contract processing to another farmer who has a small plant. Fish can also be sold live at the pond bank.

Processing techniques are fairly well standardized and have been described in some detail by Ammerman (1985). Much of the work in the older plants is done by hand, but the more modern facilities are increasingly employing automated equipment for at least part of the processing.

Various types of catfish products are being marketed by catfish processors, with common forms of the product being filets; steaks; and skinned, gutted, and deheaded whole fish. The products may be rapidly frozen in blast freezers or marketed on ice. Farm-raised catfish can be found throughout the U.S., particularly in major cities. A considerable amount of effort has gone into marketing in recent years as the industry has expanded. For several years, catfish were not particularly popular north of the Mason-Dixon line, but aggressive marketing of an excellent product by the industry has overcome much of the resistance that once existed. Quality control has played a major role in the spread of catfish throughout the country.

Off-flavors have long been a major problem with pond-reared channel catfish. Off-flavors are not unique to channel catfish (Thaysen 1936a, 1936b), but seem to occur more commonly in that species than in others sold in the U.S. Off-flavors have become more and more common as the intensity of culture has increased. One study has indicated that off-flavor tends to intensify as the amount of feed provided to the fish increases (Brown and Boyd 1982).

The off-flavor problem is commonly described as an earthy-musty flavor and odor, though flavors described as sewage, stale, rancid, metallic, moldy, weedy, and petroleum-like have also been identified by taste panelists (Lovell 1983). The primary chemical responsible for off-flavor appears to be geosmin (*trans*-1,10-dimethyl-*trans*-9-decalol) which has been isolated from a number of microorganisms, particularly blue-green algae (Gerber and Lechevalier 1965, Safferman et al. 1967, Lovell and Sackey 1973, Tabachek and Yurkowski 1976). Transfer of affected fish to geosmin-free water (for example, placing the fish in a tank that is plumbed to a well-water source) will lead to resolution of the problem within a few days as the fish metabolize the chemical (Lovell and Sakey 1973; Yurkowski and Tabachek 1974; Iredale and York 1976).

The off-flavor problem became so serious that the processing plants in Mississippi introduced a system by which fish organoleptic testing is performed up to three times before the fish are accepted. The first taste test may be conducted 2 weeks before the anticipated date of harvest; the second, 3 d before harvest; and the third, when the fish reach the processing plant but before they are unloaded. The test consists of selecting a fish at random, cutting off the tail, and cooking it in a microwave oven for about 2 min. A processing plant employee then smells and tastes the fish. If off-flavor is detected as a result of any of the three tests, the fish are rejected. Preharvest testing is highly desirable since once loaded and hauled on a truck, the fish are heavily stressed and heavy mortality may occur if they are returned to a pond for purging the off-flavor. Most farmers do not have appropriate facilities for purging large numbers of fish and are willing cooperators in the testing program. That program has greatly improved the image of channel catfish throughout the nation and is one reason the industry has continued to expand at a rapid rate.

If a pond fails the tests, it may be several weeks or even months before the problem is corrected (the bloom of algae responsible for the off-flavor will have to die back, after which a period of time will be required for the fish to metabolize the geosmin). Farmers would like to find a way to avoid the problem entirely, and some research has been conducted to determine if a way can be found to prevent or significantly reduce the occurrence of off-flavor. Tucker and van der Ploeg (1991) reported on work in Mississippi where five approaches were used. They were (1) use of the selective algicide potassium ricinoleate; (2) use of another algicide, copper sulfate; (3) polyculture with tilapia (since some tilapia species eat algae); (4) addition of dyes to the water to retard aquatic plant growth; and (5) flushing ponds with new water to reduce nutrient levels, and therefore, algal production. None of the treatments succeeded in reducing off-flavor problems.

REFERENCES

Allen, K.O., Effects of stocking density and water exchange rate on growth and survival of channel catfish *Ictalurus punctatus* (Rafinesque) in circular tanks, *Aquaculture*, 4, 29, 1974.

Allen, K.O., Effects of flow rate and aeration on survival and growth of channel catfish in tanks, *Prog. Fish-Cult.*, 38, 204, 1976.

Allen, K.O. and Avault, J.W., Jr., Effects of brackish water on ichthyophthiriasis of channel catfish, *Prog. Fish-Cult.*, 32, 227, 1970a.

Allen, K.O. and Avault, J.W., Jr., Effects of salinity on growth and survival of channel catfish, *Ictalurus punctatus, Proc. S.E. Assoc. Game Fish Comm.*, 23, 319, 1970b.

Allen, K.O. and Avault, J.W., Jr., Notes on the relative salinity tolerance of channel and blue catfish, *Prog. Fish-Cult.*, 33, 135, 1971.

Ammerman, G.R., Processing, in *Channel Catfish Culture*, Tucker, C.S., Ed., Elsevier, Amsterdam, 1985, chap. 12.

Andrews, J.W., Some effects of feeding rate on growth, feed conversion and nutrient absorption of channel catfish, *Aquaculture*, 16, 243, 1979.

Andrews, J.W. and Page, J.W., The effects of frequency of feeding on culture of catfish, *Trans. Am. Fish. Soc.*, 104, 317, 1975.

Andrews, J.W. and Stickney, R.R., Interactions of feeding rates and environmental temperature on growth, food conversion, and body composition of channel catfish, *Trans. Am. Fish. Soc.*, 101, 74, 1972.

Andrews, J.W., Knight, L.H., and Murai, T., Temperature requirements for high density rearing of channel catfish from fingerling to market size, *Prog. Fish-Cult.*, 34, 240, 1972.

Andrews, J.W., Murai, T., and Campbell, C., Effects of dietary calcium and phosphorus on growth, food conversion, bone ash and hematocrit levels of catfish, *J. Nutr.*, 103, 766, 1973a.

Andrews, J.W., Murai, T., and Gibbons, G., The influence of dissolved oxygen on the growth of channel catfish, *Trans. Am. Fish. Soc.*, 102, 835, 1973b.

Andrews, J.W., Murai, T., and Page, J.W., Effects of dietary cholecalciferol and ergocalciferol on catfish, *Aquaculture*, 19, 49, 1980.

Andrews, J.W., Knight, L., Page, J.W., Matsuda, Y., and Brown, E.E., Interactions of stocking density and water turnover on growth and food conversion of channel catfish, *Prog. Fish-Cult.*, 33, 197, 1971.

Anon., Status of world aquaculture: 1990, in *Aquaculture Magazine Buyer's Guide '90 and Industry Directory*, Aquaculture Magazine, Asheville, NC, 1990, 10.

Anon., Status of world aquaculture: 1991, in *Aquaculture Magazine Buyer's Guide '90 and Industry Directory*, Aquaculture Magazine, Asheville, NC, 1991a, 6.

Anon., Catfish industry's growth continues in 1990, *Aquacult. Mag.*, 1991b, (March/April), 6.

Awad, M.A., Nusbaum, K.E., and Brady, V.J., Preliminary studies of a newly developed subunit vaccine for channel catfish virus disease, *J. Aquat. Anim. Health*, 1, 233, 1989.

Bardach, J.E., Ryther, J.H., and McLarney, W.O., *Aquaculture,* Wiley-Interscience, New York, 1972, 1.

Beaver, J.A., Sneed, K.E., and Dupree, H.K., The difference in growth of male and female channel catfish in hatchery ponds, *Prog. Fish-Cult.*, 28, 47, 1966.

Beleau, M.H., High-density culture systems, in *Channel Catfish Culture*, Tucker, C.S., Ed., Elsevier, Amsterdam, 1985, chap. 3.

Bondari, K., Caudal fin abnormality and growth and survival of channel catfish, *Growth*, 67, 361, 1983a.

Bondari, K., Efficiency of male reproduction in channel catfish, *Aquaculture*, 35, 79, 1983b.

Bondari, K., Genetic and environmental control of fingerling size in channel catfish, *Aquaculture*, 34, 171, 1983c.

Bondari, K., Response of bidirectional selection for body weight in channel catfish, *Aquaculture*, 33, 73, 1983d.

Bondari, K., Response of channel catfish to multi-factor and divergent selection of economic traits, in *2nd Int. Symp. Genetics in Aquaculture*, Gall, G.A.E. and Busack, C.A., Eds.; *Aquaculture*, 57, 163, 1986.

Bondari, K., Reproduction and genetics of the channel catfish (*Ictalurus punctatus*), *Rev. Aquat. Sci.*, 2, 357, 1990.

Bondari, K., Washburn, K.W., and Ware, G.O., Effect of initial size on subsequent growth and carcass characteristics of divergently selected channel catfish, *Theor. Appl. Genet.*, 71, 153, 1985.

Boyd, C.E., *Water Quality in Warmwater Fish Ponds*, Alabama Agricultural Experiment Station, Auburn University, Auburn, AL, 1979, 1.

Boyd, C.E., Hydrology and pond construction, in *Channel Catfish Culture*, Tucker, C.S., Ed., Elsevier, Amsterdam, 1985, chap. 4.

Boyd, C.E., and Tucker, C.S., Emergency aeration of fish ponds, *Trans. Am. Fish. Soc.*, 108, 299, 1979.

Boyd, C.E., Preacher, J.W., and Justice, L., Hardness, alkalinity, pH, and pond fertilization, *Proc. S.E. Assoc. Fish Wildl. Agen.*, 32, 605, 1978a.

Boyd, C.E., Romaire, R.P., and Johnston, E., Predicting early morning dissolved oxygen concentrations in channel catfish ponds, *Trans. Am. Fish. Soc.*, 107, 484, 1978b.

Boyd, C.E., Romaire, R.P., and Johnston, E., Water quality in channel catfish production ponds, *J. Envir. Qual.*, 8, 423, 1979.

Brauhn, J.L., Fall spawning of channel catfish, *Prog. Fish-Cult.*, 33, 150, 1971.

Brooks, M.J., Smitherman, R.O., Chapell, J.A., and Dunham, R.A., Sex-weight relations in blue, channel, and white catfishes: implications for brood stock selection, *Prog. Fish-Cult.*, 44, 105, 1982.

Broussard, M.C., Evaluation of Four Strains of Channel Catfish, *Ictalurus punctatus*, and Intraspecific Hybrids under Aquacultural Conditions, Ph.D. dissertation, Texas A&M University, College Station, 1979, 1.

Broussard, M.C., Jr. and Stickney, R.R., Growth of four strains of channel catfish in communal ponds, *Proc. S.E. Assoc. Fish Wildl. Agen.*, 35, 541, 1981.

Brown, E.E., *World Fish Farming Cultivation and Economics*, AVI Publishing, Westport, CT, 1977, 1.

Brown, S.W. and Boyd, C.E., Off-flavor in channel catfish from commercial ponds, *Trans. Am. Fish. Soc.*, 111, 379, 1982.

Brunson, M.W., Robinette, H.R., Bowser, P.R., and Wellborn, T.L., Jr., Nutritional gill disease associated with starter feeds for channel catfish fry, *Prog. Fish-Cult.*, 45, 119, 1983.

Bryan, R.D., and Allen, K.O., Pond culture of channel catfish fingerlings, *Prog. Fish-Cult.*, 31, 38, 1969.

Busch, R.L., Channel catfish culture in ponds, in *Channel Catfish Culture*, Tucker, C.S., Ed., Elsevier, Amsterdam, 1985, chap. 2.

Carlson, A.R., Blocher, J., and Herman, L.J., Growth and survival of channel catfish and yellow perch exposed to lowered constant and diurnally fluctuating dissolved oxygen concentrations, *Prog. Fish-Cult.*, 42, 73, 1980.

Carter, R.R. and Thomas, A.E., Spawning channel catfish in tanks, *Prog. Fish-Cult.*, 39, 13, 1977.

Clady, M.D., Cool-weather growth of channel catfish held in pens alone and with other species, *Prog. Fish-Cult.*, 43, 92, 1981.

Clapp, A., Some experiments in rearing channel catfish, *Trans. Am. Fish. Soc.*, 59, 114, 1929.

Clem, L.W., Faulmann, E., Miller, N.W., Ellsaesser, C., Lobb, C.J., and Cucheus, M.A., Temperature-mediated processes in teleost immunity: differential effects of *in vitro* and *in vivo* temperature on mitogenic responses of channel catfish lymphocytes, *Dev. Comp. Immun.*, 8, 313, 1984.

Clemens, H.P., and Sneed, K.E., Spawning Behavior of Channel Catfish *Ictalurus punctatus*, Special Scientific Report-Fisheries, No. 316, U.S. Department of the Interior, 1957, 1.

Clemens, H.P. and Sneed, K.E., Bioassay and the Use of Pituitary Materials to Spawn Warmwater Fishes, Resources Report 61, U.S. Fish and Wildlife Service, Washington, D.C., 1962, 1.

Colt, J. and Tchobanoglous, G., Evaluation of the short-term toxicity of nitrogenous compounds to channel catfish, *Ictalurus punctatus, Aquaculture,* 8, 209, 1976.

Colt, J. and Tchobanoglous, G., Chronic exposure of channel catfish, *Ictalurus punctatus,* to ammonia: effects on growth and survival, *Aquaculture,* 15, 353, 1978.

Colt, J., Ludwig, R., Tchobanoglous, G., and Cech, J.J., Jr., The effects of nitrite on the short-term growth and survival of channel catfish, *Ictalurus punctatus, Aquaculture,* 24, 111, 1981.

Crawford, B., Propagation of channel catfish (*Ictalurus punctatus*) at a state fish hatchery, *Proc. S.E. Assoc. Game Fish Comm.,* 11, 132, 1957.

Dorsa, W.J., Robinette, H.R., Robinson, E.H., and Poe, W.E., Effects of dietary cottonseed meal and gossypol on growth of young channel catfish, *Trans. Am. Fish. Soc.,* 111, 651, 1982.

Dunham, R.A. and Smitherman, R.O., Growth in response to winter feeding by blue, channel, and white catfishes, *Prog. Fish-Cult.,* 43, 63, 1981.

Dunham, R.A., Smitherman, R.O., and Webber, C., Relative tolerance of channel × blue hybrid and channel catfish to low dissolved concentrations, *Prog. Fish-Cult.,* 45, 55, 1983.

Dunham, R.A., Smitherman, R.O., and Goodman, R.K., Comparison of mass selection, cross-breeding, and hybridization for improving growth of channel catfish, *Prog. Fish-Cult.,* 49, 293, 1987a.

Dunham, R.A., Brummett, R.E., Ella, M.O., and Smitherman, R.O., Genotype-environment interactions for growth of blue, channel and hybrid catfish in ponds and cages at varying densities, *Aquaculture,* 85, 143, 1990.

Dunham, R.A., Eash, J., Askins, J., and Townes, T.M., Transfer of the metallothionein-human growth hormone fusion gene into channel catfish, *Trans. Am. Fish. Soc.,* 116, 87, 1987b.

Dunham, R.A., Smitherman, R.O., Brooks, M.J., Benchakan, M., and Chappell, J.A., Paternal predominance in reciprocal channel-blue hybrid catfish, *Aquaculture,* 29, 389, 1982.

Dunseth, D. and Smitherman, N., Pond Culture of Catfish, Tilapia and Silver Carp, Highlights of Agricultural Research, Auburn University Agricultural Experiment Station, Auburn, AL, 24, 4, 1977.

Dupree, H.K. and Huner, J.V., Eds., *Third Report to the Fish Farmers,* U.S. Fish and Wildlife Service, Washington, D.C., 1984, 1.

Durve, V.S. and Lovell, R.T., Vitamin C and disease resistance in channel catfish (*Ictalurus punctatus*), *Can. J. Fish. Aquat. Sci.,* 39, 948, 1982.

Eddy, S. and Underhill J.C., *Northern Fishes,* University of Minnesota Press, Minneapolis, 1974, 1.

El-Ibiary, H.M., Joyce, J.A., Page, J.W., and Hill, T.K., Comparison between sequential and concurrent matings of two females and one male channel catfish, *Ictalurus punctatus,* in spawning pens, *Aquaculture,* 10, 153, 1977.

El-Ibiary, H.M., Andrews, J.W., Joyce, J.A., Page, J.W., and DeLoach, H.L., Sources of variations in body size traits, dress-out weight, and lipid content and their correlations in channel catfish, *Ictalurus punctatus, Trans. Am. Fish. Soc.,* 105, 267, 1976.

Gabel, S.J., Avault, J.W., Jr., and Romaire, R.P., Polyculture of channel catfish (*Ictalurus punctatus*) with all-male tilapia hybrids (*Sarotherodon mossambica* male × *Sarotherodon hornorum* female), *J. World Maricult. Soc.,* 12, 153, 1981.

Gatlin, D.M., III and Phillips, H.F., Dietary calcium, phytate and zinc interactions in channel catfish, *Aquaculture,* 79, 259, 1989.

Gatlin, D.M., III and Stickney, R.R., Fall-winter growth of young channel catfish in response to quantity and source of dietary lipid, *Trans. Am. Fish. Soc.,* 111, 90, 1982.

Gatlin, D.M., III and Wilson, R.P., Zinc supplementation of practical channel catfish diets, *Aquaculture,* 41, 31, 1984.

Gerber, N.N. and Lechevalier, H.A., Geosmin, and earthy-smelling substance isolated from actinomycetes, *Appl. Microbiol.,* 13, 935, 1965.

Giudice, J.J., Growth of a blue × channel catfish hybrid as compared to its parent species, *Prog. Fish-Cult.,* 28, 142, 1966.

Goudie, C.A., Redner, B.D., Simco, B.A., and Davis, K.B., Feminization of channel catfish by oral administrations of steroid sex hormones, *Trans. Am. Fish. Soc.*, 112, 670, 1983.

Goven, B.A., Dawe, D.L., and Gratzek, J.B., Protection of channel catfish (*Ictalurus punctatus*) against *Ichthyophthirius multifiliis* (Fouquot) by immunization with varying doses of *Tetrahymena pyriformis* (Lwoff) cilia, *Aquaculture*, 23, 269, 1981.

Green, O.L., Comparison of production and survival of channel catfish stocked alone and in combination with blue and white catfish, *Prog. Fish-Cult.*, 35, 225, 1973.

Green, O.L., Smitherman, R.O., and Pardue, G.B., Comparisons of growth and survival of channel catfish, *Ictalurus punctatus*, from distinct populations, in *Advances in Aquaculture*, Pillay, T.V.R. and Dill, W.A., Eds., Fishing News Books, Farnham, Surrey, England, 1979, 626.

Greenland, D.C., Recent developments in harvesting, grading, loading and hauling pond raised catfish, *Trans. Am. Soc. Agric. Eng.*, 17, 59, 1974.

Guerrero, R.D., III, Use of androgens for the production of all-male *Tilapia aurea* (Steindachner), *Trans. Am. Fish Soc.*, 104, 342, 1975.

Hanson, L.A. and Grizzle, J.M., Nitrite-induced predisposition of channel catfish to bacterial diseases, *Prog. Fish-Cult.*, 47, 98, 1985.

Hedrick, R.P. and McDowell, T., Passive transfer of sera with antivirus neutralizing activity from adult channel catfish protects juveniles from channel catfish virus disease, *Trans. Am. Fish. Soc.*, 116, 277, 1987.

Henderson, A., Production potential of catfish grow-out ponds supplemental stocked with silver and bighead carp, *Proc. S.E. Assoc. Fish Wildl. Agen.*, 33, 584, 1979.

Hendricks, J.D. and Bailey, G.S., Adventitious toxins, in *Fish Nutrition*, Halver, J.E., Ed., 1989, chap. 11.

Hill, T.K., Chesness, J. L., and Brown, E.E., Growing channel catfish, *Ictalurus punctatus* (Rafinesque) in raceways, *Proc. S.E. Assoc. Game Fish Comm.*, 27, 488, 1974.

Hollerman, W.D. and Boyd, C.E., Nightly aeration to increase production of channel catfish, *Trans. Am. Fish. Soc.*, 109, 446, 1980.

Huey, D.W. and Beitinger, T.L., A methemoglobin reductase system in channel catfish *Ictalurus punctatus*, *Can. J. Zool.*, 60, 1511, 1982.

Huey, D.W., Simco, B.A., and Criswell, D.W., Nitrite induced methemoglobin formation in channel catfish, *Trans. Am. Fish. Soc.*, 109, 558, 1980.

Huner, J.V., Avault, J.W., Jr., and Bean, R.A., Interactions of freshwater prawns, channel catfish fingerlings, and crayfish in earthen ponds, *Prog. Fish-Cult.*, 45, 36, 1983.

Iredale, D.G. and York, R.K., Purging a muddy-earthy flavor taint from rainbow trout (*Salmo gairdneri*) by transferring to artificial and natural holding environments, *J. Fish. Res. Bd. Can.*, 33, 160, 1976.

Johnson, S.K., Laboratory evaluation of several chemicals as preventatives of ich disease, in *Proc. 1976 Fish Farming Conf. Annu. Convention Catfish Farmers Texas*, Texas A&M University, College Station, 1976, 91.

Joint Subcommittee on Aquaculture, National Aquaculture Development Plan, Vol. 2, Washington, D.C., 1973, 1.

Jordan, D.S., *American Food and Game Fishes*, Dover Publications, New York, 1969, 1.

Kilambi, R.V., Noble, J., and Hoffman, C.E., Influence of temperature and photoperiod on growth, food consumption, and body composition of channel catfish, *Proc. S.E. Assoc. Game Fish Comm.*, 24, 519, 1970.

Kilambi, R.V., Adams, J.C., Brown, A.V., and Wickizer, W.A., Effects of stocking density and cage size on growth, feed conversion, and production of rainbow trout and channel catfish, *Prog. Fish-Cult.*, 39, 62, 1977.

Kilgen, R.H., Growth of channel catfish and striped bass in small ponds stocked with grass carp and water hyacinths, *Trans. Am. Fish. Soc.*, 107, 176, 1978.

Knepp, G.L. and Arkin, G.F., Ammonia toxicity levels and nitrate tolerance of channel catfish, *Prog. Fish-Cult.*, 35, 221, 1973.

Konikoff, M., Toxicity of nitrite to channel catfish, *Prog. Fish-Cult.*, 37, 96, 1975.

Leary, D.F. and Lovell, R.T., Value of fiber in production-type diets for channel catfish, *Trans. Am. Fish. Soc.*, 104, 328, 1975.

Lenz, G., Propagation of catfish, *Prog. Fish-Cult.*, 9, 231, 1947.

Lewis, S.D., Effect of selected concentrations of sodium chloride on the growth of channel catfish, *Proc. S.E. Assoc. Game Fish Comm.*, 25, 459, 1972.

Lewis, W.M., Cage culture of channel catfish, *Catfish Farmer*, 1(4), 511, 1969.

Lewis, W.M., Observations on the grass carp in ponds containing fingerling channel catfish and hybrid sunfish, *Trans. Am. Fish. Soc.*, 107, 153, 1978.

Limsuwan, T. and Lovell, R.T., Intestinal synthesis and absorption of vitamin B-12 in channel catfish, *J. Nutr.*, 111, 2125, 1981.

Lovell, R.T., Dietary phosphorus requirement of channel catfish (*Ictalurus punctatus*), *Trans. Am. Fish. Soc.*, 107, 617, 1978.

Lovell, R.T., New off-flavors in pond-cultured channel catfish, *Aquaculture*, 30, 329, 1983.

Lovell, R.T., Diet and fish husbandry, in *Fish Nutrition*, 2nd ed., Halver, J.E., Ed., Academic Press, New York, 1989, chap. 10.

Lovell, T., Ed., *Nutrition and Feeding of Fish*, Van Nostrand Reinhold, New York, 1989b, 1.

Lovell, T. and Li, Y.-P., Essentiality of vitamin D in diets of channel catfish (*Ictalurus punctatus*), *Trans. Am. Fish. Soc.*, 107, 809, 1978.

Lovell, R.T. and Sackey, L.A., Absorption by channel catfish of earthy-musty flavor compounds by cultures of blue-green algae, *Trans. Am. Fish. Soc.*, 102, 774, 1973.

Lovell, T. and Sirikul, B., Winter feeding of channel catfish, *Proc. S.E. Assoc. Game Fish Comm.*, 28, 208, 1974.

Loyacano, H.A., Effects of aeration in earthen ponds on water quality and production of white catfish, *Aquaculture*, 3, 261, 1974.

Maclean, N., and Penman, D., The application of gene manipulation to aquaculture, *Aquaculture*, 85, 1, 1990.

Macmillan, J.R., Infectious diseases, in *Channel Catfish Culture*, Tucker, C.S., Ed., Elsevier, Amsterdam, 1985, chap. 9.

Mandal, B.K. and Boyd, C.E., Reduction of pH in waters with high total alkalinity and low total hardness, *Prog. Fish-Cult.*, 42, 183, 1980.

Martin, M., Techniques of catfish fingerling production, in *Proc. 1967 Fish Farming Conf.*, Texas A&M University, College Station, 1967, 13.

McKee, J.E. and Wolf, H.W., Eds., Water Quality Criteria, 2nd ed., State Water Quality Control Board Publ. 3-A, State of California, Sacramento, 1963, 1.

Merkowsky, A. and Avault, J.W., Jr., Polyculture of channel catfish and hybrid grass carp, *Prog. Fish-Cult.*, 38, 76, 1976.

Meyer, F.P., The impact of diseases on fish farming, *Am. Fishes Trout News*, March-April, 1967.

Meyer, F.P. and Schnick, R.A., A review of chemicals used for the control of fish diseases, *Rev. Aquat. Sci.*, 1, 693, 1989.

Meyer, F.P., Sneed K.E., and Eschmeyer, P.T., *2nd Report to the Fish Farmers*, U.S. Fish and Wildlife Service Resource Publ. 113, Washington, D.C., 1973, 123.

Mitchell, S.J. and Cech, J.J., Ammonia-caused gill damage in channel catfish (*Ictalurus punctatus*): compounding effects of residual chlorine, *Can. J. Fish. Aquat. Sci.*, 40, 242, 1983.

Murai, T., High-density rearing of channel catfish fry in shallow troughs, *Prog. Fish-Cult.*, 41, 57, 1979.

Murai, T. and Andrews, J.W., Effects of frequency of feeding on growth and food conversion of channel catfish fry, *Bull. Jpn. Soc. Sci. Fish.*, 42, 159, 1976.

Murai, T. and Andrews, J.W., Vitamin K and anticoagulant relationships in catfish diets, *Bull. Jpn. Soc. Sci. Fish.*, 43, 785, 1977.

Murai, T. and Andrews, J.W., Channel catfish: the absence of an effect of dietary salt on growth, *Prog. Fish-Cult.*, 44, 155, 1979.

Murai, T., Andrews, J.W., and Smith, R.G., Effects of dietary copper on channel catfish, *Aquaculture*, 22, 353, 1981.

National Research Council, Nutrient Requirements of Trout, Salmon and Catfish, National Academy of Science Press, Washington, D.C., 1973, 1.

National Research Council, Nutrient Requirements of Warmwater Fishes, National Academy of Science Press, Washington, D.C., 1977, 1.

National Research Council, Nutrient Requirements of Warmwater Fishes and Shellfishes, National Academy of Science Press, Washington, D.C., 1983, 1.

Nelson, B., Spawning of channel catfish by use of hormone, *Proc. S.E. Assoc. Game Fish Comm.*, 14, 145, 1960.

Norton, V.M., Nishimura, H., and Davis, K.B., A technique for sexing channel catfish, *Trans. Am. Fish. Soc.*, 105, 460, 1976.

Pennington, C.H., Cage Culture of Channel Catfish, *Ictalurus punctatus* (Rafinesque), in a Thermally Modified Texas Reservoir, Ph.D. dissertation, Texas A&M University, College Station, 1977, 1.

Perry, W.G., Jr., Distribution and relative abundance of blue catfish, *Ictalurus punctatus*, with relation to salinity, *Proc. S.E. Assoc. Game Fish Comm.*, 21, 436, 1967.

Perry, W.G., Jr. and Avault, J.W., Jr., Culture of blue, channel and white catfish in brackish water ponds, *Proc. S.E. Assoc. Game Fish Comm.*, 23, 592, 1970.

Perry, W.G., Jr. and Avault, J.W., Jr., Comparisons of striped mullet and tilapia for added production in caged catfish studies, *Prog. Fish-Cult.*, 34, 229, 1972a.

Perry, W.G., Jr. and Avault, J.W., Jr., Polyculture studies with blue, white and channel catfish in brackish water ponds, *Proc. S.E. Assoc. Game Fish Comm.*, 25, 466, 1972b.

Perry, W.G., Jr. and Avault, J.W., Jr., Polyculture studies with channel catfish and buffalo, *Proc. S.E. Assoc. Game Fish Comm.*, 29, 91, 1975.

Piper, R.G., McElwain, I.B., Orme, L.E., McCraren, J.P., Fowler, L.G., and Leonard, J.R., Fish Hatchery Management, Fish and Wildlife Service, U.S. Department of the Interior, Washington, D.C., 1982, 1.

Plumb, J.A., Principal Diseases of Farm-Raised Catfish, Southern Cooperative Series Bull. No. 225, Auburn University, Auburn, AL, 1979, 1.

Plumb, J.A., Vaccination against *Edwardsiella ictaluri*, in *Fish Vaccination*, Ellis, A.E., Ed., Academic Press, New York, 1988, 152.

Randolph, K.N. and Clemens, H.P., Home areas and swimways in channel catfish culture ponds, *Trans. Am. Fish. Soc.*, 105, 725, 1976a.

Randolph, K.N. and Clemens, H.P., Some factors influencing the feeding behavior of channel catfish in culture ponds, *Trans. Am. Fish. Soc.*, 105, 718, 1976b.

Randolph, K.N. and Clemens, H.P., Effects of short term food deprivation on channel catfish and implications for culture practices, *Prog. Fish-Cult.*, 40, 48, 1978.

Ray, L., Channel catfish production in geothermal water, in Proc. Bio-Engineering Symposium Fish Culture, Fish Culture Section, American Fisheries Society, Bethesda, MD, 1981, 192.

Reagan, R.E., Jr., Survey of current practices and problems in the Mississippi catfish industry, *Proc. S.E. Assoc. Game Fish Comm.*, 34, 1982.

Reagan, R.E., Jr. and Conley, C.M., Effect of egg diameter on growth of channel catfish, *Prog. Fish-Cult.*, 39, 133, 1977.

Reis, L.M., Reutebuch, E.M., and Lovell, R.T., Protein-to-energy ratios in production diets and growth, feed conversion and body composition of channel catfish, *Ictalurus punctatus, Aquaculture*, 77, 21, 1989.

Rhodes, R.J., Status of world aquaculture: 1989, in *Aquaculture magazine* buyer's guide '89 and industry directory, *Aquacult. Mag.*, Asheville, NC, 1989, 6.

Robinette, H.R., Effect of selected sublethal levels of ammonia on the growth of channel catfish, *Prog. Fish-Cult.*, 38, 26, 1976.

Robinette, H.R. and Dearing, A.S., Shrimp by-product meal in diets of channel catfish, *Prog. Fish-Cult.*, 40, 39, 1978.

Robinette, H.R., Busch, R.L., Newton, S.H., Haskins, C.J., Davis, S., and Stickney, R.R., Winter feeding of channel catfish in Mississippi, Arkansas, and Texas, *Proc. S.E. Assoc. Game Fish Comm.*, 36, 162, 1982.

Robinson, E.H., Vitamin requirements, in *Nutrition and Feeding of Channel Catfish*, Robinson, E.H. and Lovell, R.T., Eds., Southern Cooperative Series Bull. 296, Texas A&M University, College Station, 1984.

Robinson, E.H., Channel Catfish Nutrition, *Rev. Aquat. Sci.*, 1, 365, 1989.

Robinson, E.H. and Wilson, R.P., Nutrition and feeding, in *Channel Catfish Culture*, Tucker, C.S., Ed., Elsevier, Amsterdam, 1985, chap. 8.

Robinson, E.H., Rawles, S.D., Brown, P.B., Yette, H.E., and Greene, L.W., Dietary calcium requirement of channel catfish, *Ictalurus punctatus*, reared in calcium-free water, *Aquaculture*, 53, 263, 1986.

Romaire, R.P. and Boyd, C.E., Effects of solar radiation on the dynamics of dissolved oxygen in channel catfish ponds, *Trans. Am. Fish. Soc.*, 108, 473, 1979.

Safferman, R.S., Rosen, A.A., Mashni, C.I., and Morris, M.E., Earthy-smelling substance from blue-green algae, *Environ. Sci. Technol.*, 1, 429, 1967.

Saksena, V.P., Yamamoto, K., and Riggs, C.D., Early development of channel catfish, *Prog. Fish-Cult.*, 23, 156, 1961.

Satoh, S., Poe, W.E., and Wilson, R.P., Studies on the essential fatty acid requirement of channel catfish, *Ictalurus punctatus*, *Aquaculture*, 79, 121, 1989.

Schmittou, H.R., Developments in the culture of channel catfish, *Ictalurus punctatus* (Rafinesque) in cages suspended in ponds, *Proc. S.E. Assoc. Game Fish Comm.*, 23, 226, 1969.

Schwedler, T.E. and Tucker, C.S., Empirical relationship between percent methemoglobin in channel catfish and dissolved nitrite and chloride in ponds, *Trans. Am. Fish. Soc.*, 112, 117, 1983.

Schwedler, T.E., Tucker, C.S., and Beleau, M.H., Non-infectious diseases, in *Channel Catfish Culture*, Tucker, C.S., Ed., Elsevier, Amsterdam, 1985, chap. 10.

Smith, B.W. and Lovell, R.T., Digestibility of nutrients in semi-purified rations by channel catfish in stainless steel troughs, *Proc. S.E. Assoc. Game Fish Comm.*, 25, 452, 1972.

Smitherman, R.O. and Dunham, R.A., Genetics and breeding, in *Channel Catfish Culture*, Tucker, C.S., Ed., Elsevier, Amsterdam, 1985, chap.7.

Smitherman, R.O., Dunham, R.A., Bice, T.O., and Horn, J.L., Reproductive efficiency in the reciprocal pairings between two strains of channel catfish, *Prog. Fish-Cult.*, 46, 106, 1984.

Smitherman, R.O., El-Ibiary, H.M., and Reagan, R.E., Genetics and Breeding of Channel Catfish, Regional Research Publ. 223, Southern Cooperative Series, Auburn, AL, 1, 1978.

Sneed, K.E. and Clemens, H.P., The use of human chorionic gonadotrophin to spawn warm-water fishes, *Prog. Fish-Cult.*, 21, 117, 1959.

Sneed, K.E. and Clemens, H.P., Use of Fish Pituitaries to Induce Spawning in Channel Catfish, U.S. Fish and Wildlife Service, Special Scientific Report — Fisheries, No. 329, 1, 1960.

Snow, J.R., Size Control of Catfish Fingerlings by Varying the Stocking Density, Highlights of Agricultural Research, Auburn University Agricultural Experiment Station, Auburn, AL, 28, 18, 1981.

Snow, J.R., Jones, R.O., and Rogers, W.A., Training Manual for Warm-Water Fish Culture, mimeo, Bureau of Sport Fisheries and Wildlife, Warm Water Inservice Training School, Marion, AL, 1964, 1.

Stickney, R.R., Rearing channel catfish fingerlings under intensive culture conditions, *Prog. Fish-Cult.*, 34, 100, 1972.

Stickney, R.R., *Principles of Warmwater Aquaculture*, Wiley-Interscience, New York, 1979, 1.

Stickney, R.R. and Andrews, J.W., Combined effects of dietary lipids and environmental temperature on growth, metabolism and body composition of channel catfish, *J. Nutr.*, 101, 1703, 1971.

Stickney, R.R. and Andrews, J.W., Effects of dietary lipids on growth, food conversion, lipid and fatty acid composition of channel catfish, *J. Nutr.*, 102, 249, 1972.

Stickney, R.R. and Hardy, R.W., Lipid requirements of some warmwater species, *Aquaculture*, 79, 145, 1989.

Stickney, R.R. and Lovell, R.T., Eds., Nutrition and Feeding of Channel Catfish, Southern Cooperative Series Bull., 218, Auburn, Alabama, 1977, 1.

Stickney, R.R. and Person N.K., An efficient heating method for recirculating water systems, *Prog. Fish-Cult.*, 47, 71, 1985.

Stickney, R.R. and Shumway, S.E., Occurrence of cellulase activity in the stomachs of fishes, *J. Fish. Biol.*, 6, 779, 1974.

Stickney, R.R. and Simco, B.A., Salinity tolerance of catfish hybrids, *Trans. Am. Fish. Soc.*, 100, 790, 1971.

Stickney, R.R., McGeachin, R.B., Lewis, D.H., and Marks, J., Response of young channel catfish to diets containing purified fatty acids, *Trans. Am. Fish. Soc.*, 112, 665, 1983.

Suppes, V.C., Jar incubation of channel catfish eggs, *Prog. Fish-Cult.*, 34, 48, 1972.

Swingle, H.S., Preliminary results on the commercial production of channel catfish in ponds, *Proc. S.E. Assoc. Game Fish Comm.*, 10, 160, 1957.

Swingle, H.S., Experiments on growing fingerling channel catfish to marketable size in ponds, *Proc.S.E. Assoc. Game Fish Comm.*, 12, 63, 1958.

Tabachek, J.A.L. and Yurkowski, M., Isolation and identification of blue-green algae producing muddy odor metabolites, gesomin, and 2-methyliso-orneol, in saline lakes in Manitoba, *J. Fish. Res. Bd. Can.*, 33, 25, 1976.

Tackett, D.L., Carter, R.R., and Allen, K.O., Daily variation in feed consumption by channel catfish, *Prog. Fish-Cult.*, 50, 107, 1988.

Tackett, D.L., Carter, R.R., and Allen, K.O., Winter feeding of channel catfish based on maximum air temperature, *Prog. Fish-Cult.*, 49, 290, 1987.

Tave, D. and Smitherman, R.O., Spawning success of reciprocal hybrid pairings between blue and channel catfishes with and without hormone injection, *Prog. Fish-Cult.*, 44, 73, 1982.

Tave, D., Ramsey, J.S., and Grizzle, J.M., Channel catfish with a rayed adipose fin, *J. Aquat. Anim. Health*, 2, 71, 1990.

Thaysen, A.C., The origin of an earthy or muddy taint in fish. I. The nature and isolation of the tains, *Ann. Appl. Biol.*, 23, 99, 1936a.

Thaysen, A.C., The origin of an earthy or muddy taint in fish. II. The effect on fish of the taint produced by an odoriferous species of *Actinomyces*, *Ann. Appl. Biol.*, 23, 105, 1936b.

Thune, R.L. and Plumb, J.A., Effect of delivery method and antigen preparation on the production of antibodies against *Aeromonas hydrophila* in channel catfish, *Prog. Fish-Cult.*, 44, 53, 1982.

Tilton, J.E. and Kelley J.E., Experimental cage culture of channel catfish in heated discharge water, *Proc. World Maricult. Soc.*, 1, 73, 1970.

Tomasso, J.R., Simco, B.A., and Davis, K.B., Chloride inhibition of nitrite-induced methemoglobinemia in channel catfish *(Ictalurus punctatus)*, *J. Fish. Res. Bd. Can.*, 36, 1141, 1979a.

Tomasso, J.R., Simco, B.A., and Davis, K.B., Inhibition of ammonia and nitrite toxicity to channel catfish, *Proc. S.E. Assoc. Fish Wildl. Agen.*, 33, 600, 1979b.

Tomasso, J.R., Goudie, C.A., Simco, B.A., and Davis, K.B., Effects of environmental pH and calcium on ammonia toxicity in channel catfish *(Ictalurus punctatus)*, *Trans. Am. Fish. Soc.*, 109, 229, 1980a.

Tomasso, J.R., Wright, M.I., Simco, B.A., and Davis, K.B., Inhibition of nitrite-induced toxicity by channel catfish by calcium chloride and sodium chloride, *Prog. Fish-Cult.*, 42, 144, 1980b.

Toole, M., Channel catfish culture in Texas, *Prog. Fish, Cult.*, 13, 3, 1951.

Tucker, C.S., Variability of percent methemoglobin in pond populations of nitrite-exposed channel catfish, *Prog. Fish-Cult.*, 45, 108, 1983.

Tucker, C.S., Fertilization of catfish fry ponds, in *For Fish Farmers*, Mississippi Cooperative Extension Service Newsletter, July, 1984, 1.

Tucker, C.S. and Boyd, C.E., Effects of simazine treatment on channel catfish and bluegill production in ponds, *Aquaculture*, 15, 345, 1978.

Tucker, C.S. and Schwedler, T.E., Variability of percent methemoglobin in pond populations of nitrite-exposed channel catfish, *Prog. Fish-Cult.*, 45, 108, 1983.

Tucker, C.S., Boyd, C.E., and McCoy, E.W., Effects of feeding rate on water quality, production of channel catfish and economic returns, *Trans. Am. Fish. Soc.*, 108, 389, 1979.

Tucker, C.S., MacMillan, J.R., and Schwedler, T.E., Influence of *Edwardsiella ictaluri* septicemia on nitrite-induced methemoglobinemia in channel catfish *(Ictalurus punctatus)*, *Bull. Environ. Contam. Toxicol.*, 32, 669, 1984.

Tucker, C.S. and van der Ploeg, M., Evaluation of five techniques to reduce incidence of off-flavor in channel catfish, in *For Fish Farmers,* Mississippi Agricultrual and Forestry Experiment Station, Stoneville, 1991.

Tuten, J.S. and Avault, J.W., Jr., Growing red swamp crayfish *(Procambarus clarkii)* and several North American fish species together, *Prog. Fish-Cult.,* 43, 97, 1981.

Westerman, A.G. and Birge, W.J., Accelerated rate of albinism in channel catfish exposed to metals, *Prog. Fish-Cult.,* 40, 143, 1978.

Williamson, J. and Smitherman, R.O., Food habits of hybrid buffalofish, tilapia, Israeli carp and channel catfish in polyculture, *Proc. S.E. Assoc. Game Fish Comm.,* 29, 86, 1975.

Wilson, J.L. and Hilton, L.R., Effects of tilapia densities on growth of channel catfish, *Prog. Fish-Cult.,* 44, 207, 1982.

Wilson, J.L. and Robinson, E.H., Protein and Amino Acid Nutrition for Channel Catfish, Information Bull. 25, Mississippi Agricultural and Forestry Experiment Station, Mississippi State University, Starkville, 1982, 1.

Wilson, R.P. and Poe, W.E., Choline nutrition of fingerling channel catfish, *Aquaculture,* 68, 65, 1988.

Winfree, R.A. and Stickney, R.R., Starter diets for channel catfish: effects of dietary protein on growth and carcass composition, *Prog. Fish-Cult.,* 46, 79, 1984a.

Winfree, R.A. and Stickney, R.R., Formulation and processing of hatchery diets for channel catfish, *Aquaculture,* 41, 311, 1984b.

Wolters, W.R., Libey, G.S., and Chrisman, C.L., Induction of triploidy in channel catfish, *Trans. Am. Fish. Soc.,* 110, 310, 1981.

Wolters, W.R., Libey, G.S., and Chrisman, C.L., Effect of triploidy on growth and gonad development in channel catfish, *Trans. Am. Fish. Soc.,* 111, 102, 1982.

Worsham, R.L., Nitrogen and phosphorus levels in water associated with a channel catfish *(Ictalurus punctatus)* feeding operation, *Trans. Am. Fish. Soc.,* 104, 811, 1975.

Yant, D.R., Smitherman, R.O., and Green, O.L., Production of hybrid (blue × channel) catfish and channel catfish in ponds, *Proc. S.E. Assoc. Game Fish Comm.,* 29, 82, 1975.

Yingst, W.L., III and Stickney, R.R., Growth of caged channel catfish fingerlings reared on diets containing various lipids, *Prog. Fish-Cult.,* 42, 24, 1980.

Yurkowski, M. and Tabachek, J.L., Identification, analysis, and removal of geosmin from muddy-flavored fish, *J. Fish. Res. Bd. Can.,* 31, 1851, 1974.

Chapter 3

TILAPIA

Robert R. Stickney

TABLE OF CONTENTS

I. INTRODUCTION

Tilapia are members of the family Cichlidae. In body configuration, they are, in general, similar to sunfishes (Figure 1). Tilapia have become perhaps the most important warmwater aquaculture fish group in the world over the past few decades (FAO 1980).

There have been some attempts to change the classification of the tilapia species in recent years. The most recent reclassification scheme, proposed by Trewavas (1982), would place species of culture interest into three genera: *Tilapia*, *Sarotherodon*, and *Oreochromis*. The species of most interest to aquaculturists are all members of the genus *Tilapia*. Many of those fishes were reclassified as species within the genus *Sarotherodon* for a brief period (Trewavas 1973). The later classification scheme (Trewavas 1982) changed members of the genus *Sarotherodon* to *Oreochromis*. That taxonomic scheme has been widely adopted throughout the world, but it has not been accepted by the American Fisheries Society (Robins 1991), so the traditional genus name, *Tilapia*, is used in this text. The common name tilapia is still used with respect to all of the classification schemes that are currently in use.

While the family Cichlidae is widely found throughout tropical and subtropical latitudes, the original distribution of tilapia was from south central Africa northward into Syria, with populations also occurring on the island of Madagascar (Boulenger 1915, Copley 1952). Tilapia first appeared outside their native range when they were introduced to Java by persons unknown in 1939 (Atz 1954). Further dispersion around the islands of the Pacific Basin was carried out by the Japanese during World

FIGURE 1.   Fingerling and harvestable *Tilapia aurea.*

War II (Chimits 1955). The ease with which tilapia can be cultured and their high level of productivity prompted development of the belief that the fish could contribute significant amounts of animal protein to people living in developing countries. As a result, tilapia were widely distributed in the tropical Indo-Pacific and into both North and South America (Chimits 1957).

Information on the distribution and habitat preference of various species of tilapia was presented by Philippart and Ruwet (1982). Reviews of tilapia and their culture have been prepared by Balarin and Hatton (1979), Chervinski (1982), Hepher and Pruginin (1982), and Coche (1982). Schoenen (1982) prepared an extensive bibliography on selected species of tilapia. According to Pullin (1983), four species or hybrids are superior for culture, another three have likely potential, and five more may have limited potential. Those considered to be superior were *T. aureus*, *T. nilotica*, monosex male hybrids, and red hybrids.

Red hybrid tilapia, also sometimes called golden tilapia, are touted as being more readily marketable than specimens of normal color, since the normal fish often are dark and may have a black peritoneum which is considered to be unattractive. Red tilapia have been produced in several places, including Taiwan, The Philippines, Guam, and the U.S. (reviewed by Galman and Avtalion 1983). Red hybrids have been produced from combinations of *T. aurea*, *T. hornorum*, *T. mossambica*, and *T. nilotica*.

The natural distribution of *T. aurea* overlaps that of *T. nilotica* in Senegal and Chad (Philippart and Ruwet 1982), so some physical or biological requirement leading to ecological isolation in the wild is indicated. Both species were historically present in the Nile River, and *T. aurea* is native to portions of Israel (Payne and Collinson 1983).

T. mossambica was widely introduced into Asia and has become well-established in a number of countries. While *T. mossambica* continues be be widely cultured, in recent years, species such as *T. aurea* and *T. nilotica* have become increasingly popular. They tend to grow more rapidly than *T. mossambica*, reproduce at a later age and larger size, have a more silvery color, and have a lower head to body ratio. Many regard *T. mossambica* as an undesirable species, though it still has some proponents (DeSilva and Senaratne 1988).

Because of their intolerance to cold (discussed in Section II.A), the culture of tilapia occurs primarily in tropical environments. Some tilapia production occurs in the U.S., with pond culture largely restricted to Hawaii and extreme southern regions of the mainland, with the exception of isolated situations where warm water is available throughout the year. Through proper overwintering of broodstock and/or fingerlings, summer production of tilapia is possible in temperate climates. Israel, for example, has a considerable tilapia culture industry, most of which depends on overwintering of broodstock and fingerlings in greenhouses.

II. WATER QUALITY REQUIREMENTS

Tilapia as a group are among the most hardy fishes known. They are highly tolerant of crowding and degraded water quality, features which make them excellent candidates for culture. Other important attributes which the primary cultured species of tilapia share include their fast growth and excellent flavor. Tilapia feed low on the food web, making culture in fertilized waters feasible and greatly simplifying the culture technology relative to feeding. That is of particular importance to subsistence culturists.

A. TEMPERATURE

While tilapia are able to tolerate high environmental water temperatures extremely well, all the species being cultured are tropical and have only a limited ability to survive cold. When exposed to cold water, disease resistance is severely impaired and death may result in a matter of hours or days following exposure, depending on the species and temperature (Table 1). Under conditions when temperature is a few degrees above the lower lethal limit, tilapia rapidly develop such problems as fungal and bacterial infections which may contribute to high mortality even if the temperature does not reach the lower lethal limit.

As a general rule, tilapia do not grow well at temperatures below about 16°C and cannot usually survive for more than a few days below about 10°C (Chervinski 1982). The normal range of temperature over which the representative species *T. aurea* can be found in nature is from about 13 to 32°C (Philippart and Ruwet 1982).

Growth of such species as *T. mossambica* is three times more rapid at 30°C (approximately the optimum temperature for growth) than as 22°C (Chimits 1955). The optimum temperature for feeding *T. zillii* (a species sometimes used in aquatic vegetation control) is from 28.8 to 31.4°C (Platt and Hauser 1978), which may be representative of other species as well.

Tilapia can tolerate temperatures of over 40°C in some instances, though there are also reports of mortality occurring at temperatures above 38°C (Allanson and Noble 1964; Denzer 1968; Gleastine 1974; Chervinski 1982). Acclimation time and temperature may have a bearing on subsequent upper lethal temperature (Chung 1983).

T. aurea seems to be one of the most cold tolerant of the cultured tilapia species (Table 1), with a generally accepted lower lethal temperature of between 8 and 9°C (McBay 1961, Sarig 1969), though one report indicated that the species may be able to tolerate temperatures as low as 6°C (Shafland and Pestrak 1982). *T. mossambica*

TABLE 1
Minimum Temperature Tolerances for
Some Species of Tilapia

Species	Lower lethal temperature (°C)
Tilapia aurea	8–9
Tilapia mossambica	8–10
Tilapia rendalli	11

Adapted from Chervinski 1982.

is perhaps typical in terms of its response to cold. That species ceases feeding at about 16°C, and mortalities begin at between 11 and 14°C. Total mortality occurs in the range of 8 to 10°C (Kelly 1956).

In tropical environments where ambient water temperature may vary only a few degrees even on an annual basis, few problems with temperature are encountered by tilapia culturists. However, in subtropical and temperate climates, winter conditions are typically associated with water temperatures that reach or surpass the lethal minimum temperature tolerated by tilapia. To prevent winterkill, some source of warm water must be provided. Geothermal water sources and the thermal effluent from fossil fuel and nuclear power plants have been utilized commercially or on a research basis for tilapia rearing, but their use has not been very well developed.

Given a temperature gradient such as that which occurs in the discharge canal of a power plant, tilapia will move to the warmest water and remain therein throughout the winter, moving into the reservoir that supplies the power plant with cooling water only after there has been a general and significant increase in ambient water temperature the following spring. Thus, power plant lakes can support self-sustaining populations of tilapia even in regions which have relatively cold winters, provided the power plant is not shut down during cold weather.

When abundant supplies of geothermal water are not available and the culturist does not have access to the heated effluent from a power plant, other options for overwintering the fish must be employed. The problem is simply one of providing suitable thermal conditions to carry the fish through the winter. For reasons of economics, it makes the most sense to overwinter brood stock, small fingerlings, or both, though in Israel it has been standard practice to overwinter intermediate size fingerlings for final growout the following spring (Hepher and Pruginin 1981).

Tilapia can be held in covered ponds that receive a constant supply of well water of sufficient temperature to keep them alive and healthy or they may be held in ponds, tanks, or raceways constructed under greenhouses (Chervinski and Stickney 1981; Stickney and Winfree 1982). Well water of suitable temperature should be constantly flowed into the culture chambers if no other source of supplemental heat is available, otherwise diel temperature fluctuations may lead to fish mortality. A properly designed greenhouse system can provide for maintenance and even spawning of broodstock during the winter, allowing the culturist to have a supply of fingerlings available for stocking as soon as ambient pond water temperatures are sufficiently high in the spring (Henderson-Arzapalo 1984).

Speculation as to the potential for breeding tilapia which are more cold tolerant or utilizing molecular biological techniques to introduce an antifreeze gene into tilapia has developed over the past several years. Behrends and Smitherman (1984) reported that they had been able to develop a strain of red tilapia which exhibited cold tolerance similar to that of *T. aurea*.

B. SALINITY

While tilapia do not appear to occur in salt water over their historical range, tolerance to salinity is, at least in some instances, remarkable (reviewed by Stickney 1986). Reports of large populations of tilapia becoming established in coastal waters following their introduction as exotic species have been appearing in recent years. Perhaps the most striking example is in Tampa Bay, FL, where *T. melanopleura* is now the dominant fish species present in the upper bay. The impact of such population explosions on native fishes is of concern to marine ecologists and environmentalists, though such problems appear to be rare at present. Studies to determine the conditions required for the establishment of tilapia in areas where they would not normally occur are warranted. As tilapia become increasingly introduced into the marine environment by culturists, problems that occur because of their competition with native species can only be expected to increase.

Growth of such species at *T. aurea* in natural and artificial seawater (in ponds and in the laboratory, respectively) is similar to that obtained in freshwater environments (Chervinski and Yashov 1971, Rahimaldin 1983). The ability of *T. zillii* to acclimate and grow in seawater is even better developed than that of *T. aurea* (Chervinski and Zorn 1974). One of the tilapia hybrids (cross between a *T. aurea* male and an *T. nilotica* female) has been shown to perform well in brackish water ponds (Fishelson and Popper 1968, Loya and Fishelson 1969).

According to Chervinski (1982), neither *T. zillii* nor *T. aurea* is able to reproduce in seawater. While no nest building has been observed in seawater ponds containing *T. aurea* (Chervinski and Yashouv 1971), nesting behavior by that species has been observed at various salinities in aquaria (Rahimaldin 1983). In a pond study, the gonadosomatic index (relationship between gonad weight and fish body weight) of *T. aurea* dropped in fish reared in seawater (Chervinski and Zorn 1974), but in aquaria, developing ovaries were found at salinities from 0 to 35 ppt (the experiment was conducted at 5 ppt salinity intervals) with the exception of 25 ppt (Rahimaldin 1983).

At least one species, *T. mossambica*, will reproduce over a salinity range from fresh to full-strength seawater, with one report indicating that reproduction occurred at 49 ppt salinity (Popper and Lichatowich 1975), but there have been no reports of *T. mossambica* becoming widely established in marine environments.

Research on the potential of producing red hybrid tilapia in seawater has developed in recent years, particularly in the Caribbean. While some studies were conducted with such species as *T. aurea* (McGeachin et al. 1987, Watanabe et al. 1989a), red hybrids have become the fish of choice because they are salt tolerant and their appearance more closely resembles that of some of the desirable marine species.

Florida red tilapia hybrids (*T. urolepis hornorum* × *T. mossambica*) can be spawned in brackish water. Watanabe et al. (1989b) found that sex-reversed (all-male) red hybrids spawned at 18 ppt grew better in the laboratory at 28°C and a 12:12 light to dark photoperiod at salinities of both 18 and 36 ppt than fish spawned in 4 ppt salinity water. The same authors found that fish spawned and sex-reversed at 2 ppt and 18 ppt did not differ in growth or survival when reared in outdoor tanks under ambient temperature conditions until the water temperature fell below 25°C at which time the growth and survival of fish spawned at 18 ppt were significantly higher.

Fry and juveniles of Florida strain red tilapia were shown to survive direct transfer from freshwater to 19 ppt salinity without suffering mortality or apparent stress. Adults did not begin to suffer mortality until they were exposed to salinities at or above 29 ppt (Pershbacher and McGeachin 1988).

The effects of salinity and temperature on growth of *T. spilurus*, *T. mossambica*, and red hybrids were investigated by Payne et al. (1988). Red hybrids grew best at what was considered an optimum salinity of 12 ppt. Watanabe et al. (1988a) reported that Florida red tilapia hybrids grew more rapidly at salinities above 10 ppt (up to 36 ppt) than at 1 ppt. The latter study employed Florida strain red hybrids (*T. urolepis hornorum* × *T. mossambica*), while the hybrids used in the study of Payne et al. (1988) appeared to be a combination of *T. urolepis hornorum* × *T. aurea*). Watanabe et al. (1988b) attributed the improved growth of Florida red hybrids at elevated salinities to a reduction in agonistic behavior.

C. DISSOLVED OXYGEN

The ability of various species of tilapia to survive low levels of dissolved oxygen (DO) is widely recognized. The lowest DO level at which survival has been reported is 0.1 mg/l. That level has been recorded for *T. mossambica* by Maruyama (1958) and *T. nilotica* by Magid and Babiker (1975). A slightly higher value of 0.3 mg/l was reported by Lovshin et al. (1977) for hybrid tilapia (*T. hornorum* male × *T. nilotica* female).

In situations where high levels of primary productivity are present in association with tilapia-rearing activities, occurrences of extremely low DO levels are not unusual. DO may frequently approach 0.0 mg/l during the predawn hours. That situation may recur over a period of days or even weeks under the proper conditions, and while it would prove quickly fatal to most species, tilapia are often able not only to survive, but may actually continue to grow at a relatively normal rate. This is particularly true when the oxygen depletion is remedied by photosynthetic oxygen input once sufficient light becomes available.

Tilapia have the ability to obtain oxygen from the saturated microlayer that exists at the air-water interface. During periods when the water column is depleted in oxygen, the fish will skim along the water surface with their mouths open. That activity causes oxygen-rich water to pass over the gills. Mortality in *T. aurea* has been reported as a result of low DO in one instance where a pond was covered with a heavy growth of duckweed (*Lemna* sp.) which precluded the tilapia from obtaining oxygen at the air-water interface (Stickney et al. 1977a).

D. AMMONIA AND PH

Few studies on the tolerance of tilapia to environmental ammonia have been conducted, but it is generally felt that these fishes are more tolerant than many others. Redner and Stickney (1979) found that *T. aurea* which had not been chronically exposed to elevated ammonia levels had a 48-h LC_{50} (concentration which was lethal to 50% of the exposed fish in 48 h) of 2.4 mg/l un-ionized ammonia. When the same species was exposed to sublethal un-ionized ammonia concentrations of 0.4 to 0.5 mg/l for 35 d, the LC_{50} value increased to 3.4 mg/l. That study showed that acclimation conditions affect the response of the fish to environmental ammonia concentration, a concept which may apply to other contaminants.

Daud et al. (1988) exposed red hybrid tilapia (*T. mossambica* × *T. nilotica*) to un-ionized ammonia and determined LC_{50} values for 48, 72, and 96 h of 6.6, 4.1, and 2.9 mg/l. Those authors reported that the threshold lethal concentration was 0.24 mg/l. The study indicates that the red hybrid is more tolerant of ammonia than *T. aurea*.

The pH of freshwater fish ponds is generally within the range of 6.5 to 8.5 and is controlled by the carbonate-bicarbonate buffer system as described in Chapter 1. In poorly buffered waters, relatively broad fluctuations in pH can occur when primary productivity is high since carbon dioxide levels will vary considerably as a function of photosynthetic activity which removes that chemical from the water leading to an increase in pH when the hydrogen donor bicarbonate ion has been depleted. Conversely, respiration by both plants and animals at night releases carbon dioxide into the water and the pH will fall once the available carbonate has been converted to bicarbonate allowing free hydrogen ions to occur in solution. Tilapia commonly occur in environments that have high levels of primary produc-tivity; thus, the fish may be exposed to relatively high diurnal pH fluctuations. The pH tolerance of tilapia, in general, has been reported to range from 5 to 11 (Chervinski 1982).

III. PRODUCTION

Tilapia can be grown in a variety of culture systems, the most common of which is ponds. Because of their tolerance to crowding and degraded water quality, tilapia are most commonly reared in monoculture, though polyculture has also been practiced, particularly when tilapia are a species of secondary importance.

Yields of tilapia in ponds have been shown to positively correlate with phytoplankton numbers, Secchi disk transparency, gross primary productivity, and the level of chlorophyll *a* in the water (Almazan and Boyd 1978). The best correlation among those variables was for chlorophyll *a* ($r^2 = 0.89$). Since tilapia feed low on the food chain (most species consume phytoplankton, rooted aquatic vegetation, zooplankton, or some combination of those food items), the positive correlation between fish production and the indicated variables seems logical.

Stocking rates for tilapia are highly variable depending on the type of culture technique employed. In regions where subsistence culture is practiced and little or no supplemental feed is provided, stocking densities may be as low as a few hundred fish per hectare. More intensive pond culture, in which heavy fertilization is used or prepared feeds are provided, is commonly associated with stocking rates of one or more fish per square meter. If water exchange is provided within a pond, several fish can be stocked per square meter of pond surface. Representative stocking densities for tilapia in ponds range from 3000 to 5000 fish per hectare in polyculture to from 10,000 to 60,000 fish per hectare in monoculture (Sarig and Marek 1974; Lovshin et al. 1977; Hepher and Pruginin 1981).

Pond culture of tilapia is highly developed in various parts of the world, with Israeli technology being among the most advanced. Nursery ponds are typically used as an intermediate stage for holding tilapia between spawning and stocking into production ponds (Hepher and Pruginin 1981). In many regions, fry or early fingerlings are reared in small net-pens (happas) for a few weeks before final stocking. Both spawning and early fry rearing are commonly conducted in happas in the Philippines. Elsewhere, fry are collected from spawning ponds and stocked immediately into production ponds.

Intensive tank, raceway, and cage culture can lead to production levels higher than those obtainable in ponds. In one cage study where the growth of hybrid tilapia (*T. mossambica* × *T. hornorum*) was examined, estimated production was as high as 50,000 kg/ha per year (Suffern et al. 1978). Coche (1982) compared the production of tilapia in cages with that of other warmwater fish species and showed that there were no striking differences among the fishes with respect to production and food conversion ratios (Table 2). Christensen (1989) also reviewed the cage culture literature and reported that stocking densities in tropical regions commonly range from 50 to 300 fish per cubic meter and production typically ranges from 50 to 150 kg/m³. Examination of published literature on *T. nilotica* by Carro-Anzalotta and McGinty (1986) showed that maximum carrying capacity of that species in cages is lower than for various other species, ranging from 10 to 70 kg/m³. Those authors concluded that the range may be more closely related to the total number of fish being reared in a particular water body than to the actual density of fish per unit volume of cage.

Growth, condition factor, and food conversion ratio (measured as dry weight of feed offered divided by weight gain for a given period of time) are better for *T. aurea* than *T. mossambica* as demonstrated in laboratory studies. Various densities of both species have been reared in flow-through tank culture systems with the results indicating that growth of *T. aurea* continued to be good even when their biomass reached 66.2 g/l (Henderson-Arzapalo and Stickney 1980). The particular strain of *T. mossambica* used in that study began to die when their biomass reached about 20 g/l, with mortality being attributed to an autoimmune response (Henderson-Arzapalo et al. 1980).

TABLE 2
Production Levels of Tilapia, Common Carp, and Channel Catfish Produced in Cages

Species	Estimated annual production (kg/m³)	Protein level (% in feed)	Feed conversion ratio
Tilapia aurea	200 to >300	40	1.0–1.8
Tilapia nilotica	200 to >300	25–30	1.9–2.2
Cyprinus carpio	Over 240	30–35	1.6–2.3
Ictalurus punctatus	Over 240	36	1.4–2.5

Adapted from Coche 1982.

Tilapia have been stocked in polyculture with a variety of species, including Chinese carps (various species), common carp (*Cyprinus carpio*), mullet (*Mugil cephalus*), channel catfish (*Ictalurus punctatus*), freshwater shrimp (*Macrobrachium rosenbergii*), and tiger shrimp (*Penaeus monodon*), as a means of increasing overall pond production (Yashouv 1969, Spataru and Hepher 1977, Brick and Stickney 1979, Hepher and Pruginin 1981, Wilson and Hilton 1982, Rouse and Stickney 1983, Meriwether et al. 1984, Williams et al. 1987, Gonzales-Corre, 1988, Shen and Xu 1988). If *T. aurea* are stocked with common carp at a density of no more than 5000 fish per hectare, they will not adversely affect, and might even stimulate carp growth (Spataru and Hepher 1977). The same study demonstrated that tilapia growth was not affected by the presence of either carp or mullet at densities of 2500 to 3000 fish per hectare. Part of the success of polyculture relates, at least in some cases, to the fact that other fish species may prey upon tilapia fry as they are produced, thus reducing the competition for food by the products of unwanted reproduction. This subject is considered in more detail under Section V below.

Polyculture can be practiced by having the various species intermingle within a pond, or one of the species may be placed in cages to provide physical separation. For example, Meriwether et al. (1984) reared caged red tilapia hybrids in ponds containing freshwater or marine shrimp. Freshwater shrimp and tilapia can be allowed to intermingle in ponds without any apparent negative impact due to agonistic behavior (Brick and Stickney 1979; Rouse and Stickney 1983).

As is true with the culture of various other species, tilapia culture has sometimes been integrated with hydroponics to produce both fish and vegetables. Tomatoes and cucumbers have been successfully produced in such systems which employed fish waste as a source of nutrients for the plants (Watten and Busch 1984; Wren 1984). As interest in closed recirculating water systems for producing tilapia in temperate regions has grown, such integrated systems are actively being considered or put into place in conjunction with private sector production units. Watten and Busch (1984) described a water system which they indicated could produce tilapia and tomatoes economically in the U.S. Virgin Islands.

IV. FERTILIZATION

Since the cultured species of tilapia feed low on the food chain and most are filter feeders, the promotion of plankton blooms by fertilizing ponds is a logical culture practice and one that has been adopted in many regions. Inorganic fertilizers, organic wastes (agricultural byproducts, manures), and combinations of the two have been widely utilized to promote natural food production in tilapia ponds. Organic fertilization has long been utilized as a primary method of providing food not only for tilapia but also for various species of carp and other fishes in Asia. The technique is also employed in commercial aquaculture in Israel (Hepher and Pruginin 1981) and other countries in the Middle East, Europe, and elsewhere.

In the U.S., pond fertilization aimed at the promotion of algal blooms has depended primarily on inorganic fertilizers, though a considerable amount of research has been conducted in conjunction with organic fertilization of tilapia ponds and some commercial interest in organic fertilization has developed.

Boyd (1976) examined the effect of 15 biweekly applications of 3 inorganic fertilizers at the rate of 22.5 kg/ha on the production of *T. aurea* in Alabama ponds. The fertilizers were 0-20-5, 5-20-5, and 20-20-5 (N-P-K percentage) compounds. Increased fish production occurred as the level of nitrogen in the fertilizer was increased because of the influence of that nutrient on phytoplankton production. In general, the types of fertilizer utilized in tilapia ponds are similar to those used in channel catfish ponds, though the frequency and rates of application may be higher in tilapia culture as compared with the standard fertilization rates employed in the catfish culture industry.

With respect to organic fertilizers, the most commonly utilized materials are poultry and swine manures, though cattle waste has also been evaluated as a fertilizer for tilapia ponds (Schroeder 1974). In addition, commercial organic fertilizers are available and were evaluated in at least one study (Mang-Umphan and Arce 1988). Organic fertilization with animal manures has been reviewed by Edwards (1980).

While most of the emphasis has been placed on establishing fish culture systems to which organic fertilizers are added in the interest of producing fish for human consumption, the use of tilapia in removing algae from sewage treatment ponds is also a possibility. With the latter type of system, sale of the fish for human consumption in the U.S. may be prohibited, but the use of fish could actually be less expensive than other forms of secondary waste treatment in certain instances, even if the fish were discarded. The removal of microalgae from sewage systems has received some attention by researchers (Edwards and Sinchumpasak 1981; Edwards et al. 1981a, 1981b).

Several studies have been conducted in recent years in Texas where tilapia were reared in ponds that received the waste from either poultry (Figure 2) or swine (Stickney et al. 1977a; Stickney and Hesby 1978; Stickney et al. 1979; Burns and Stickney 1980; McGeachin and Stickney 1982). Conclusions drawn from those studies were that the waste from the equivalent of 50 growing-finishing hogs per hectare or between 70 and 140 kg/ha/d of poultry manure will produce high levels of primary and secondary productivity that lead to excellent tilapia growth. Similar figures were developed relative to swine waste by Aguilar (1983), though Hopkins et al. (1981) recommended 103 hogs per hectare for ponds in the Philippines that were stocked with tilapia (*T. nilotica*), common carp (*C. carpio*), and snakeheads (*Ophicephalus striatus*). Broussard et al. (1983) recommended fertilization of *T. nilotica* broodfish ponds in the Philippines with dried chicken manure at the rate of 3000 kg/ha per month (about 100 kg/ha/d) in conjunction with 100 kg/ha per month of 16-20-0 inorganic fertilizer.

Tilapia production in saline waters has also been conducted in conjunction with organic fertilization. Tamse et al. (1985), working in the Philippines, introduced swine waste into brackishwater ponds containing *T. mossambica*. They found that application of swine waste effluent from biogas digesters produced better fish production when added five times weekly as compared with twice weekly applications. In another study in the Philippines, *T. nilotica* were reared along with milkfish (*Chanos chanos*) and shrimp (*Penaeus indicus*) in experimental ponds that received the waste from broiler chickens (Pudadera et al. 1986).

FIGURE 2.   Laying hens housed over a pond containing *Tilapia aurea*. Manure and spilled chicken feed fertilize the pond creating a dense plankton bloom that provides feed for the fish.

Mixtures of manure and sorghum have been evaluated in Israel (Spataru, 1976b), where it was determined that the natural food supplies were better in ponds which received the grain than in those receiving only manure. It was felt that the *T. aurea* in the ponds were feeding on benthic animals and detritus to a large extent and that the production of those types of food organisms was improved with the addition of the sorghum.

V. REPRODUCTION AND GENETICS

Most of the commercially cultured tilapia species are mouthbrooders. The strategy involves nest-building activity by the male (Figure 3), after which he attempts to attract a gravid female. Following a brief period of courtship, the female deposits her eggs in the nest where they are fertilized and then picked up by one of the adults for incubation in the mouth. In most species, and true for all of the cultured fish within the genus *Tilapia*, the female incubates the eggs. Once the fry have hatched, they remain in the mouth of the female through yolk-sac absorption. Swim-bladder inflation will also generally occur during the posthatch period when the fry are under the protection of the female (Doroshov and Cornacchia 1979). For a few days after the fry are released, they will remain in close proximity to the mother and seek refuge in her mouth should they perceive a threat of danger.

FIGURE 3. Circular tilapia nests create a moonscape in the bottom of a drained brood pond. The diameter of the nests provide an indication of the size of the male that constructed each.

While the fecundity of an individual tilapia female is relatively low (for example from 69 to 302 eggs per female in *Sarotherodon galilaeus* [Blay 1981]), the onset of spawning at an early age, coupled with a short refractory period, lead to sexual maturity at a size as small as 10 cm (Babiker and Ibrahim 1979) and the production of as many as 6 to 11 broods within a year under proper environmental conditions (Chimits 1955). Onset of spawning may occur when the fish are only 3 months old in the case of *T. mossambica*. Other species, such as *T. aurea* and *T. nilotica*, generally begin spawning at an age of about 6 months. *T. nilotica* females mature at 11.4 cm and males at 14.3 on an average (Babiker and Ibrahim 1979). Once the females of any of the mouthbrooding species begin to mature, energy is diverted from growth to egg production. In addition, females are unable to eat for periods in excess of 2 weeks during which eggs and fry are held in the mouth. Thus, there is a distinct divergence with respect to size between the males and females once the fish become sexually mature.

Another spawning strategy involves substrate spawning where the eggs and fry remain within the nest during incubation and yolk-sac absorption. Fecundity in such substrate spawning species as *T. zillii* often exceeds 1000 eggs per spawn, but the fish are initially smaller than those of the mouthbrooding species, and growth tends to be slower, so mouthbrooders are more popular among culturists. *T. zillii* have been produced in the U.S. for use as biological weed control agents, for example, in California irrigation canals.

In addition to growth reduction during gonadal development and following maturity in mouthbrooding tilapia females, a secondary problem of overpopulation and consequent stunting in ponds where reproduction is not controlled has been reported. Reproduction leads to increased competition for food within a pond and can lead to degradation of water quality. Strategies have been developed, as discussed below, for reduction or elimination of spawning in production ponds.

A. CAPTIVE BREEDING

Spawning will occur in ponds, tanks, cages, and other types of culture environments. Photoperiod does not seem to be a significant controlling factor, but temperature is critical. Tilapia will not spawn if the water temperature falls below 20 to 23°C (Uchida and King 1962; Huet 1970), thus in temperate climates, spawning begins in the spring and may continue until early fall. In the tropics, year-round spawning may occur.

The simplest technique for spawning tilapia involves stocking broodfish in an open pond and later collecting fry or fingerlings. Once the fry permanently leave the mouth of the female, they swim in schools for several days, often appearing near the edge of the pond where the water is warmest. Such schools can be easily netted. As the spawning season progresses, individuals and schools of young fish that escape capture will reach sufficient size that they can cannibalize newly released fry. Cannibalism may be the reason for the apparent disappearance of fry in brood ponds after several weeks of spawning have occurred. The phenomenon has been documented in both aquaria and tanks (Uchida and King 1962; Silvera 1978; Pierce 1980; Berrios-Hernandez 1983; Pantastico et al. 1988) as well as in ponds. While it is generally thought that overpopulation and stunting are significant problems, cannibalism of fry as described above may actually limit the problem significantly. There is not much documentation of stunting problems in the literature though the perception that stunting is of concern is frequently mentioned.

Pond spawning has been widely used in the production of hybrid tilapia. Such crosses as *T. mossambica* × *T. nilotica* can be readily produced (Avault and Shell 1967) which may grow more rapidly than either of the parents. A number of other hybrid crosses have been produced, and as discussed below, some of them result in high percentages of male offspring. The golden or red tilapia (a hybrid between *T. mossambica* and *T. hornorum*) has recently become popular in many parts of the world. The obvious problem with hybridization is that the species used in the cross must be maintained separately until paired for spawning, and a high degree of precision must occur in conjunction with sex determination to ensure that both sexes of one of the species are not stocked in the same brood pond.

The happa spawning method is widely employed in the Philippines. Happas are small net-pens with mesh size sufficiently small to retain eggs and fry. Several happas stocked with one or more pairs of broodfish can be placed within a small pond, and fry can be removed for stocking as they are produced. A modification on the happa technique involves the use of larger mesh netting on one side of the net-pen. The larger-mesh side will allow the fry to escape but will retain the adults. Once sufficient numbers of fry have escaped to stock a pond, the broodfish happas can be moved to another pond. So long as the bottom of such happas have sufficient fine mesh to retain the eggs upon their deposition, the adults should be able to perform their reproductive function satisfactorily.

Tilapia can be spawned with relative ease in aquaria (Rothbard and Pruginin 1975, Rothbard 1979), but that technique is not generally suitable for commercial culturists because of the requirement for special facilities and trained personnel which are often not available in developing nations where tilapia culture is most extensively practiced. Aquarium studies have been important in outlining the spawning behavior (Rothbard 1979) and environmental conditions (Rothbard and Pruginin 1975) required by various tilapia species.

Stripping of females after egg development has been accomplished, followed by external incubation of the eggs. Eggs may also be obtained from the mouths of brooding females and incubated under artificial conditions. Shaker tables (Rothbard and Pruginin 1975) and conical-shaped incubation chambers with updraft water-flow (Rothbard and Hulata 1980) have been successfully used as egg incubators. While there may be no particular difference in hatching success for eggs taken from the mouths of females and hatched in incubation chambers as compared with collection of fry after hatching (Berrios-Hernandez 1983), externally incubated eggs taken from the mouths of *T. aurea* females in one study had relatively low survival rates, particularly when removed during the early stages of development (Henderson-Arzapalo 1984).

Ratios of males to females in spawning chambers may vary considerably, though it would seem to make sense to stock larger numbers of females than males. A male will quickly begin to court secondary females once he has spawned and the original female has picked up the eggs and abandoned the nest site. In tanks, a ratio of three females to each male has been effective (Uchida and King 1962, Berrios-Hernandez 1983).

B. CONTROL OF REPRODUCTION

As reviewed by Guerrero (1982), reproduction in tilapia can be controlled by (1) ensuring that fertilized eggs do not survive, (2) stocking predators that will harvest fry, or (3) rearing unisex populations. In the latter case, it is generally desirable to rear all-male populations because of their more rapid growth as compared with females (Chimits 1955; Lowe-McConnell 1958; Van Someren and Whitehead 1960a, 1960b; Avault and Shell 1967; Fryer and Iles 1972; Guerrero and Guerrero 1975). If reproduction is not controlled, stunted populations may occur as has been demonstrated in one tilapia species in West Africa (Eyeson 1983).

Early studies indicated that tilapia reared in cages could not reproduce. While disruption of normal spawning behavior was attributed to the lack of reproduction in cages (Pagan 1969), instances of successful fry production in cages were reported when the mesh size was 0.3 cm or smaller (Pagan-Font 1975; Rifai 1980). The dependence on mesh size for successful reproduction is related to the size of the eggs being produced; if the mesh is sufficiently large, the eggs will fall through before the female can pick them up. The use of small mesh for egg retention is the key to production of tilapia fry in happas.

Various species of fish have been stocked in culture ponds as predators on tilapia fry (Table 3). It is important that the growth of the tilapia originally stocked is rapid enough to outpace that of the predator; otherwise, the crop may be decimated. A

TABLE 3
Predatory Fish Species Used to Assist in Control of Tilapia Reproduction and Regions Where Species Have been Employed

Predator	Location	Ref.
Channa striata (Snakehead)	Thailand	Balasuriva (1988)
Cichla ocellaris (Peacock bass)	Puerto Rico	McGinty (1984)
Cichlosoma manguense (Oscar)	El Salvador	Dunseth and Bayne (1978)
Dicentrarchus labrax (Sea bass)	Israel	Chervinski (1974)
Lates niloticus (Nile perch)	Africa	Pruginin (1965, 1967)
		Meschkat (1967)
		Ofori (1988)
Megalops cyprinoides (Tarpon)	Philippines	Fortes (1985)
Micropterus salmoides (Largemouth bass)	U.S.	Swingle (1960)
		McGinty (1985)
Mudfish (species not given)	Thailand	Chimiits (1957)

properly selected predator should not only consume tilapia fry as they are produced, but could also be a secondary cash crop. In most instances, the predator is selected from among locally available species; it is not generally considered wise to introduce an exotic predatory species.

Numbers of predators to be stocked will vary depending upon its species, the number of tilapia stocked, and the availability of other types of food in addition to tilapia fry. McGinty (1985) found that 143 largemouth bass per hectare effectively decreased recruitment while promoting increased yields in *T. nilotica* stocked at 9500 per hectare (90% of the tilapia were males).

Perhaps the most popular means of controlling tilapia reproduction involves stocking all-male fish. This can be accomplished by hand sexing or by producing all-male stocks. Hand sexing may be effective with fish as small as 40 g (Sarig and Arieli 1980), but fish from 50 to 70 g seem to be preferred (Hepher and Pruginin 1982). The distinction between the sexes is more easily made if the genital papilla is stained with a water-soluble die such as Azorubin prior to inspection (Chervinski and Rothbard 1982.) In any case, the technique is time-consuming and not 100% effective, as errors are almost inevitable.

The production of all-male populations or even all-female populations can be achieved from certain hybrid crosses (Table 4) and by feeding hormones to tilapia fry. Hybridization has been popular in Israel and Asia, but as previously indicated, the accidental mixing of both sexes of the same species in the same spawning pond can lead to problems. Further, the hybrids must be maintained separate from the parental fish to preserve the integrity of the technique. Finally, the crosses which are reputed to lead to all-male hybrids are not always 100% effective (reviewed by Wohlfarth and Hulata 1983). One reason may be that because tilapia hybridize so freely, genetically pure representatives of the various species may be difficult to obtain. Many stocks appear to be hybrids themselves, though meristics show that they conform very well to a particular species.

The most recent technique that has been developed for sex-reversing tilapia involves feeding hormones to fry before sexual differentiation takes place. While

TABLE 4
Results of Tilapia Hybridization
for Production of Unisex
or Predominantly Unisex Populations

Species used in cross	Ref.
Crosses leading to predominantly unisex populations	
T. niloticus × *T. hornorum*	Hickling (1960); Chen (1969)
T. mossambica × *T. aurea*	Avault and Shell (1967); Guerrero and Gaguan (1979); Pierce (1980); Hsiao (1980)
T. hornorum × *T. aurea*	Pruginin (1965); Lee (1979); Pinto (1982)
T. niloticus female × *T. aurea* male	Pruginin et al. (1975)
T. vulcani female × *T. aurea* male	Pruginin et al. (1975)
Crosses leading to all-male offspring	
T. mossambica female × *T. hornorum* male	Hickling (1960); Chen (1969)
T. nilotica female × *T. hornorum* male	Pruginin (1967)
T. nilotica × *T. macrochir*	Jalabert et al. (1971)
T. nilotica female × *T. aurea* male	Fishelson (1962); Pruginin (1967)
T. nilotica × *T. variabilis*	Pruginin (1967)
T. spilurus niger × *T. hornorum*	Pruginin (1967)
Cross leading to all-female offspring	Edwards and Sinchumpasak
T. melanotheron female × *T. mossambica* male	(1981); Edwards et al. (1981a)

most of those who have investigated the technique were interested in producing all-male populations and therefore fed androgens, some studies have focused on the production of all-female populations by feeding estrogens (Hopkins et al. 1979; Jensen and Shelton 1979). Successful production of all-male tilapia by feeding hormones has been reported for *T. mossambica* (Clemens and Inslee 1968; Guerrero 1979; Varadaraj 1990), *T. nilotica* (Jalabert et al. 1974; Guerrero and Abella 1976; Tayamen and Shelton 1978), and *T. aurea* (Guerrero 1975; Shelton et al. 1981; Henderson-Arzapalo 1984).

The technique described by Guerrero (1975) is representative. He found that when *T. aurea* fry were fed for about 3 weeks, beginning at first feeding, a prepared diet containing 60 mg/kg of 17-α-ethynyltestosterone, only male fish were produced. The feed is prepared by dissolving the appropriate amount of androgen in ethyl alcohol and mixing the alcohol with the feed at a ratio of about two parts alcohol to one part feed. The mixture is then placed in a drying oven at about 60°C until the alcohol evaporates. During the evaporation process, the androgen is adsorbed on the surface of the feed particles. When large numbers of fish are being sex-reversed, the success rate may decline slightly below 100% because there is an increased chance that any given fish will not obtain sufficient amounts of hormone to affect sex reversal. Also, the level of hormone and duration of feeding that work well on *T. aurea* may not apply as well to other species.

Torrans et al. (1988) exposed *T. aurea* to various concentrations of the synthetic steroid mibolerone in the water for 5 weeks. Exposure of tilapia to 0.6 ppm of the hormone appeared to be an effective method of achieving sex-reversal. In a related study, Meriwether and Torrans (1986) found that 10 ppm mibolerone was lethal to tilapia fry within 24 h.

Public health concerns relating to the use of hormones in sex-reversal of fish center around whether traces of those hormones remain in the body of the fish at the time of harvest. Studies have been conducted in which the elimination of radio-labeled methyltestosterone was evaluated in tilapia. Johnstone et al. (1983), who worked with *T. mossambica* and rainbow trout (*Oncorhynchus mykiss*, formerly *Salmo gairdneri*) found that 100 h after the hormone was withdrawn, less than 1% of the initial radioactivity was present, indicating that the hormone was rapidly eliminated. Goudie et al. (1986) worked with *T. aurea* and found that more than 90% of the radioactivity was in the viscera during 21 d of feeding labelled hormone. Once the hormone was withdrawn the radioactivity decreased exponentially. The radio-activity level after 21 d of withdrawal was less than 1% of that present when the fish were being fed the hormone. The conclusion that can be drawn from these experiments is that the use of hormones to sex-reverse tilapia poses no public health problem to consumers as the fish will have eliminated the hormone long before reaching market size.

Gynogenetic and triploid fish have been produced in conjunction with at least three species of tilapia: *T. mossambica*, *T. nilotica*, and *T. aurea* (Penman et al. 1987). Such fish offer another option for controlling reproduction.

C. GENETICS

Improvements of tilapia stocks through genetic control have not been widely addressed as yet, though there are areas where selective breeding and genetic engineering might be expected to have significant impacts. Breeding for improved growth in *T. nilotica* was of only limited success in one study because of low genetic variation (Tave and Smitherman 1980). Low heritabilities in age and size at maturation have also been demonstrated in the same species (Uraiwan 1988). The subject has been reviewed by Tave (1988), who reported that limited success in using selection to improve tilapia growth has been achieved.

Since many culturists are working with relatively small gene pools and there has been so much hybridization among the various stocks around the world, selective breeding may not be an effective means of improving performance. Electrophoretic stock identification with respect to tilapia is being developed (McAndrew and Majumdar 1983). Comparisons of presently cultured fish with museum specimens dating back to the period before tilapia were almost indiscriminately moved from one location to another would be very useful.

Research on the inheritance of body color in the Taiwanese strain of the red tilapia hybrid seems to indicate that color is under the control of a single gene. The red color appears to be dominant and the black recessive (Huang et al. 1988a, 1988b).

Genetic engineering may provide the means by which improved strains of tilapia can be produced. For example, gene splicing to meet that objective may be on the horizon.

One intriguing concept would be to introduce one or more genes which would improve the cold tolerance of tilapia (so-called antifreeze genes). While desirable from a culture standpoint, cold-tolerant tilapia could cause problems in areas where they would compete with native fishes considered to be more desirable. Such competition is reduced, at present, in temperate climates where tilapia die during winter. Year-round survival of tilapia may not be desirable in many regions outside the tropics. For example, in the U.S., *T. aurea* have been shown to outcompete largemouth bass (*Micropterus salmoides*) for spawning sites in a Texas power plant cooling lake where the tilapia were able to overwinter (Noble et al. 1975). If tilapia were genetically engineered to become tolerant to temperate waters, similar competition could conceivably develop in a variety of aquatic habitats, leading to significant reductions or even the elimination of largemouth bass and perhaps other species.

VI. NUTRITION

Even though tilapia are among the most widely cultured fishes in the world, relatively little attention has been directed toward outlining their nutritional requirements until recently. The predominance of culture systems that rely on frequent fertilization as a means of providing natural food has certainly taken the emphasis away from prepared rations and, thus, the need to clearly define the nutritional requirements of these fishes. Over the past few years, some fish nutritionists have turned their attention toward tilapia, and significant advances in our knowledge have been made. Tilapia nutrition was recently reviewed by El-Sayed and Teshima (1991).

A. NATURAL FOODS
One way to begin evaluating the nutritional requirements of a fish or group of fishes is to examine their food habits. The composition of the preferred foods can provide some indication of how a prepared diet should be formulated relative to the protein, carbohydrate, lipid, and energy levels.

The ability of such species as *T. aurea* to grow rapidly on blue-green algae (Burns and Stickney 1980) and other primary producers has led to the view in some circles that tilapia of commercial aquaculture importance are herbivorous as juveniles in nature, though most species appear, in reality, to be omnivores (Table 5). Even species such as *T. zillii*, which has been introduced throughout much of the southwestern U.S. as a weed-control species in irrigation canals, consumes some animal matter in nature, though it could be contended that animals are incidentally ingested as the fish feed on aquatic vegetation.

B. PROTEIN AND ENERGY REQUIREMENTS
Studies to evaluate protein requirements and protein utilization by tilapia have employed a variety of feedstuffs, some commonly utilized in commercial fish feeds for other species, and others of an exotic nature. Studies with *T. aurea*, *T. mossambica*, *T. nilotica*, *T. zillii*, and hybrid tilapia have indicated that each performs best when dietary protein is in the 30 to 40% range (Davis and Stickney

TABLE 5
Food Habits of Selected Tilapia Species

Species	Primary food ingested	Ref.
Tilapia aurea	Zooplankton, detritus	Spataru and Zorn (1978) Spataru and Gophen (1983)
T. nilotica	Phytoplankton	Moriarty and Moriarty (1973)
T. mossambica	Macrophytes, benthic algae, phytoplankton, periphyton, zooplankton, fish larvae, fish eggs, detritus	Munro (1967); Naik (1973); Man and Hodgkiss (1977); Weatherly and Cogger (1977); Bowen (1979, 1980)
Sarotherodon galilaeus	Phytoplankton	Spataru (1976a); Spataru and Gophen (1983)
Tilapia zillii	Macrophytes, benthos	Abdel-Malek (1972); Buddington (1979)
Red tilapia hybrids	Microalgae, phytoplankton, zooplankton	Grover et al. (1988, 1989)

1978; Mazid et al. 1979; Newman et al. 1979; Jauncey 1982, Viola and Zohar 1984; Teshima et al. 1985; Siddiqui et al. 1988; Wee and Tuan 1988). The studies cited used various protein sources, including casein-gelatin (Teshima et al. 1985), and also various energy levels so additional work to determine precise protein requirements may be required.

High protein diets have been recommended for brood tilapia. Chang et al. (1988) found that a 44%-protein eel-ration led to better fry production and broodfish growth in Taiwanese red hybrids than rations with protein levels of 22 and 24%. Similar results were reported with respect to *T. nilotica* by Santiago et al. (1985) who compared feeds with 20 and 40% crude protein.

The protein to energy requirements of *T. aurea* and *T. mossambica* have been examined (Jauncey 1982; Winfree and Stickney 1981) and appear to be similar, though more work in that area should also be conducted.

Most of the effort on protein and amino acid utilization has been centered on *T. mossambica*. That species has been shown capable of using free amino acids (Jackson and Capper 1982). The quantitative indispensable amino acid requirements of *T. mossambica* have also been determined (Table 6).

It is possible to substitute such plant proteins as soybean meal and cottonseed meal for fish meal in tilapia diets (Davis and Stickney 1978; Jackson et al. 1982; Viola and Arieli 1983; Tacon et al. 1984), though even some exotic plant proteins can be used. Jackson et al. (1982) evaluated copra, groundnut, sunflower, rapeseed, and leucaena meals in addition to soybean and cottonseed meal.

In most of the studies to date, growth has been depressed if excessive levels of plant protein are used as substitutes for fish meal. In general, 25% replacement of animal protein with plant meals appears to be acceptable. At a relatively low protein level (25%), half of the fish meal can be replaced by soybean meal without negative impacts on growth or food conversion (Viola and Arieli 1983). Other sources of animal protein, such as meat and bone meal and feather meal, have also been used as substitutes for fish meal in tilapia rations (Tacon et al. 1984).

There remains some confusion as to the appropriateness of using plant proteins as animal protein replacements in tilapia feeds. While in one study, digestibility studies on various feedstuffs have shown that protein and energy in animal-based foodstuffs are more available to *T. nilotica* than those from plants (Hanley 1987), another study reported that soybean protein concentrate was more efficiently utilized than anchovy meal by the same species (Abdelghany 1987). In yet another study, Viola et al. (1988a) found that soybean meal could be used as a complete replacement for fish meal in feeds provided to hybrid tilapia (*T. nilotica* × *T. aurea*). The soybean meal diet was also altered from the original commercial feed formulation by the addition of 3% dicalcium phosphate and 2% oil.

Antinutritional factors are sometimes present in plant proteins and treatment of those protein sources prior to their incorporation into fish feeds is one factor that can influence the result of feeding trials. Proper heating of soybean meal, for example, will eliminate trypsin inhibitor activity, and studies have shown that full-fat soybean meal containing trypsin inhibitor leads to decreased performance in *T. nilotica*, as compared with the same meal after proper heat treatment (Wee and Shu 1989). It is not always clear from published reports whether trypsin inhibitor has been taken into consideration in studies where soybean meal is substituted for animal protein.

Cottonseed meal may contain gossypol which can lead to poor growth and toxicity in fish. Robinson et al. (1984) fed *T. aurea* diets containing cottonseed meal with and without gossypol and found that, in both cases, growth was depressed

TABLE 6
Amino Acid Requirements of
Tilapia mossambicas

Amino acid	Minimum requirement (% of diet)
Arginine	0.93
Histidine	0.37
Isoleucine	0.75
Leucine	1.28
Lysine	1.62
Methionine	0.16
Phenylalanine	0.79
Threonine	0.95
Tryptophan	0.16
Valine	0.75

Data from Jackson, A. J. and Capper, B. S., *Aquaculture*, 29, 289, 1982; Jauncey, K., Tacon, A. G. J., and Jackson, A. J., in *Int. Symp. Tilapia in Aquaculture*, Fishelson, L. and Yaron, Z., Eds., Tel Aviv University, Tel Aviv, 1984, 328.

compared with diets formulated with either soybean or peanut meal. The presence of gossypol did not appear to be related to the growth depression.

While some plant proteins may lead to slower growth and lower feed-conversion efficiency, their lower cost may mean that they are more economical on a cost to benefit basis than fish meal-based diets (El-Sayed 1990). The degree to which plants can replace animal protein in tilapia diets, both with respect to the effect on performance and economics, should be investigated on a case-by-case basis. Certain plant proteins may not be suitable. For example, when the alga *Cladophora glomerata* was used to replace fish meal in diets fed to *T. nilotica*, growth and protein utilization were depressed when replacement exceeded 5% (Appler and Jauncey 1983).

In many countries where tilapia culture is popular, such high quality plant proteins as soybean, corn, peanut, and cottonseed meal are either not available or are given priority for use in human diets and livestock feed, leaving none for incorporation into fish feeds. Another problem of particular importance in developing countries is that animal proteins and certain plant proteins are very expensive. Thus, there have been many attempts to find suitable alternative protein and energy sources for tilapia. Some have shown promise (a few are mentioned in earlier paragraphs in this section), while others have not promoted good growth in the fish. Such exotic feed ingredients as, ipil-ipil meal (*Leucaena leucocephala*), sesbania seed meal (*Sesbania grandiflora*), duckweed (*Lemna gibba*), *Azolla pinnata* meal, mulberry leaf meal, tapioca, sweet lupin, coffee pulp, copra, cassava, groundnut, distillation solubles, pharmaceutical wastes, spent beer waste, Pito brewery waste, chicken feed, mosquito and soldier fly larvae, zooplankton, lettuce, and an industrial single cell protein have been incorporated into tilapia diets with various degrees of success (Moriarty 1973; Bayne et al. 1976; Cruz and Laudencia 1978; Kohler and Pagan-Font 1978; Gophen 1980; Guerrero 1980; Jackson et al. 1982; Appler and Jauncey 1983; Gaigher et al. 1984; Wee and Ng 1986; Bondari and Sheppard 1987; Wee and Wang 1987; Davies and Wareham 1988; Oduro-Boateng and Bart-Plange 1988; Olvera et al. 1988; Santiago et al. 1988a, 1988b; Viola et al. 1988b). Besides having been used as organic fertilizers, poultry and cattle manure have been incorporated into prepared feeds for tilapia (Stickney et al. 1977b; Degani et al. 1982).

C. CARBOHYDRATES

Few studies on the carbohydrate requirements of tilapia have been conducted, though it has been demonstrated that there is considerable potential for carbohydrate to spare protein in the diet of *T. mossambica* (Anderson et al. 1983). Diets high in fiber cannot be expected to have as much value for tilapia as those containing less complex carbohydrates since it has been established that *T. aurea* do not contain cellulase activity within their intestinal tracts (Stickney 1976).

D. LIPIDS

Interest in determining the lipid requirements of tilapia has developed over the past several years. In 1980, it was determined that *T. zillii*, unlike many other fishes, may have a dietary requirement for fatty acids in the linoleic acid family (Kanazawa et al. 1980). Since then, research with various species of *Tilapia* has led to the conclusion that linolenic acid may not be required or is required at relatively low levels. There may even be interference with *de novo* fatty acid synthesis when linolenic acid is present at high dietary levels (Stickney and McGeachin 1984). Total n-3 dietary fatty acid composition seems less important than the level of linolenic acid, which if it is below 1% of the total diet, appears to not have a detrimental effect on growth (Stickney and Wurts 1986). The lipid requirements of various fishes, including tilapia, were reviewed by Stickney and Hardy (1989). From that review, it is clear that additional research is required, and indicates that tilapia as a group may differ from trout and catfish in requiring dietary n-6 fatty acids.

The difference between tilapia and other fishes was demonstrated in one study in which *T. aurea* grew as well on diets supplemented with soybean oil as fish oil. Beef tallow, which has produced good growth in channel catfish, does not appear to be a very good lipid source for tilapia (Stickney and McGeachin 1983).

E. VITAMINS AND MINERALS

Diets containing vitamin supplements developed from other species seem to perform well when provided in tilapia diets. Deficiency signs can be induced in tilapia, at least in the case of vitamin C, vitamin E, and pantothenic acid (Stickney et al. 1984; Roem et al. 1990a, 1990b). Complete diets for *T. aurea* should contain 50 mg/kg of vitamin C (Stickney et al. 1984) and 6 mg/kg of pantothenic acid (Roem et al. 1990b). Vitamin E in the form of *dl*-α-tocopherol acetate is required at a level of about 10 mg/kg of diet when the feed contains 3% dietary lipid and 25 mg/kg of diet when the lipid level is 6% (Roem et al. 1990a).

No choline requirement has been demonstrated for tilapia, even though diets lacking that vitamin have been fed in conjunction with other diets containing graded levels of the vitamin up to 500 mg/kg of diet (Roem et al. 1990b). In addition, there appears to be no dietary vitamin B_{12} requirement by *T. nilotica* since that vitamin can be synthesized in the intestine (Lovell and Limsuwan 1982).

The mineral requirements of tilapia are also sketchy. Viola et al. (1986) found that the phosphorus requirement of tilapia hybrids varied with fish size and dietary protein source. For small fish, the requirement appeared to be about 1% available dietary phosphorus, while larger fish grew well when a herring meal-sorghum feed supplied 0.7% available phosphorus. Among the trace metals necessary for good tilapia growth, the zinc requirement has been investigated by McClain and Gatlin (1988). That study demonstrated the requirement, and showed that zinc bioavailability is reduced by dietary phytate.

F. BODY COLOR

Commercial tilapia producers have become interested in the red tilapia hybrid and report that consumer acceptance is increased as compared with fish having normal coloration. A wide range in color occurs with respect to red tilapia hybrids, even when fish from the same hatch are compared, though as more and more generations of hybrids are produced, the color morphs breed more true with each generation. Carotenoid pigments have been found to enhance color in various species of fish, and tilapia are not an exception. When lutein, rhodoxanthin, and the alga *Spirulina* sp. were fed to red tilapia, there was color enhancement as compared with controls (Matsuno et al. 1980). As yet, color-enhancing chemicals are not being widely used by tilapia culturists as they depend more heavily on breeding to produce the external red coloration of the fish and appear to be satisfied with the white flesh that is typical of all the commercially produced tilapia species.

G. GROWTH ENHANCEMENT

As described above, hormones have been effectively used to produce all-male tilapia. Hormones have also been fed to tilapia as a means of improving growth rate. Rothbard et al. (1988) found that they could improve the growth rate of *T. aurea* × *T. nilotica* hybrids by adding 17-α-ethynyltestosterone to the feed at rates of a few milograms per Kilogram. They felt that utilizing the hormone at low concentrations and limited periods of time might be a good management technique, but cautioned that studies to determine the elimination rate of the hormone from the fish and fate of the hormone in the culture environment need to be conducted before the technique is applied in the industry.

Improved growth of *T. spilurus* treated with 17-α-methyltestosterone was reported by Ridha and Lane (1990). A growth advantage was obtained during the 38-d period when the fish were fed the hormone to induce sex-reversal. Treated fish also showed improved specific growth rates and food conversion efficiencies as compared with controls, though control survival was higher. Howerton et al. (1988) found that growth improved in *T. mossambica* fed increasing levels of testosterone from 0.1 to 10 mg/kg of feed. Those same authors also fed T_3 (triiodothyronine) at levels of from 1 to 25 mg/kg of feed. While the T_3-treated fish grew more rapidly than the controls, the lowest level of dietary T_3 produced the greatest growth response.

VII. DISEASE

Tilapia rarely demonstrate signs of disease except after exposure to cold temperatures. That does not mean, however, that tilapia are immune from diseases when reared within their optimum temperature range. Various diseases have been reported for tilapia under culture conditions, including the virus *Lymphocystis*; bacterial problems with *Aeromonas*, myxobacteria, and *Edwardsiella tarda*; and such parasites as trematodes, *Trichodina*, *Epistylis*, *Apiosoma*, *Ambiphrya*, *Haplorchis pumilio*, *Ichthybodo*, and *Ichthyophthirius* (Sommerville 1982; Roberts and Sommerville 1982; Lightner et al. 1988). In closed systems, gill hyperplasia has also been reported, apparently in conjunction with chronic ammonia and nitrite toxicity (Lightner et al. 1988). *T. aurea* was shown to be only slightly susceptible to the bacterium *Edwardsiella ictaluri* (Plumb and Sanchez 1983).

Treatment of diseases in tilapia closely parallels that for channel catfish. A vaccine developed against *Aeromonas hydrophila* has been shown to protect *T.*

nilotica which received intraperitoneal injections (Ruangpan et al. 1986). Terramycin resistance in *E. tarda* has been reported from *T. aurea* (Hilton and Wilson 1980).

T. mossambica has been shown to display a hypersensitivity response to some component in the mucus under high-density culture (Henderson-Arzapalo et al. 1980). The unidentified component induced cutaneous anaphylactic reactions, not only in *T. mossambica,* but also in three other species (*T. aurea, T. nilotica,* and *T. zillii*). No reaction was elicited from channel catfish.

REFERENCES

Abdelghany, A.E., Optimum Protein Requirements and Optimum Ratio Between Animal Protein to Plant Protein in Formulated Diets for Nile Tilapia (*Tilapia niloticus* L.), Ph.D. Dissertation, University of Idaho, Boise, 1, 1987.

Abdel-Marek, S.A., Food and feeding habits of some Egyptian fishes in Lake Quarun. I. *Tilapia zillii* (Gerv.). B. According to different length groups, *Bull. Inst. Oceanogr. Fish. Cairo,* 2, 204, 1972.

Aguilar, V.L., Feasibility of growing tilapia fish in swine waste lagoons, (Abstr.), in *Int. Symp. on Tilapia in Aquaculture,* Nazareth, Israel, May 8 to 13, 1983, 78.

Allanson, B.R. and Noble, R.G., The tolerance of *Tilapia mossambica* (Peters) to high temperature, *Trans. Am. Fish. Soc.,* 93, 323, 1964.

Almazan, G. and Boyd, C.E., Plankton production and tilapia yield in ponds, *Aquaculture,* 15, 75, 1978.

Anderson, J., Jackson, A.J., and Matty, J.A., Effects of purified carbohydrates and fibre on the growth of the *Oreochromis* (*Tilapia*) *niloticus,* (Abstr.), in *Int. Symp. on Tilapia in Aquaculture,* Nazareth, Israel, May 8 to 13, 1983, 80.

Appler, H.N. and Jauncey, K., The utilization of a filamentous green alga (*Cladophora glomerata* (L.) Kutzin) as a protein source in pelleted feeds for *Sarotherodon* (*Tilapia*) *niloticus* fingerlings, *Aquaculture,* 30, 21, 1983.

Atz, J.W., The peregrinating tilapia, *Anim. Kingdom,* 57, 148, 1954.

Avault, J.W. and Shell, E.W., Preliminary studies with hybrid *Tilapia nilotica* and *T. mossambica, Food Agric. Organ. Fish. Rep.,* 44, 237, 1967.

Babiker, M.M. and Ibrahim, H., Studies on the biology of reproduction in the cichlid *Tilapia nilotica* (L.): gonadal maturation and fecundity, *J. Fish Biol.* 14, 437, 1979.

Balarin, J.D. and Hatton, J.P., *Tilapia—a Guide to the Biology and Culture in Africa,* University of Stirling, Stirling, KS, 1979, 1.

Balasuriva, C., Snakehead (*Channa striata*) as a Controlling Predator in Culture of Nile Tilapia (*Oreochromis niloticus*), Master's thesis, Asian Institute of Technology, Bangkok, Thailand, 1988, 1.

Bayne, D.R., Dunseth, D., and Ramirios, C.G., Supplemental feeds containing coffee pulp for rearing *Tilapia* in Central America, *Aquaculture,* 7, 133, 1976.

Behrends, L.L. and Smitherman, R.O., Development of a cold-tolerant population of red tilapia through introgressive hybridization, *J. World Maricult. Soc.,* 14, 172, 1984.

Berrios-Hernandez, J.M., Comparison of methods for reducing fry losses to cannibalism in tilapia production, *Prog. Fish-Cult.,* 45, 116, 1983.

Blay, J., Jr., Fecundity and spawning frequency of *Sarotherodon galilaeus* in a concrete pond, *Aquaculture,* 25, 95, 1981.

Bondari, K. and Sheppard, D.C., Soldier fly, *Hermetia illucens* L. larvae as feed for channel catfish, *Ictalurus punctatus* (Rafinesque), and blue tilapia, *Oreochromis aureus* (Steindachner), *Aquacult. Fish. Man.,* 18, 209, 1987.

Boulenger, G.A., Catalogue of the freshwater fishes of Africa, III, *Br. Mus. Nat. Hist. (London)*, 1, 1915.

Bowen, S.H., A nutritional constraint in detritivory by fishes: the stunted population of *Sarotherodon mossambicus* in Lake Sigaya, South Africa, *Ecol. Monogr.*, 49, 17, 1979.

Bowen, S.H., Detrital nonprotein amino acids are the key to rapid growth of tilapia in Lake Valencia, *Science*, 207, 1216, 1980.

Boyd, C.E., Nitrogen fertilizer effects on production of *Tilapia* in ponds fertilized with phosphorus and potassium, *Aquaculture*, 7, 385, 1976.

Brick, R.B. and Stickney, R.R., Polyculture of *Tilapia aurea* and *Macrobrachium rosenbergii* in Texas, *Proc. World Maricult. Soc.*, 10, 222, 1979.

Broussard, M.C., Jr., Reyes, R., and Raguindin, F., Evaluation of hatchery management schemes for large scale production of *Oreochromis niloticus* in Central Luzon, Philippines, in *Int. Symp. Tilapia in Aquaculture*, Nazareth, Israel, May 8 to 13, Tel Aviv University Press, Tel Aviv, 1983, 414.

Buddington, R.K., Digestion of an aquatic macrophyte by *Tilapia zillii*, *J. Fish Biol.*, 15, 449, 1979.

Burns, R.P. and Stickney, R.R., Growth of *Tilapia aurea* in ponds receiving poultry wastes, *Aquaculture*, 20, 117, 1980.

Carro-Anzalotta, A.E. and McGinty, A.S., Effects of stocking density on growth of *Tilapia nilotica* cultured in cages in ponds, *J. World Aquacult. Soc.*, 17, 52, 1986.

Chang, S.-L., Huang, C.-M., and Liao, I.-C., The effect of various feeds on seed production by Taiwanese red tilapia, in *2nd Int. Symp. Tilapia in Aquaculture*, Pullin, R.S.V., Bhukaswan, T., Tonguthai, K., and Maclean, J.L., Eds., Department of Fisheries, Bangkok, Thailand; International Center for Living Aquatic Resources Management, Manila, Philippines, 1988, 319.

Chen, F.Y., Preliminary studies on the sex-determining mechanism of *Tilapia mossambica* Peters and *T. hornorum* Trewavas, *Verh. Int. Ver. Limnol.*, 17, 719, 1969.

Chervinski, J., Sea bass, *Dicentrarchus labrax* (Pisces, Serranidae) a "policefish" in freshwater ponds and its adaptability to saline conditions, *Bamidgeh*, 26, 110, 1974.

Chervinski, J., Environmental physiology of tilapias, in *The Biology and Culture of Tilapia*, Pullin, R.S.V. and Lowe-McConnell, R.H., Eds., International Center for Living Aquatic Resources Management, Manila, Philippines, 1982, 119.

Chervinski, J. and Rothbard, S., An aid in manually sexing *Tilapia*, *Aquaculture*, 26, 389, 1982.

Chervinski, J. and Yashouv, A., Preliminary experiments on the growth of *Tilapia aurea* (Steindachner) (Pisces, Cichlidae) in sea water ponds, *Bamidgeh*, 23, 125, 1971.

Chervinski, J. and Stickney, R.R., Overwintering facilities for tilapia in Texas, *Prog. Fish-Cult.*, 43, 20, 1981.

Chervinski, J. and Zorn, M., Note on the growth of *Tilapia aurea* (Steindachner) and *Tilapia zillii* (Gervais) in sea-water ponds, *Aquaculture*, 4, 249, 1974.

Chimits, P., Tilapia and its culture, *FAO Fish. Bull.*, 8, 1, 1955.

Chimits, P., The tilapias and their culture, a second review and bibliography, *FAO Fish. Bull.*, 10, 1, 1957.

Christensen, M.S., The intensive cultivation of freshwater fish in cages in tropical and subtropical regions, *Anim. Res. Dev.*, 29, 7, 1989.

Chung, K.S., Effects of temperature on growth, survival, acclimation rate and body temperature in *Oreochromis* (*Tilapia mossambicus*), (Abstr.) in Int. Symp. Tilapia in Aquaculture, Nazareth, Israel, May 8 to 13, 1983, 30.

Clemens, H.P. and Inslee, T., The production of unisexual broods of *Tilapia mossambica* sex-reversed with methyltestosterone, *Trans. Am. Fish. Soc.*, 97, 18, 1968.

Coche, A.G., Cage culture of tilapias, in *The Biology and Culture of Tilapia*, Pullin, R.S.V. and Lowe-McConnell, R.H., Eds., International Center for Living Aquatic Resources Management, Manila, Philippines, 1982, 205.

Copley, H., The tilapia of Kenya colong, *East Afr. Agric. J.*, 18, 30, 1952.

Cruz, E.M. and Laudencia, I.L., Screening of feedstuffs as ingredients in the rations of Nile tilapia, *Kalkasan, Philipp. J. Biol.*, 7, 159, 1978.

Daud, S.K., Hasbollah, D., and Law, A.T., Effects of unionized ammonia on red tilapia (*Oreochromis mossambicus/O. niloticus* hybrid, in *2nd Int. Symp. Tilapia in Aquaculture,* Pullin, R.S.V., Bhukaswan, T., Tonguthai, K., and Maclean, J.L., Eds., Department of Fisheries, Bangkok, Thailand, and International Center for Living Aquatic Resources Management, Manila, Philippines, 1988, 411.

Davies, S.J. and Wareham, H., A preliminary evaluation of an industrial single cell protein in practical diets for tilapia (*Oreochromis mossambicus* Peters), *Aquaculture,* 73, 189, 1988.

Davis, A.T. and Stickney, R.R., Growth responses of *Tilapia aurea* to dietary protein quality and quantity, *Trans. Am. Fish. Soc.,* 107, 479, 1978.

Degani, G., Dosoretz, C., Levanon, D., Marchaim, U., and Perach, Z., Feeding *Sarotherodon aureus* with fermented cow manure, *Bamidgeh,* 34, 119, 1982.

Denzer, H.W., Studies on the physiology of young *Tilapia, FAO Fish. Rep.,* 44, 356, 1968.

DeSilva, S.S. and Senaratne, K.A.D.W., *Oreochromis mossambicus* is not universally a nuisance species: the Sri Lankan experience, in *2nd Int. Symp. Tilapia in Aquaculture,* Pullin, R.S.V., Bhukaswan, T., Tonguthai, K., and Maclean, J.L., Eds., Department of Fisheries, Bangkok, Thailand, and International Center for Living Aquatic Resources Management, Manila, Philippines, 1988, 445.

Doroshov, S.I. and Cornacchia, J.W., Initial swim bladder inflation in the larvae of *Tilapia mossambica* (Peters) and *Morone saxatilis* (Walbaum), *Aquaculture,* 16, 57, 1979.

Dunseth, D.R. and Bayne, D.R., Recruitment control and production of *Tilapia aurea* (Steindachner) with the predator, *Cichlasoma managuense* (Gunther), *Aquaculture,* 14, 383, 1978.

Edwards, P., A review of recycling organic wastes in fish, with emphasis on the tropics, *Aquaculture,* 21, 261, 1980.

Edwards, P. and Sinchumpasak, O.-A., The harvest of microalgae from the effluent of a sewage fed high rate stabilization pond by *Tilapia nilotica.* I. Description of the system and the study of the high pond, *Aquaculture,* 23, 83, 1981.

Edwards, P., Sinchumpasak, O.-A., and Tabucanon, M., The harvest of microalgae from the effluent of a sewage fed high rate stabilization pond by *Tilapia nilotica.* II. Studies of the fish ponds, *Aquaculture,* 23, 107, 1981a.

Edwards, P., Sinchumpasak, O.-A., Labhsetwar, V.K., and Tabucanon, M., The harvest of microalgae from the effluent of a sewage fed high rate stabilization pond by *Tilapia nilotica.* III. Maize cultivation experiment, bacteriological studies, and economic assessment, *Aquaculture,* 23, 149, 1981b.

El-Sayed, A.-F. M., Long-term evaluation of cotton seed meal as a protein source for Nile tilapia, *Oreochromis niloticus* (Linn.), *Aquaculture,* 84, 315, 1990.

El-Sayed, A.-F. M., and Teshima, T.-I., Tilapia nutrition in aquaculture, *Rev. Aquat. Sci.,* 1991, in press.

Eyeson, K.N., Stunting and reproduction in pond-reared *Sarotherodon melanotheron, Aquaculture,* 31, 257, 1983.

FAO, Report of the Ad Hoc Consultation on Aquaculture Research, FAO Fish. Rep. 238, Food and Agriculture Organization, U.N., Rome, 1, 1980.

Fishelson, L., Hybrids of two species of fishes in the genus *Tilapia* (Cichlidae, Teleostei), *Fishermen's Bull.,* 4(2), 14, 1962.

Fishelson, L. and Popper, D., Experiments on rearing fish in salt water near the Dead Sea, Israel, *FAO Fish. Rep.,* 44, 244, 1968.

Fortes, R.D., Tarpon as biological control in milkfish-tilapia polyculture, *Fish. J. Coll. Fish. Univ. Philipp. Visayas,* 1, 47, 1985.

Fryer, G. and Iles, T.D., *The Cichlid Fishes of the Great Lakes of Africa: Their Biology and Evolution,* T.F.H. Publications, Neptune City, NJ, 1972, 1.

Gaigher, I.G., Porath, D., and Granoth, G., Evaluation of duckweed (*Lemna gibba*) as feed for tilapia (*Oreochromis niloticus × O. aureus*) in a recirculating system, *Aquaculture,* 41, 235, 1984.

Galman, O.R. and R.R. Avtalion, A preliminary investigation on the characteristics of red tilapias from the Philippines and Taiwan, in *Int. Symp. Tilapia in Aquaculture,* Nazareth, Israel, May 8 to 13, Tel Aviv University Press, Tel Aviv, 1983, 291.

Gleastine, B.W., A Study of the Cichlid *Tilapia aurea* (Steindachner) in a Thermally Modified Texas Reservoir, Master's thesis, Texas A&M University, College Station, 1974, 1.

Gonzales-Corre, K., Polyculture of the tiger shrimp (*Penaeus monodon*) with Nile tilapia (*Oreochromis niloticus*) in brackishwater fishponds, in *2nd Int. Symp. Tilapia in Aquaculture*, Pullin, R.S.V., Bhukaswan, T., Tonguthai, K., and Maclean, J.L., Eds., Department of Fisheries, Bangkok, Thailand; International Center for Living Aquatic Resources Management, Manila, Philippines, 1988, 15.

Gophen, M., Food sources, feed behaviour and growth rates of *Sarotherodon galilaeus* (Linnaeus) fingerlings, *Aquaculture*, 20, 101, 1980.

Goudie, C.A., Shelton, W.L., and Parker, N.C., Tissue distribution and elimination of radiolabelled methyltestosterone fed to sexually undifferentiated blue tilapia, *Aquaculture*, 58, 215, 1986.

Grover, J.J., Olla, B.L., O'Brien, M., and Wicklund, R.I., Food habits of Florida red tilapia fry in manured seawater pools, in *2nd Int. Symp. Tilapia in Aquaculture*, Pullin, R.S.V., Bhukaswan, T., Tonguthai, K., and Maclean, J.L., Eds., Department of Fisheries, Bangkok, Thailand; International Center for Living Aquatic Resources Management, Manila, Philippines, 1988, 595.

Grover, J.J., Olla, B.L., O'Brien, M., and Wicklund, R.I., Food habits of Florida red tilapia fry in manured seawater pools in the Bahamas, *Prog. Fish-Cult.*, 51, 152, 1989.

Guerrrero, R.D., III, Use of androgens for the production of all-male *Tilapia aurea* (Steindachner), *Trans. Am. Fish. Soc.*, 104, 342, 1975.

Guerrrero, R.D., Culture of male *Tilapia mossambica* produced through artificial sex-reversal, in *Advances in Aquaculture*, Pillay, T.V.R. and Dill, W.A., Eds., Fishing News Books, Farnham, Surrey, England, 1979, 166.

Guerrero, R.D., III, Studies on the feeding of *Tilapia nilotica* in floating cages, *Aquaculture*, 20, 169, 1980.

Guerrrero, R.D., Control of tilapia reproduction, in *The Biology and Culture of Tilapia*, Pullin, R.S.V. and Lowe-McConnell, R.H., Eds., International Center for Living Aquatic Resources Management, Manila, Philippines, 1982, 309.

Guerrero, R.D. and Abella, T.A., Induced sex-reversal of *Tilapia nilotica* with methyltestosterone, *Fish. Res. J. Philipp.*, 1, 46, 1976.

Guerrero, R.D. and Caguan, A.G., Culture of male *Tilapia mossambica* produced through artificial sex reversal, in *Advances in Aquaculture*, Pillay, T.V.R. and Dill, W.A., Eds., Fishing News Books, Farnham, Surrey, England, 1979, 166.

Hanley, F., The digestibility of foodstuffs and the effects of feeding selectivity on digestibility determinations in tilapia, *Oreochromis niloticus* (L.), *Aquaculture*, 66, 163, 1987.

Henderson-Arzapalo, A., Design and Operational Characteristics of a Tilapia Rearing System for Temperate Regions, Ph.D. dissertation, Texas A&M University, College Station, 1984, 1.

Henderson-Arzapalo, A. and Stickney, R.R., Effects of stocking density on two tilapia species raised in an intensive culture system, *Proc. S.E. Assoc. Fish Wildl. Agen.*, 34, 379, 1980.

Henderson-Arzapalo, A., Stickney, R.R., and Lewis, D.H., Immune hypersensitivity in intensively cultured tilapia species, *Trans. Am. Fish. Soc.*, 109, 244, 1980.

Hepher, B. and Pruginin, Y., *Commercial Fish Farming*, John Wiley & Sons, New York, 1981, 1.

Hepher, B., and Pruginin, Y., Tilapia culture in ponds under controlled conditions, in *The Biology and Culture of Tilapia*, Pullin, R.S.V. and Lowe-McConnell, R.H., Eds., International Center for Living Aquatic Resources Management, Manila, Philippines, 1982, 185.

Hickling, C.F., The Malacca *Tilapia* hybrids, *J. Genet.*, 57, 1, 1960.

Hilton, L.R. and Wilson, J.L., Terramycin-resistant *Edwardsiella tarda* in channel catfish, *Ictalurus punctatus*, *Prog. Fish-Cult*, 42, 159, 1980.

Hopkins, K.D., Shelton, W.L., and Engle, C.R., Estrogen sex-reversal of *Tilapia aurea*, *Aquaculture*, 18, 263, 1979.

Hopkins, K.D., Cruz, E.M., Hopkins, M.L., and Chjong, K.-C., Optimum manure loading rates in tropical freshwater fish ponds receiving untreated piggery wastes, in The ICLARM-CLSU Integrated Animal-Fish Farming Project: Poultry-Fish and Pig-Fish Trials, International Center for Living Aquatic Resources Management, Central Luzon State University, Philippines, Tech. Rep. 2, 1981, 15.

Howerton, R.D., Okimoto, D.K., and Grau, E.G., Changes in the growth rate of *Oreochromis mossambicus* following treatment with the hormones, triiodothyronine and testosterone (Abstr.), in *2nd Int. Symp. Tilapia in Aquaculture*, Pullin, R.S.V., Bhukaswan, T., Tonguthai, K., and Maclean, J.L., Eds., International Center for Living Aquatic Resources Management, Manila, Philippines, 1988, 598.

Hsiao, S.M., Hybridization of *Tilapia mossambica, T. nilotica, T. aurea* and *T. zillii* — a preliminary report, *China Fish. Mon. (Taipei)*, 332, 3, 1980.

Huang, C.-M., Chang, S.-L., Cheng, H.-J., and Liao, I.-C., Single gene inheritance of red body coloration in Taiwanese red tilapia, *Aquaculture*, 74, 227, 1988.

Huang, C.-M., Cheng, H.-J., Chang, S.-L., and Liao, I.-C., Inheritance of body color in Taiwanese red tilapia, in *2nd Int. Symp. Tilapia in Aquaculture*, Pullin, R.S.V., Bhukaswan, T., Tonguthai, K., and Maclean, J.L., Eds., International Center for Living Aquatic Resources Management, Manila, Philippines, 1988b, 593.

Huet, M., *Traite de Pisciculture*, Editions Ch. de Wyngaert, Brussels, 1970, 1.

Jackson, A.J. and Capper, B.S., Investigations into the requirements of the tilapia *Sarotherodon mossambicus* for dietary methionine, lysine and arginine in semi-synthetic diets, *Aquaculture*, 29, 289, 1982.

Jackson, A.J., Capper, B.S., and Matty, J.A., Evaluation of some plant proteins in complete diets for the tilapia *Sarotherodon mossambicus*, *Aquaculture*, 27, 97, 1982.

Jalabert, B., Kammacher, P., and Lessent, P., Sex determination in *Tilapia machrochir* × *Tilapia nilotica* hybrids. Investigations of sex ratios in first generation × parent crossings, *Biochim. Biophys. Acta*, 11, 155, 1971.

Jalabert, B., Moreau, J., Planquette, P., and Billard, R., Determination de sexe chez *Tilapia macrochir* et *Tilapia nilotica*. Action de la methyltestosterone dans l'alimentation des alevins sur la differentiation sexuelle; proportion des sexes dans la descendance des males "inverses," *Ann. Biol. Anim. Biochim. Biophys.*, 14, 729, 1974.

Jauncey, K., The effects of varying dietary protein level on the growth, food conversion, protein utilization and body composition of juvenile tilapias (*Sarotherodon mossambicus*), *Aquaculture*, 27, 43, 1982.

Jauncey, K., Tacon, A.G.J., and Jackson, A.J., The quantitative essential amino acid requirements of *Oreochromis* (=*Sarotherodon*) *mossambicus*, in *Int. Symp. Tilapia in Aquaculture*, Fishelson, L. and Yaron, Z., Eds., Tel Aviv University, Tel Aviv, 1984, 328.

Jensen, G.L. and Shelton, W.L., Effects of estrogens on *Tilapia aurea*: implications for production of monosex genetic male tilapia, *Aquaculture*, 16, 233, 1979.

Johnstone, R., MacIntosh, D.J., and Wright, R.S., Elimination of orally administered 17α-methyltestosterone by *Oreochromis mossambicus* (tilapia) and *Salmo gairdneri* (rainbow trout) juveniles, *Aquaculture*, 35, 249, 1983.

Kanazawa, A., Teshima, S., Sakamoto, M., and Awai, M.A., Requirements of *Tilapia zillii* for essential fatty acids, *Bull. Jpn. Soc. Sci. Fish.*, 46, 1353, 1980.

Kelly, H.D., Preliminary studies on *Tilapia mossambica* Peters relative to experimental pond culture, *Proc. S. E. Assoc. Game Fish Comm.*, 10, 139, 1956.

Kohler, C.C. and Pagon-Font, F.A., Evaluations of rum distillation wastes, pharmaceutical wastes and chicken feed for rearing *Tilapia aurea* in Puerto Rico, *Aquaculture*, 14, 339, 1978.

Lee, J.C., Reproduction and Hybridization of Three Cichlid Fishes, *Tilapia aurea* (Steindachner), *T. hornorum* (Trewavas) and *T. nilotica* (Linnaeus) in Aquaria and Plastic Pools, Ph.D. dissertation, Auburn University, Auburn, AL, 1979, 1.

Lightner, D., Redman, R., Mohney, L., Dickenson, G., and Fitzsimmons, K., Major diseases encountered in controlled environment culture of tilapias in fresh- and brackishwater over a three-year period in Arizona, in *2nd Int. Symp. Tilapia in Aquaculture*, Pullin, R.S.V., Bhukaswan, T., Tonguthai, K., and Maclean, J.L., Eds., International Center for Living Aquatic Resources Management, Manila, Philippines, 1988, 111.

Lovell, R.T. and Limsuwan, T., Intestinal synthesis and dietary nonessentiality of vitamin B_{12} in *Tilapia nilotica*, *Trans. Am. Fish. Soc.*, 111, 485, 1982.

Lovshin, L.L., da Silva, A.B., and Fernandes, J.A., The intensive culture of all male hybrids of *Tilapia hornorum* (male) and *T. nilotica* (female) in northeast Brazil, *FAO Fish. Rep.,* 159, 162, 1977.

Lowe-McConnell, R.H., Observations on the biology of *Tilapia nilotica* Linné in east Africa waters, *Rev. Zool. Bot. Afr.,* 57, 131, 1958.

Loya, L. and Fishelson, L., Ecology of fish breeding in brackish water ponds near the Dead Sea (Israel), *J. Fish Biol.,* 1, 261, 1969.

Magid, A. and Babiker, M.M., Oxygen consumption and respiratory behaviour of three Nile fishes, *Hydrobiology,* 46, 359, 1975.

Man, H.S.H. and Hodgkiss, I.J., Studies on the ichthyo-fauna in Plover Cove Reservoir, Hong Kong: feeding and food relations, *J. Fish Biol.,* 11, 1, 1977.

Mang-Umphan, K. and Arce, R.G., Culture of Nile tilapia (*Oreochromis niloticus*) in a rice-fish culture system using chemical and commercial organic fertilizers, in *2nd Int. Symp. Tilapia in Aquaculture,* Pullin, R.S.V., Bhukaswan, T., Tonguthai, K., and Maclean, J.L., Eds., Department of Fisheries, Bangkok, Thailand; International Center for Living Aquatic Resources Management, Manila, Philippines, 1988, 59.

Maruyama, T., An observation on *Tilapia mossambica* in ponds referring to the diurnal movement with temperature change, *Bull. Freshw. Fish. Res. Lab. Tokyo,* 8, 25, 1958.

Matsuno, T., Katsuyama, M., Iwahashi, M., Koike, T., and Okada, M., Intensification of color of red *Tilapia* with lutein, rhodoxanin and sprulina, *Bull. Jpn. Soc. Sci. Fish.,* 46, 479, 1980.

Mazid, M.A., Tanaka, Y., Katayama, T., Rahman, M.A., Simpson, K.L., and Chichester, C.O., Growth response of *Tilapia zillii* fingerlings fed iso-caloric diets with variable protein levels, *Aquaculture,* 18, 115, 1979.

McAndrew, B.J. and Majumdar, K.C., Tilapia stock identification using electrophoretic markers, *Aquaculture,* 30, 249, 1983.

McBay, L.G., The biology of *Tilapia nilotica* Linnaeus, *Proc. S. E. Assoc. Game Fish Comm.,* 15, 208, 1961.

McClain, W.R. and Gatlin, D.M., III, Dietary zinc requirement of *Oreochromis aureus* and effects of dietary calcium and phytate on zinc bioavailability, *J. World Aquacult. Soc.,* 19, 103, 1988.

McGeachin, R.B. and Stickney, R.R., Manuring rates for production of blue tilapia in simulated sewage lagoons receiving laying hen waste, *Prog. Fish-Cult.,* 44, 25, 1982.

McGeachin, R.B., Wicklund, R.I., Olla, B.L., and Winton, J.R., Growth of *Tilapia aurea* in seawater cages, *J. World Aquacult. Soc.,* 18, 31, 1987.

McGinty, A.S., Effects of predation by largemouth bass in fish production ponds stocked with *Tilapia nilotica, Aquaculture,* 46, 269, 1985.

McGinty, A.S., Population dynamics of peacock bass, *Cichla ocellaris* and *Oreochromis* (*Tilapia*) *niloticus* in fertilized ponds, in *Int. Symp. Tilapia in Aquaculture,* Fishelson, L., and Yaron, Z., Eds., Tel Aviv University, Tel Aviv, 1984, 86.

Meriwether, F.H. and Torrans, E.L., Evaluation of a new androgen (mibolerone) and procedure to induce functional sex reversal in tilapia, in *Proc. First Asian Fisheries Forum,* Manila, Philippines, May 26 to 31,1986, Maclean, J.L., Dizon, L.B., and Hosillos, L.V., Eds., Asian Fisheries Forum, Manila, Philippines, 1986, 675.

Meriwether, F.H., II, Scura, E.D., and Okamura, W.Y., Cage culture of red tilapia in prawn and shrimp ponds, *J. World Maricult. Soc.,* 14, 254, 1984.

Meschkat, A., The status of warm-water fish culture in Africa, *FAO Fish. Rep.,* 44, 88, 1967.

Moriarty, D.J.W., The physiology of digestion of blue-green algae in the cichlid fish, *Tilapia niloticus,* in *Int. Symp. Tilapia in Aquaculture,* Fishelson, F. and Yaron, Z., Eds., Tel Aviv University, Tel Aviv, 1984, 336.

Moriarty, C.M. and Moriarty, D.J.W., Quantitative estimation of the daily ingestion of phytoplankton by *Tilapia nilotica* and *Haplochromis nigripinnis* in Lake George, Uganda, *Proc. R. Soc. London Ser. B.,* 184, 299, 1973.

Munro, J.L., The food of a community of East African freshwater fishes, *J. Zool.,* 151, 389, 1967.

Naik, I.U., Studies on *Tilapia mossambica* Peters in Pakistan, *Agric. Pak.,* 24, 47, 1973.

Newman, M.W., Huezo, H.E., and Hughes, D.G., The response of all-male tilapia hybrids to four levels of protein in isocaloric diets, *Proc. World Maricult. Soc.,* 10, 788, 1979.

Noble, R.L., Germany, R.D., and Hall, C.R., Interactions of blue tilapia and largemouth bass in a power plant cooling reservoir, *Proc. S. E. Assoc. Game Fish Comm.*, 29, 247, 1975.

Oduro-Boateng, F. and Bart-Plange, A., Pito brewery waste as an alternative protein source to fishmeal in feeds for *Tilapia busumana*, in *2nd Int. Symp. Tilapia in Aquaculture*, Pullin, R.S.V., Bhukaswan, T., Tonguthai, K., and Maclean, J.L., Eds., Department of Fisheries, Bangkok, Thailand; International Center for Living Aquatic Resources Management, Manila, Philippines, 1988, 357.

Ofori, J.K., The effect of predation by *Lates niloticus* on overpopulation and stunting in mixed sex culture of tilapia species in ponds, in *2nd Int. Symp. Tilapia in Aquaculture*, Pullin, R.S.V., Bhukaswan, T., Tonguthai, K., and Maclean, J.L., Eds., Department of Fisheries, Bangkok, Thailand; International Center for Living Aquatic Resources Management, Manila, Philippines, 1988, 69.

Olvera, N.M.A., Martinez, P.C.A., Galvan, C.R., and Chaves, S.C., The use of seed of the leguminous plant *Sesbania grandiflora* as a partial replacement for fish meal in diets for tilapia (*Oreochromis mossambicus*), *Aquaculture*, 71, 51, 1988.

Pagan, F.A., Cage culture of tilapia, *FAO Fish Cult. Bull.*, 2, 6, 1969.

Pagan-Font, F.A., Cage culture as a mechanical method for controlling reproduction of *Tilapia aurea* (Steindachner), *Aquaculture*, 6, 243, 1975.

Pantastico, J.B., Danlan, M.M.A., and Equia, R.V., Cannibalism among different sizes of tilapia (*Oreochromis niloticus*) fry/fingerlings and the effect of natural food, in *2nd Int. Symp. Tilapia in Aquaculture*, Pullin, R.S.V., Bhukaswan, T., Tonguthai, K., and Maclean, J.L., Eds., Department of Fisheries, Bangkok, Thailand; International Center for Living Aquatic Resources Management, Manila, Philippines, 1988, 465.

Payne, A.I. and Collinson, R.I., A comparison of the biological characteristics of *Sarotherodon niloticus* (L.) with those of *S. aureus* (Steindachner) and other tilapia of the delta and lower Nile, *Aquaculture*, 30, 335, 1983.

Payne, A.I., Ridgway, J., and Hamer, J.L., The influence of salt (NaCl) concentration and temperature on the growth of *Oreochromis spilurus spilurus*, *O. mossambicus* and a red tilapia hybrid, in *2nd Int. Symp. Tilapia in Aquaculture*, Pullin, R.S.V., Bhukaswan, T., Tonguthai, K., and Maclean, J.L., Eds., Department of Fisheries, Bangkok, Thailand; International Center for Living Aquatic Resources Management, Manila, Philippines, 1988, 481.

Penman, D.J., Shah, M.S., Beardmore, J.A., and Skibinski, D.O.F., Sex ratios of gynogenetic and triploid tilapia, in *Selection, Hybridization and Genetic Engineering in Aquaculture*, Tiews, K., Ed., Schr. Bundesforschungsanst. Fisch., Hamburg, 18-19, 1987, 267.

Perschbacher, P.W. and McGeachin, R.B., Salinity tolerances of red hybrid tilapia fry, juveniles and adults, in *2nd Int. Symp. Tilapia in Aquaculture*, Pullin, R.S.V., Bhukaswan, T., Tonguthai, K., and Maclean, J.L., Eds., International Center for Living Aquatic Resources Management, Manila, Philippines, 1988, 415.

Philippart, J.-Cl. and Ruwet, J.-Cl., Ecology and distribution of tilapias, in *The Biology and Culture of Tilapia*, Pullin, R.S.V. and Lowe-McConnell, R.H., Eds., International Center for Living Aquatic Resources Management, Manila, Philippines, 1982, 15.

Pierce, B.A., Production of hybrid tilapia in indoor aquaria, *Prog. Fish-Cult.*, 42, 233, 1980.

Pinto, L.G., Hybridization between species of tilapia, *Trans. Am. Fish. Soc.*, 111, 481, 1982.

Platt, S. and Hauser, W.J., Optimum temperature for feeding and growth of *Tilapia zillii*, *Prog. Fish-Cult.*, 40, 105, 1978.

Plumb, J.A. and Sanchez, D.J., Susceptibility of five species of fish to *Edwardsiella ictaluri*, *J. Fish Dis.*, 6, 261, 1983.

Popper, D. and Lichatowich, T., Preliminary success in predator control of *Tilapia mossambica*, *Aquaculture*, 5, 213, 1975.

Pruginin, Y., Report to the Government of Uganda on the Experimental Fish Culture Project in Uganda, 1962-1964, FAO/UNDP (Technical Assistance) Reports on Fisheries, TA Reports, Food and Agriculture Organization, Rome, 1960, 1965, 1.

Pruginin, Y., Report to the Government of Uganda on the Experimental Fish Culture Project in Uganda, 1965-1966, FAO/UNDP (Technical Assistance) Reports on Fisheries, TA Reports, Food and Agriculture Organization, Rome, 2446, 1967, 1.

Pruginin, Y., Rothbard, S., Wohlfarth, G., Halevy, A., Moav, R., and Hulata, G., All-male broods of *Tilapia nilotica* × *T. aurea* hybrids, *Aquaculture*, 6, 11, 1975.

Pudadera, B.J., Jr., Corre, K.C., Coniza, A., and Taleon, G.A., Integrated farming of broiler chickens with fish and shrimp in brackishwater ponds, in *Proc. First Asian Fisheries Forum*, Maclean, J.L., Dizon, L.B., and Hosdillos, L.V., Eds., Manila, Philippines, 1986, 141.

Pullin, R.S.V., Choice of tilapia species for aquaculture, in *Int. Symp. Tilapia in Aquaculture*, Nazareth, Israel, May 8 to 13, Tel Aviv University Press, Tel Aviv, 1983, 64.

Rahimaldin, S.A., The Effect of Salinity on the Growth of Blue Tilapia (*Tilapia aurea*), Master's thesis, Texas A&M University, College Station, 1983, 1.

Redner, B.D. and Stickney, R.R., Acclimation to ammonia by *Tilapia aurea*, *Trans. Am. Fish. Soc.*, 108, 383, 1979.

Ridha, M.T. and Lane, K.P., Effect of oral administration of different levels of 17a-methyltestosterone on sex reversal, growth and food conversion efficiency of the tilapia *Oreochromis spilurus* (Guenther) in brackish water, *Aquacult. Fish. Man.*, 21, 391, 1990.

Rifai, S.A., Control of reproduction of *Tilapia nilotica* using cage culture, *Aquaculture*, 20, 177, 1980.

Roberts, R.J. and Sommerville, C., Diseases of tilapias, in *The Biology and Culture of Tilapias*, Pullin, R.S.V. and Lowe-McConnell, R.H., Eds., International Center for Living Aquatic Resources Management, Manila, Philippines, 1982, 247.

Robins, R., A List of Common and Scientific Names of Fishes from the U.S. and Canada, American Fisheries Society, Bethesda, MD, 1991, 1.

Robinson, E.H., Rawles, S.D., Oldenburg, P.W., and Stickney, R.R., Effects of feeding glandless and glanded cottonseed products and gossypol to *Tilapia aurea*, *Aquaculture*, 38, 145, 1984.

Roem, A.J., Kohler, C.C., and Stickney, R.R., Vitamin E requirement of the blue tilapia, *Oreochromis aureus* (Steindachner), in relation to dietary lipid level, *Aquaculture*, 87, 155, 1990a.

Roem, A.J., Stickney, R.R., and Kohler, C.C., Vitamin requirements of blue tilapia in a recirculating water system, *Prog. Fish-Cult.*, 52, 15, 1990b.

Rothbard, S., Observations on the reproductive behavior of *Tilapia zillii* and several *Sarotherodon* spp. under aquarium conditions, *Bamidgeh*, 31, 35, 1979.

Rothbard, S. and Hulata, G., Closed-system incubator for cichlid eggs, *Prog. Fish-Cult.*, 42, 203, 1980.

Rothbard, S. and Pruginin, Y., Induced spawning and artificial incubation of *Tilapia*, *Aquaculture*, 5, 315, 1975.

Rothbard, S., Yaron, Z., and Moav, B., Field experiments on growth enhancement of tilapia (*Oreochromis niloticus* × *O. aureus* F$_1$ hybrids) using pellets containing an androgen (17α-ethynyltestosterone), in *2nd Int. Symp. Tilapia in Aquaculture*, Pullin, R.S.V., Bhukaswan, T., Tonguthai, K., and Maclean, J.L., Eds., Department of Fisheries, Bangkok, Thailand; International Center for Living Aquatic Resources Management, Manila, Philippines, 1988, 367.

Rouse, D.B. and Stickney, R.R., Evaluation of the production potential of *Macrobrachium rosenbergii* in monoculture and in polyculture with *Tilapia aurea*, *Proc. World Maricult. Soc.*, 13, 73, 1983.

Ruangpan, L., Kitao, T., Yoshida, T., Protective efficacy of *Aeromonas hydrophila* vaccines in Nile tilapia, *Vet. Immunol. Immunopathol.*, 12, 346, 1986.

Santiago, C.B., Aldaba, M.B., Abuan, E.F., and Laron, M.A., The effects of artificial diets on fry production and growth of *Oreochromis niloticus* breeders, *Aquaculture*, 47, 193, 1985.

Santiago, C.B., Aldaba, M.B., Reyes, O.S., and Laron, M.A., Response of Nile tilapia (*Oreochromis niloticus*) fry to diets containing Azolla meal, in *2nd Int. Symp. Tilapia in Aquaculture*, Pullin, R.S.V., Bhukaswan, T., Tonguthai, K., and Maclean, J.L., Eds., Department of Fisheries, Bangkok, Thailand; International Center for Living Aquatic Resources Management, Manila, Philippines, 1988a, 377.

Santiago, C.B., Aldaba, M.B., Laron, M.A., and Reyes, O.S., Reproductive performance and growth of Nile tilapia (*Oreochromis niloticus*) broodstock fed diets containing *Leucaena leucocephala* leaf meal, *Aquaculture*, 70, 53, 1988b.

Sarig, S., Winter storage of *Tilapia*, *FAO Fish Cult. Bull.*, 2, 8, 1969.

Sarig, S. and Arieli, Y., Growth capacity of tilapia in intensive culture, *Bamidgeh*, 32, 57, 1980.

Sarig, S. and Marek, M., Results of intensive and semi-intensive fish breeding techniques in Israel in 1971-1973, *Bamidgeh*, 26, 28, 1974.

Schoenen, P., *A Bibliography of Important Tilapias (Pisces:Cichlidae) for Aquaculture,* Bibliographies 3, International Center for Living Aquatic Resources Management, Manila, Philippines, 1982, 1.

Schroeder, G., Use of cowshed manure in fish ponds, *Bamidgeh*, 26, 84, 1974.

Shafland, P.L. and Pestrak, J.M., Lower lethal temperatures for fourteen nonnative fishes in Florida, *Environ. Biol. Fish.*, 7, 149, 1982.

Shelton, W.L., Rodriquez-Guerrero, D., and Lopez-Marcias, J., Factors affecting androgen sex reversal of *Tilapia aurea, Aquaculture*, 25, 59, 1981.

Shen, P. and Xu, B., Pen culture in Wuli Lake, Jiangsu, China, *Aquaculture*, 71, 301, 1988.

Siddiqui, A.Q., Howlader, M.S., and Adam, A.A., Effects of dietary protein on growth, feed conversion and protein utilization in fry and young Nile tilapia, *Oreochromis niloticus, Aquaculture*, 70, 63, 1988.

Silvera, P.A.W., Factors Affecting Fry Production in *Sarotherodon niloticus* (Linnaeus), Master's thesis, Auburn University, Auburn, AL, 1978, 1.

Sommerville, C., The pathology of *Haplorchis pumilio* (Looss, 1896) infections in cultured tilapias, *J. Fish Dis.*, 5, 243, 1982.

Spataru, P., The feeding habits of *Tilapia galilaea* (Artedi) in Lake Kinneret (Israel), *Aquaculture*, 9, 47, 1976a.

Spataru, P., Natural feed of *Tilapia aurea* Steindachner in polyculture, with supplementary feed and intensive manuring, *Bamidgeh*, 28, 57, 1976b.

Spataru, P. and Gophen, M., A comparative study of the food and feeding habits of *Sarotherodon galilaeus* and *Oreochromis aureus* (Cichlidae) from Lake Kinneret, (Abstr.), in *Int. Symp. Tilapia in Aquaculture*, Nazareth, Israel, May 8 to 13, 1983, 28.

Spataru, P. and Hepher, B., Common carp predating on tilapia fry in a high density polyculture fish pond system, *Bamidgeh*, 29, 1, 1977.

Spataru, P. and Zorn, M., Food and feeding habits of *Tilapia aurea* (Steindachner) (Cichlidae) in Lake Kinneret (Israel), *Aquaculture*, 13, 67, 1978.

Stickney, R.R., Cellulase activity in the stomachs of freshwater fishes from Texas, *Proc. S. E. Assoc. Game Fish Comm.*, 26, 282, 1976.

Stickney, R.R., Tilapia tolerance of saline waters: a review, *Prog. Fish-Cult.*, 48, 161, 1986.

Stickney, R.R. and Hardy, R.W., Lipid requirements of some warmwater species, *Aquaculture*, 79, 145, 1989.

Stickney, R.R. and Hesby, J.H., Tilapia production in ponds receiving swine wastes, in *Symp. Culture of Exotic Fishes*, Fish Culture Section, American Fisheries Society, Auburn, AL, 1978, 90.

Stickney, R.R. and McGeachin, R.B., Effects of dietary lipid quality on growth and food conversion of tilapia, *Proc. S.E. Assoc. Fish Wildl. Agen.*, 37, 352, 1983.

Stickney, R.R. and McGeachin, R.B., Responses of *Tilapia aurea* to semipurified diets of differing essential fatty acid composition, in *Proc. Conf. Tilapia in Aquaculture*, Fishelson, F. and Yaron, Z., Eds., Tel Aviv University, Tel Aviv, 1984, 346.

Stickney, R.R. and Winfree, R.A., Tilapia overwintering systems, in Proc. 1982 Fish Farming Conf. Annu. Convention Fish Farmers of Texas, 1982, 58.

Stickney, R.R. and Wurts, W.A., Growth response of blue tilapia to selected levels of dietary menhaden and catfish oils, *Prog. Fish-Cult.*, 48, 107, 1986.

Stickney, R.R., Rowland, L.O., and Hesby, J.H., Water quality — *Tilapia aurea* interactions in ponds receiving swine and poultry wastes, *Proc. World Maricult. Soc.*, 8, 55, 1977a.

Stickney, R.R., Simmons, H.B., and Rowland, L.O., Growth responses of *Tilapia aurea* to feed supplemented with dried poultry waste, *Tex. J. Sci.*, 28, 93, 1977b.

Stickney, R.R., Hesby, J.H., McGeachin, R.B., and Isbell, W.A., Growth of *Tilapia nilotica* in ponds with differing histories of organic fertilization, *Aquaculture*, 17, 189, 1979.

Stickney, R.R., McGeachin, R.B., Lewis, D.H., Marks, J., Roggs, A., Sis, R., Robinson, E.H., and Wurts, W., Response of *Tilapia aurea* to dietary vitamin C, *Proc. World Maricult. Soc.*, 15, 179, 1984.

Suffern, J.S., Adams, S.M., Blaylock, B.G., Coutant, C.C., and Guthrie, C.A., Growth of monosex hybrid tilapia in the laboratory and sewage oxidation ponds, in *Symp. Culture of Exotic Fishes*, Fish Culture Section, American Fisheries Society, Auburn, AL, 65, 1978.

Swingle, H.S., Comparative evaluation of two tilapias as pondfishes in Alabama, *Trans. Am. Fish. Soc.*, 89, 142, 1960.

Tacon, A.G.J., Jauncey, K., Falaye, A.E., and Pantha, M.B., The use of meat and bone meal, hydrolyzed feathermeal and soybean meal in practical fry and fingerling feeds for *Oreochromis niloticus*, in *Int. Symp. Tilapia in Aquaculture*, Fishelson, F. and Yaron, Z., Eds., Tel Aviv University, Tel Aviv, 1984, 356.

Tamse, A.F., Fortes, N.R., Catedrilla, L.C., and Yuseco, J.E.H., The effect of using piggery wastes in brackishwater fishponds on fish production, *Fish. J. Coll. Fish. Univ. Philipp. Visayas*, 1, 69, 1985.

Tave, D., Genetics and breeding of tilapia: a review, in *2nd Int. Symp. Tilapia in Aquaculture*, Pullin, R.S.V., Bhukaswan, T., Tonguthai, K., and Maclean, J.L., Eds., International Center for Living Aquatic Resources Management, Manila, Philippines, 1988, 285.

Tave, D., and Smitherman, R.O., Predicted response to selection for early growth in *Tilapia nilotica*, *Trans. Am. Fish. Soc.*, 109, 439, 1980.

Tayamen, M.M. and Shelton, W.L., Inducement of sex reversal in *Sarotherodon niloticus* (Linnaeus), *Aquaculture*, 14, 349, 1978.

Teshima, S., Kanazawa, A., and Uchiyama, Y., Optimum protein levels in casein-gelatin diets for *Tilapia nilotica, Mem. Fac. Fish. Kagoshima Univ*, 34, 45, 1985.

Torrans, L., Meriwether, F., Lowell, F., Wyatt, B., and Swinup, P.D., Sex reversal of *Oreochromis aureus* by immersion in mibolerone, a synthetic steroid, *J. World Aquacult. Soc.*, 19, 97, 1988.

Trewavas, E., On the cichlid fishes of the genus *Pelmatochromis* with proposal of a new genus for *P. congicus*; on the relationship between *Pelmatoachromis* and *Tilapia* and the recognition of *Sarotherodon* as a distinct genus, *Bull. Br. Mus. Nat. Hist. Zool.*, 25, 1, 1973.

Trewavas, E., Tilapias: taxonomy and speciation, in *The Biology and Culture of Tilapia*, Pullin, R.S.V. and Lowe-McConnell, R.H., Eds., International Center for Living Aquatic Resources Management, Manila, Philippines, 1982, 3.

Uchida, R.N. and King, J.E., Tank culture of tilapia, *Fish. Bull.*, 14, 21, 1962.

Uraiwan, S., Direct and indirect responses to selection for age at first maturation of *Oreochromis niloticus*, in *2nd Int. Symp. Tilapia in Aquaculture*, Pullin, R.S.V., Bhukaswan, T., Tonguthai, K., and Maclean, J.L., Eds., International Center for Living Aquatic Resources Management, Manila, Philippines, 1988, 295.

Van Someren, V.D. and Whitehead, P.J., The culture of *Tilapia nigra* (Gunther) in ponds. III. The early growth of males and females at comparable stocking rates, and the length/weight relationship, *East Afr. Agric. For. J.*, 25, 169, 1960a.

Van Someren, V.D. and Whitehead, P.J., The culture of *Tilapia nigra* (Gunther) in ponds. IV. The seasonal growth of male *T. nigra, East Afr. Agric. For. J.*, 26, 79, 1960b.

Varadaraj, K., Production of monosex male *Oreochromis mossambicus* (Peters) by administering 19-norethisterone acetate, *Aquacult. Fish. Man.*, 21, 133, 1990.

Viola, S. and Arieli, Y., Nutrition studies with tilapia (*Sarotherodon*). I. Replacement of fishmeal by soybeanmeal in feeds for intensive tilapia culture, *Bamidgeh*, 35, 9, 1983.

Viola, S. and Zohar, G., Nutrition studies with market size hybrids of tilapia (*Oreochromis*) in intensive culture. III. Protein levels and sources, *Bamidgeh*, 36, 3, 1984.

Viola, S., Arieli, Y., and Zohar, G., Animal-protein-free feeds for hybrid tilapia (*Oreochromis niloticus ¥ O. aureus*) in intensive culture, *Aquaculture*, 75, 115, 1988a.

Viola, S., Arieli, Y., and Zohar, G., Unusual feedstuffs (tapioca and lupin) as ingredients for carp and tilapia feeds in intensive aquaculture, *Bamidgeh*, 40, 29, 1988b.

Viola, S., Zohar, G., and Arieli, Y., Phosphorus requirement and its availability from different sources for intensive pond culture species in Israel. Part 1. Tilapia, *Bamidgeh*, 38, 3, 1986.

Watanabe, W.O., Ellingson, L.J., Wicklund, R.I., and Olla, B.L., The effects of salinity on growth, food consumption and conversion in juvenile monosex male Florida red tilapia, in *2nd Int. Symp. Tilapia in Aquaculture,* Pullin, R.S.V., Bhukaswan, T., Tonguthai, K., and Maclean, J.L., Eds., Department of Fisheries, Bangkok, Thailand, and International Center for Living Aquatic Resources Management, Manila, Philippines, 1988a, 515.

Watanabe, W.O., French, K.E., Ellingson, L.J., Wicklund, R.I., and Olla, B.L., Further investigations on the effects of salinity on growth in Florida red tilapia: evidence for the influence of behavior, in *2nd Int. Symp. Tilapia in Aquaculture,* Pullin, R.S.V., Bhukaswan, T., Tonguthai, K., and Maclean, J.L., Eds., Department of Fisheries, Bangkok, Thailand, and International Center for Living Aquatic Resources Management, Manila, Philippines, 1988b, 525.

Watanabe, W.O., Wicklund, R.I., Olla, B.L., Ernst, O.H., and Ellingson, L.J., Potential for saltwater tilapia culture in the Caribbean, *Proc. Gulf Caribb. Fish. Inst.,* 39, 435, 1989a.

Watanabe, W.O., French, K.E., Ernst, D.H., Olla, B., and Wicklund, R.I., Salinity during early development influences growth and survival of Florida red tilapia in brackish and seawater, *J. World Aquacult. Soc.,* 20, 134, 1989b.

Watten, B.J. and Busch, R.L., Tropical production of tilapia (*Sarotherodon aurea*) and tomatoes (*Lycopersicon esculentum*) in a small-scale recirculating water system, *Aquaculture,* 41, 271, 1984.

Weatherly, A.H. and Cogger, B.M.G., Fishculture: problems and prospects, *Science,* 197, 427, 1977.

Wee, K.L. and Ng, L.T., Use of cassava as an energy source in a pelleted feed for the tilapia *Oreochromis niloticus* L., *Aquacult. Fish. Man.,* 17, 129, 1986.

Wee, K.L. and Shu, S.-W., The nutritive value of boiled full-fat soybean in pelleted feed for Nile tilapia, *Aquaculture,* 81, 303, 1989.

Wee, K. and Tuan, N.A., Effects of dietary protein level on growth and reproduction in Nile tilapia (*Oreochromis niloticus*), in *2nd Int. Symp. Tilapia in Aquaculture,* Pullin, R.S.V., Bhukaswan, T., Tonguthai, K., and Maclean, J.L., Eds., Department of Fisheries, Bangkok, Thailand, and International Center for Living Aquatic Resources Management, Manila, Philippines, 1988, 401.

Wee, K.L. and Wang, S.S., Nutritive value of *Leucaena* leaf meal in pelleted feed for Nile tilapia, *Aquaculture,* 62, 97, 1987.

Williams, K., Gebhart, G.E., and Maughan, O.E., Enhanced growth of cage cultured channel catfish through polyculture with blue tilapia, *Aquaculture,* 62, 207, 1987.

Wilson, J.L. and Hilton, L.R., Effects of tilapia densities on growth of channel catfish, *Prog. Fish-Cult.,* 44, 207, 1982.

Winfree, R.A. and Stickney, R.R., Effects of dietary protein and energy on growth, feed conversion efficiency and body composition of *Tilapia aurea, J. Nutr.,* 111, 1001, 1981.

Wohlfarth, G.W., and Hulata, G., *Applied Genetics of Tilapias,* ICLARM Studies and Reviews, Vol. 6, International Center for Living Aquatic Resources Management (ICLARM), Manila, Philippines, 1983, 1.

Wren, S.W., Comparison of Hydroponic Crop Production Techniques in a Recirculating Fish Culture System, Master's Thesis, Texas A&M University, College Station, 1984, 1.

Yashouv, A., Mixed fish culture in ponds and the role of tilapia in it, *Bamidgeh,* 21, 75, 1969.

Chapter 4

CARP AND BUFFALO

Robert B. McGeachin

TABLE OF CONTENTS

8633-9/93/$0.00 + $.50

I. INTRODUCTION AND HISTORY OF CARP CULTURE

The common carp, *Cyprinus carpio* (Figure 1), is a hardy species, tolerant of handling and poor water-quality conditions, which is widely cultured around the world, though not popular in the U.S. It is especially tolerant of low dissolved-oxygen (DO) concentrations and has the capacity for anaerobic respiration for short periods of time (Johnston 1977, Christiansen et al. 1982). The common carp is omniverous and readily accepts prepared diets. It grows relatively rapidly in warm water, can be easily spawned in captivity, and has high fecundity. All these traits make the species an excellent one for aquaculture. The problem of intercillary bones can be mitigated by various cooking methods, and off-flavors can be eliminated by feeding commercial diets and maintaining good culture conditions.

According to Balon (1974), the species is native to central Asia. There are three probable subspecies: (1) the wild rheophilic ancestral form; (2) the European, *Cyprinus carpio carpio*; and (3) the "big belly" eastern Asian, *Cyprinus carpio haematopterus*. Since the fish are geologically young — no older than the Pleistocene — there are few differences among the subspecies. Within domesticated populations, however, there are a number of strains that have been highly selected and that appear and perform differently.

FIGURE 1.   The mirror carp form of the common carp is characterized by having only a small number of scales, a trait developed through many generations of selective breeding.

The common carp was one of the first fish species cultured and domesticated. A Chinese treatise on fish culture by Fan Li, written in the fifth century B.C., provided recommendations for carp culture (Ling 1977). Common carp continues to be widely cultured in China, Japan, and southeast Asia.

A second independent effort at domestication began in Europe in the first century A.D. (Balon 1974.). Wild carp from the Danube region were captured and shipped live (probably in wet moss) back to Rome where they were maintained in special holding ponds called "piscinae" until eaten. This practice continued into the sixth century with the Visigoths and spread north with monks during the development of monasteries. The keeping of fish ponds was especially important for monks since the fish could be consumed during the many meatless fasting days that were observed. Eventually, breeding was conducted in the holding ponds, and by the late middle ages, carp culture had spread throughout eastern and central Europe.

The common carp was first brought to North America in 1831 (Balon 1974), and the first domestic carp were successfully introduced by Rudolph Hessel in 1877 for the newly created U.S. Fish Commission (Bowen 1970; Doughty 1980). The 227 leather-and-mirror carp and 118 scaled carp in the shipment were temporarily kept in Baltimore and then moved to Washington, D.C., where they were bred in the reflecting ponds of the Washington Monument. The U.S. Fish Commission started a program to rejuvenate the rapidly depleting inland fisheries of the U.S. by stocking carp in public waters and fish ponds on the request of private individuals. The intent was to provide inexpensive food in the face of depleted wildlife and fish stocks and rapidly rising livestock prices. Carp was particularly popular with, and requested by, European immigrants and their descendants who were familiar with the fish and its culture. The first carp were distributed in 1879, and by 1882, some 79,000 had been stocked in 298 of the 301 congressional districts (Bowen 1970; Doughty 1980).

By the end of the program in 1896, 2.2 million fingerlings had been distributed to about 50,000 individuals throughout the U.S., and about 250,000 fish had been stocked directly into public waters (Doughty 1980). Individuals who received carp were instructed to keep the fish in clean water that contained a wholesome aquatic food supply and to supplementally provide good quality feeds. In practice, many persons stocked the fish in poor quality waters, fed them refuse and offal, or left them to scavenge. The result was off-flavor, which helped the species fall into disrepute.

The grass carp or white amur, *Ctenopharyngodon idella* (Figure 2), is another species that has been introduced into the U.S. Native to the Amur River system on the border between The People's Republic of China and Siberia (Brown 1977), the species now inhabits many of the large river systems of East and Southeast Asia (George 1982). The adults feed on aquatic macrophytes, a trait that led to the introduction of the species in the U.S. for weed control in the 1960s. Common and grass carp are among the species utilized in Chinese polyculture ponds, a style of culture that dates back at least several hundred years.

Other carp species are under culture around the world, but to date, interest in carp culture in the U.S. has been centered on common and grass carp; therefore, our discussion will focus on them.

FIGURE 2.   Grass carp were introduced into the U.S. as an aquatic weed control agent.

II. STATUS OF THE COMMERCAL CARP INDUSTRY

The estimation of the Food and Agriculture Organization of the United Nations (FAO) for common carp production worldwide in 1988 was 1,194,603 tons (FAO 1990). By species it ranked 12th out of about 750 in size of catch. Although the FAO statistics do not differentiate aquaculture production from capture fisheries, a large portion of the total can be attributed to fish-farming activity.

Table 1 lists common carp production data for 1988 for the major producing countries. The production from China, Israel, Japan, and the European countries was primarily from aquaculture (Doughty 1980; Krupauer 1973; Hepher and Pruginin 1981; Lovell 1977; Debeljak et al. 1979; Kafuku and Ikenoue 1983). Common carp represent a component of the Chinese polyculture system, which generally employs 3 to 6 species (Tapidor et al. 1977, FAO 1979), and in 1988, China produced 584,600 tons of common carp (FAO 1990). In 1988, China also produced 1,481,000 tons of silver carp (*Hypophthalmichthys molitrix*), 701,500 tons of bighead carp, (*Hypophthalmichthys nobilis*), 584,600 tons of grass carp, and 77,900 tons of mud carp (*Cirrhinus mulitorella*).

In the U.S., common carp are obtained only from capture fisheries. The stocking effort of the U.S. Fish Commission led to a significant inland fishery, but the relatively low market price and great abundance of wild carp has left little room in the market place for cultured carp. Presently, there is no commercial food production

TABLE 1
Production of Common and Grass Carp
in Various Nations in 1988

Country	Common carp (tons)	Grass carp (tons)
China	584,600	584,600
U.S.S.R.	252,851	158
Indonesia	108,070	—
Mexico	27,056	—
Japan	25,043	—
Poland	21,824	2
Iran	19,800	—
Turkey	19,684	—
Germany	19,360	64
Romania	17,044	9,100
Czeckoslovakia	16,993	—
Yugoslavia	13,094	—
Hungary	12,542	383
Bulgaria	9,527	—
Israel	8,215	36
U.S.A.	1,232	—
World Total	1,194,603	596,499

Data from FAO, Yearbook of Fishery Statistics Catches
and Landings 1988, Food and Agriculture Organization,
Rome, 1990, 1.

of common carp in the U.S., though some carp are produced for the ornamental and aquarium markets, as well as for trotline bait (Martin 1984).

The grass carp was first introduced to the U.S. in 1963 for evaluation as a biological control for aquatic vegetation by the Fish Farming Experimental Station of the U.S. Fish and Wildlife Service in Stuttgart, AR (Stevenson 1965). The fish was found to be an effective tool in the control of aquatic macrophytes in fish culture ponds and has been widely employed at low densities in states where use of the fish is not prohibited by law. Since it is an exotic species, the grass carp has been prohibited in many states pending the outcome of research to determine potential environmental problems associated with its introduction.

Production of fingerling grass carp for weed control began in Arkansas during 1972 (Bailey et al. 1973) and continues to be centered in that state. But grass carp hatcheries are now located in many other midwestern and southeastern states. Some of the grass carp stocked for weed control in polyculture with such species as channel catfish are eventually harvested and marketed as food fish. In Arkansas during 1978, 19 tons of food-size grass carp were produced, and 18 tons of fingerlings were reared for stocking (Henderson and Freeze 1979).

The grass carp has been introduced to Israel — also primarily for its role in aquatic vegetation control — and small numbers have reached the human food market (Brown 1977). Data on commercial production of grass carp in 1988 are listed in Table 1 (FAO 1990). The fish has also been introduced to Japan, where a reproducing population has reportedly been established on the Tone River. However, commercial production of grass carp has not as yet been initiated in Japan (Kafuku and Ikenouke 1983). Finally, grass carp have been introduced to the Sudan (George 1982) and Egypt on an experimental basis to control aquatic vegetation in the irrigation canal systems that have been developed in those countries.

III. CULTURE SYSTEMS

A. PONDS

Common carp are cultured primarily in earthen ponds. The degree of culture intensity ranges from highly extensive (low stocking densities with no supplemental feeding or fertilization) to relatively highly intensive (high density stocking, fertilization, provision of complete diets, mechanical aeration, management of phytoplankton blooms). Pond production in Europe is mostly extensive in nature. In Austria, Belgium, France, and the former Federal Republic of Germany, pond production involves no feeding or fertilization. Production averages about 300 kg/ ha (Brown 1977). In those countries, about 3 years are required to produce carp of 1.0 kg.

Traditional carp culture in Czechoslovakia is also extensive, but fertilization with manure is practiced as a means of increasing natural food supplies. There is also some supplemental feeding with grain (Krupauer 1973). Recent emphasis has been to integrate carp culture with the production of ducks and geese. The birds provide fertilizer and spilled feed to the water, thus increasing carp production.

In Poland, traditional pond culture of carp leads to production levels of around 500 kg/ha (Lovell 1977). The addition of grain as supplemental feed has been shown to increase production to levels of 1000 to 1200 kg/ha. Experiments in which pelleted feeds have been provided led to further increases in production to 4000 kg/ ha, while the use of continuous aeration doubled the latter production figure.

Carp pond production in Hungary averages about 1000 kg/ha (Brown 1977). An experimental system of 10-year crop rotation has been initiated, which consists of 5 years of carp and duck culture in ponds, followed by 2 years of alfalfa, then 3 years of rice.

Intensive carp culture is practiced in The People's Republic of China, Japan, and Israel (Brown 1977; Hepher and Pruginin 1981; Kafuku and Ikenoue 1983). Carp production in those countries is enhanced by the moderate climates that can be found in many regions; thus, carp production requires only 2 years.

In The People's Republic of China, the stocking density of various carp species in polyculture is high, and ponds are heavily fertilized with manure and agricultural wastes. The ponds are closely monitored for water color and clarity, so DO problems can be anticipated and manuring rates adjusted accordingly. Average pond production ranges from 1500 kg/ha in northern China to 3750 kg/ha in the south (Tapidor et al. 1977).

About half of the common carp produced in Japan are reared in ponds. Production ranges from 3700 to 5200 kg/ha (Brown 1977; Kafuku and Ikenoue 1983). Ponds are heavily stocked, complete diets are provided, mechanical aeration is available, and plankton blooms are carefully managed.

In Israel, carp production, which was initiated by eastern European immigrants, has gradually shifted over the past 50 years from extensive monoculture to intensive monoculture with the use of high-protein pelleted feeds and mechanical aeration or intensive polyculture with daily manure applications and the feeding of supplemental grains (Brown 1977; Chervinski 1979). Carp production has been as high as 3700 kg/ha in polyculture and up to 30,000 kg/ha in monoculture (Chervinski 1979).

B. RACEWAYS

In Japan, 14 to 17% of carp production has been accomplished in stone or concrete raceways (Brown 1977; Kafuku and Ikenoue 1983). The water flow at the most productive of the farms employing raceways ranges from 91 to 455 l/sec, and the carp are fed pelleted diets containing silkworm pupae. The yield from such systems is the equivalent to 2,203,000 kg/ha (Brown 1977).

C. CAGES

Culture of common carp in floating cages in lakes and reservoirs accounts for from 28 to 37% of production in Japan (Brown 1977; Kafuku and Ikenoue 1983). Square cages, 9 m on a side and 2 m deep, are constructed which float to a depth of 1.5 m. An additional meter of netting around the top of the cages helps retain the fish that are subject to jumping. Pelleted feed with a protein content of 39% is provided 4 times daily. The fish are stocked at no more than 75 fish per square meter and require 2 years to reach market size of 0.8 to 1.0 kg. Production averages are equivalent to 431,000 kg/ha (Brown 1977).

Experiments with carp fingerling rearing in cages utilizing the thermal effluent from a power plant to increase the length of the effective growing season have been conducted in Poland. The technique led to a doubling in normal annual growth (Trzebiatowski 1979). Cage culture has also been initiated in irrigation canals in Egypt.

D. RIVERS, LAKES, AND CANALS

In The People's Republic of China many lakes, reservoirs, rivers, and canals have been developed into extensive fish-culture facilities (Tapidor et al. 1977; FAO, 1979). The inlets and outlets of lakes and reservoirs have been blocked off with nets, predators controlled, and carp stocked and supplementally fed. Coves are isolated with nets and used as nursery grounds for fry and fingerling carp which are later stocked into the main rearing areas. In 1987, about 600,000 ha of lakes were being farmed with an average production of 269 kg/ha (Shen and Xu 1988). Experiments with more intensive pen polyculture in lakes have been conducted with yields as high as 11,900 kg/ha.

Sections of rivers and canals may be fenced off with flexible gates which allow the passage of boats while retaining fish (FAO 1979). Predators are removed by fishing, and fry are stocked subsequent to rearing in adjacent pools on the banks of the rivers and canals. The waterways are fertilized with agricultural runoff from adjacent lands.

E. RICE PADDIES

Carp polyculture is carried out in about 0.5 million ha of rice paddies in The People's Republic of China, although culture potentially could be conducted in up to 25 million ha of paddies (Li 1988). Trenches or sumps occupying 5 to 8% of the total area are dug in the paddies as a refuge or harvest point for fish when the paddies are drained for harvest of the rice and before insecticides are applied. When the paddies are flooded, the fish live off natural algal, aquatic weed, and insect production. The fish greatly benefit rice production by increasing soil aeration, nutrient mineralization, and by eating weeds and insect pests. Fish production generally ranges from 225 to 750 kg/ha, but has been reported as high as 2250 kg/ha.

IV. STOCKING DENSITIES

A. POLYCULTURE

Common carp and grass carp are often reared with other species of fish (Brown 1977; Hepher and Pruginin 1981, Tapidor et al. 1977; FAO 1979; Chadhuri et al. 1975; Sinha and Vijaya 1975; Malacha et al. 1981). There are a number of variables that influence the species and stocking density used by polyculturists. These include

1. Economics of culture and market demand for the fish
2. Ecological niche to be filled
3. Natural food and prepared ration availability
4. Water quality, availability of water, and aeration equipment
5. Size of fish at stocking
6. Size of fish desired at harvest
7. Climate and length of growing season
8. Energy and labor available for stocking, harvesting, and processing

Polyculture in Israel generally consists of a mixture of common carp, tilapia (*Tilapia aurea*), mullet (*Mugil cephalus*), and silver carp (Hepher and Pruginin 1981). Stocking densities for common carp at some high-yield farms ranges from 2500 to 6500 fish/ha. That species may account for from 32 to 59% of total production. In many instances, 2-year classes of carp are stocked simultaneously.

In The People's Republic of China, polyculture involves a mixture of common carp, grass carp, silver carp, bighead carp, black carp (*Mylopharyngodon piceus*), mud carp, goldfish (*Carassius auratus*), and Wuchan fish (*Megalobrama amblycephala*), as described by Tapidor et al. (1977) and FAO (1979). In ponds with a total stocking density of 15,000 fish per hectare, common carp comprise 3 to 5% of the total, grass carp, which is sometimes the major species cultured, comprise from 12 to 55%. In the stocking of rivers, common carp commonly represent 20% of the total, and grass carp, 35%.

B. MONOCULTURE

The extensive monoculture practiced in most of Europe involves the stocking of common carp at densities of 300 to 600 fish per hectare (Brown 1977). Experiments aimed at intensifying the culture of carp employed 300 fish per hectare in unfed unfertilized ponds; 900 fish per hectare in fertilized ponds; and 4000 fish per hectare in ponds that received pelleted feeds (Lovell 1977).

Monoculture ponds in Israel may range in density from 1000 to 14,750 carp per hectare (Hepher and Pruginin 1981). Research in Israel has shown that at densities of 4000 to 20,000 fish per hectare, carp growth is inverse to density (Rappaport and Sarig 1979). Economic analysis of carp production showed that the most profitable stocking density for monoculture was 16,000 fish per hectare.

The 81 m^2 floating cages used in Japan are initially stocked with 43 to 75 carp fingerlings per square meter. Final production yields of from 100 to 200 kg/m^2 are achieved (Brown 1977; Kafuku and Ikenoue 1983). Stocking densities in raceways vary with water flow which ranges from 91 to 840 l/s. Average stocking in raceways is from 3 to 11 kg/m^2 of 70- to 150-g fingerlings. Production averages from 100 to 200 kg/m^2. When stocked at 181 fish per square meter, yields of 220 kg/m^2 have been obtained (Brown 1977).

V. REPRODUCTION AND GENETICS

A. NATURAL SPAWNING

The natural spawning season for common carp is during the spring when water temperatures reach 18 to 24°C and remain fairly stable (Hepher and Pruginin 1981; Bieniarz et al. 1979). Common carp lay adhesive eggs on plant substrates, so spawning ponds are supplied with mats of conifer branches or plastic artificial substrates (Hepher and Pruginin 1981). Spawning ponds range in size from a few square meters to several hectares. In smaller ponds, once egg laying is complete, spawning mats are covered with a wet cloth to prevent egg desiccation and are transferred to fry nursery ponds. Larger ponds are usually utilized both for spawning and fry nursing. After the eggs hatch, the spawning mats are removed and the broodfish are seined from the pond with large mesh nets.

During early spring, the sexes are segregated into separate ponds to prevent premature spawning. Females of 2 to 5 kg, and slightly smaller males, are utilized as broodstock. They are stocked, at a ratio of 1:2 or 2:3 (females to males), into spawning ponds that have been allowed to dry for elimination of disease organisms and are filled immediately prior to stocking. Densities of 10 females per hectare or less are utilized (FAO 1979; Hepher and Pruginin 1981).

The final environmental spawning cue for carp is spring flooding. Transportation of ripe carp and their introduction into the newly filled ponds appears to mimic that cue and leads to stimulation of spawning. Spawning usually occurs the morning after the fish are stocked into the spawning ponds. If spawning has not occurred within a few days, fresh water is flowed through the pond as a stimulus.

Fry nursery ponds are also dried prior to use and are fertilized with 60 kg/ha superphosphate and 60 kg/ha ammonium sulfate when filled. Chicken manure is added at the rate of 100 kg/ha every 2 weeks to stimulate zooplankton production which is the initial food of carp fry (Hepher and Pruginin 1981; Rothbard 1982). Fry ponds are treated every 2 or 3 d with vegetable or mineral oil or insecticides to kill predatory insects until the fry are large enough to avoid predation (Hepher and Pruginin 1981, Rothbard 1982). Ponds are also commonly treated at 3-d intervals with 0.2 ppm Bromex® to prevent the establishment of monogenetic trematodes (Hepher and Pruginin 1981).

Fry are reared in the initial spawning or nursery ponds until they reach from 0.2 to 0.5 g. They are harvested and counted to establish survival and then stocked into secondary ponds at known rates.

Secondary nursery ponds are employed for rearing carp fingerlings of 10 to 50 g by the end of the first growing season (Hepher and Pruginin 1981). Fry ponds should be stocked at densities that will lead to final standing crops at harvest of not more than 1000 kg/ha. The secondary nursery ponds are fertilized, and the fish may receive supplemental feedings of ground grains.

B. INDUCED SPAWNING

Techniques for the induction of spawning in common carp are well-established. With respect to grass carp, spawning can only be achieved in the hatchery as the fish will not produce viable fry in ponds (Jalbert et al. 1977; FAO 1979; Hepher and Pruginin 1981; Rothbard 1981, 1982; Epler et al. 1982; Rabelahatra 1982). Ripe broodfish are brought into the hatchery where the females are placed in individual containers and two to five males are placed together. The females should be weighed to provide an estimate of initial fecundity, with the average being 120 eggs per gram of body weight (Bisshai et al. 1974, Tomita et al. 1980).

Ripe broodfish are induced to spawn by injection with either carp pituitary extract, human chorionic gonadotropin, or leuteinizing hormone-releasing hormone at normal spawning temperatures of $18°C$ or higher (Hepher and Pruginin 1981; Rothbard 1981; Peter et al. 1988). At low temperatures (13 to $15°C$), spawning can be induced with carp pituitary extract and $17-\alpha$-hydroxy-$20-\beta$-dihydroprogesterone (Jalbert et al. 1977). When carp pituitary extract is employed, the fish receive 2 injections 8 h apart for common carp and 24 h apart for grass carp (Rothbard 1981). Males are injected once at the time the female receives her second injection. Spawning occurs 10 to 14 h after the final injection depending upon water temperature. The total injection rate for females is about 3 mg of dry weight pituitary per kilogram of broodfish, with 10 to 15% in the first injection and the remainder in the second.

When eggs are found to flow easily from the female, or a few released eggs are observed in the holding container, she is netted, wrapped in a towel, and the lower portion of the abdomen is dried (Rothbard 1981). One person holds the female and strips the eggs into a round basin held by a second person (Figure 3). Milt from two

FIGURE 3.   Eggs being stripped from a common carp female at the fish hatchery of Kibbutz Gan Schmuel, Israel.

males handled in a similar manner is added to the dry eggs and the mixture is stirred. Milt that has been obtained in advance and stored under refrigeration may also be successfully utilized (Hulata and Rothbard 1979). In the case of grass carp, water is added to the eggs after about 1 min of stirring to harden them (Rothbard 1981).

Incubation of the normally adhesive eggs of common carp in hatching containers requires that the eggs be treated with a salt and urea (Hepher and Pruginin 1981; Rothbard 1981; Rabelahatra 1982), salt and talc, or salt and milk solution (Soin 1977) to keep them from sticking. Egg volume for each spawn is recorded so estimates of egg numbers can be obtained. The eggs are then placed in upwelling hatching containers to incubate for 18 to 28 h for grass carp and 43 to 55 h for common carp, depending on water temperature (Rothbard 1981; Figure 4).

When eggs hatch, the larvae swim to the surface and are carried by water currents into larval rearing tanks. Three to five d after hatching, the yolk is totally absorbed, the fry surface to gulp air and inflate their swimbladders, and feeding begins. Food such as decapsulated brine shrimp nauplii, rotifers, and yeast are fed on a continuous basis (Hepher and Pruginin 1981; Rothbard 1982). After about 10 to 14 d in the hatchery, the fry are removed from the tanks and placed in nursery ponds.

C. GENETICS
Since the common carp has been domesticated longer than any other fish, it has been possible for culturists to undertake genetic studies developing several different strains and two major races (Hines et al. 1974; Wohlfarth et al. 1975, 1980, 1983; Brody et al. 1976, 1980, 1981; Hepher and Pruginin 1981; Suzuki et al. 1977; Hulata

FIGURE 4.   Carp eggs in incubators at the fish hatchery of Kibbutz Gan Schmuel, Israel.

et al. 1980; Suzuki and Yamaguchi 1980; Hulata et al. 1982; Sin, 1982; Jhingran and Pullin 1985).

The European race has been selected over the centuries for fast growth rate, late sexual maturity, ease of capture by seining, and scale patterns. The European race of common carp consists of four main strains characterized by different scale patterns: (1) the "scaly" carp is totally covered with scales, (2) the "mirror" carp has only a few scales under the dorsal fin and ventrally, (3) the "line" carp has scales only along the lateral line, and (4) the "leather" carp has no scales.

The leather and line carp varieties carry recessive lethal genes and are characterized by slow growth and low fecundity. Many highly selected and inbred lines of mirror and scaly carp have been developed in Europe and Israel and are usually named for the place or region where they originated. Some examples include the Aischgrunder, Franken, and Galacian mirror carps, the Lausitzer scaled carp, and the Bohemian leather carp (Jhingran and Pullin 1985; Michaels 1988).

The process of domestication of the common carp in Europe has led to the development of highly inbred lines with concomitant increases in deformities and reduced growth rates. The most common practice in Israel to counter these undesirable traits is to cross the line known as the "Dor-70" with a Yugoslavian line to restore hybrid vigor (Wohlfarth et al. 1975, 1980, 1983; Hepher and Pruginin 1981). Many of the carps produced in Hungary and Yugoslavia are crosses of Aischgrunder and Galacian carps, and Hungarian, Aischgrunder, Bohemian, and Lausitzer carps have been crossed to produce the "Militsch" carps (Jhingran and Pullin 1985).

The Asian race, known as the "big belly" carp, was domesticated under totally different conditions than the European carp. The former is a minor component of intensively manured polyculture systems, as compared with the European race which is usually reared in monoculture. The Asian or Chinese variety of common carp has not been highly selected or inbred and retains the wild characteristics of high fecundity, early sexual maturity, high disease resistance, and relatively slow growth after the first year once sexual maturity is reached (Suzuki et al. 1977; Wohlfarth et al. 1983; Jhingran and Pullin 1985). The high fecundity of these fish is demonstrated by the large gonads of both sexes which give rise to the common name for the race.

The big belly carp is more difficult to seine than other carps because of its ability to escape under nets. Crosses between the European and Asian carp have produced hybrids that grow more rapidly than the Asian race, have higher disease resistance, are easier to seine, yet compete well in intensive polyculture situations (Wohlfarth et al. 1975, 1983; Suzuki et al. 1977; Suzuki and Yamaguchi 1980; Hulata et al. 1982; Sin 1982).

Research to produce gynogenetic monosex female and sterile triploid common carp has led to populations with 5 to 10% faster growth compared with normal populations (Nagy et al. 1978; Gervai et al. 1980). Gynogenetic grass-carp production has also been studied in attempts to produce monosex female fish that would, when stocked, pose a reduced environmental threat since spawning could not occur (Stanley et al. 1975; Stanley and Jones 1976; Shelton 1986). In addition, female grass carp have been crossed with bighead carp males to produce sterile triploid fish, though possibly fertile diploids also occur as a result of such crosses (Beck and Biggers 1982, 1983a, 1983b). Blood cells from individual fish are now checked with a Coulter Counter to certify the animals as triploid fish.

To produce gynogenetic carp, sperm are inactivated by destruction of their DNA with either ultraviolet or gamma radiation, mixed with eggs to induce "fertilization". Then the eggs are shocked to induce retention of the second polar body to result in a diploid organism (Hollebecq et al. 1986; Linhart et al. 1986). Sperm inactivation is accomplished with either an absorbed dose of 1000 Gy of gamma radiation from cobalt-60 (Linhart et al. 1986) or 2200 J/m^2/min ultraviolet radiation for 60 to 65 min (Komen et al. 1991). Shocks that have been reported include: cold of 0 to 4°C for 60 min beginning 15 min after fertilization (Linhart et al. 1986); for eggs starting at 24°C, cold of 0°C for 45 min starting 1 to 2 min after fertilization (Komen et al. 1988); heat of 39 to 40°C for 1.33 to 2 min 3 to 5 min after fertilization (Hollebecq et al. 1986); or heat of 40°C for 2 min 28 to 30 min after fertilization (Komen et al. 1991).

For commercial production of all female carp, some of the gynogenetic females are treated with testosterone when young to produce phenotypic males which are used to breed back to the gynogenetic females to produce all female progeny. In the production of phenotypic male grass carp from gynogenetic females 17-α-methyltestosterone in silastic tubes is implanted in the body cavity of young fish (Shelton 1986). Verification of true gynogenetic broodstock is done by progeny testing, which involves spawning a pair and determining if the resulting fish are all female.

Sterile triploid grass carp have been produced by normal fertilization followed by shocking to induce second polar-body retention. Successful shocking techniques include: cold of 5 to 7°C for 25 to 30 min starting 2 to 4.5 min after fertilization; heat of 40°C for 1 min 4.75 min following fertilization (Cassani and Caton 1985); heat of 42 °C for 1 min applied 4 min following fertilization; and hydrostatic pressure of 7000 to 8000 psi for 1 to 2 min initiated 4 min after fertilization (Cassani and Caton 1986).

Research on feeding hormones to produce phenotypic changes to influence growth rates has been initiated. Komen et al. (1989) found that feeding common carp fry 50 ppm of 17-α-methyltestosterone during weeks 6 through 11 after hatching produced 92.7% males. Monzoor Ali and Rao (1989) investigated overdosing with male hormone to produce sterile common carp. They found that feeding 400 ppm 17-α-methyltestosterone for 30 d starting 1 d after hatch produced 98.25% sterile and 1.75% male fish which had 47% increased growth over control fish at 1 year of age.

A new area of genetic research is transfer of genetic material from one species to another which results in "transgenic" organisms. In the hopes of increasing growth rates, the gene for growth hormone from rainbow trout (*Salmo mykiss*) has been transferred into common carp by researchers at Auburn University, and the first pond growth studies of offspring from transgenetic carp are currently underway (Anonymous 1990, 1991). Initial aquarium studies of transgenic carp showed a 20 to 40% increase in growth rate.

VI. FEEDING AND NUTRITION

A. NATURAL AND LIVE FOODS

Common carp fingerlings and adults are primarily benthic omnivores which feed on chironomid larvae and adults, oligochaetes, and gastropods (Williamson and Smitherman 1975; Spataru and Hepher 1977; Zur 1979, 1980; Zur and Sarig 1980). In cases where food is limited, carp have been found to eat tilapia fry, helping to reduce unwanted reproductive success of tilapia in polyculture ponds (Spataru and Hepher 1977).

Carp are fed silkworm pupae in Japan where those organisms are available in large quantities as a by-product of the silk industry (Brown 1977; Kafuku and Ikenoue 1983). Carp farming in Japan started in the silk producing areas where pupae were readily available and inexpensive.

Larval and fry carp feed on zooplankton in nature (Tamas and Horvath 1976; Tamas 1979; Debeljak et al. 1979). Larval carp require small food organisms such as rotifers for about their first week of feeding, after which they consume copepods and cladocera. Fertilized fry ponds can be managed to produce primarily rotifers initially by treatment with 1.0 ppm of organophosphorus esters such as Flibol E®, Ditrifon®, or Dipterex® which selectively kill arthropods at the recommended concentrations (Tamas and Horvath 1976; Tamas 1979). Those compounds degrade within a week allowing the copepods and cladocerans to become reestablished at the time the switch in primary food organisms is made by the fish. Under intensive hatchery conditions, larval and fry carp are often fed brine shrimp nauplii (Rothbard 1982; Bryant and Matty 1980).

Adult grass carp are herbivorous. They feed on submerged vascular plants and terrestrial plants that come in contact with the water (Hepher and Pruginin 1981; Venkatesh and Shetty 1978). Grass carp have been found to have both endogenous and bacterial cellulase production in the intestinal tract (Das and Tripathi 1991). In pond polyculture, terrestrial plants are sometimes cut and thrown into the water as supplemental feed for grass carp. Studies have been done on the digestibility by grass carp of some aquatic and terrestrial plants (Hajra 1985; Law 1986; Naskar et al. 1986; Chapman et al. 1987; Hajra 1987).

Young grass carp, like common carp, initially feed on zooplankton (Grygierek 1973; Singh et al. 1977; Watkins et al. 1981; DeSilva and Weerakoon 1981). They begin feeding on rotifers and then alter their food habits to take crustacean and chironomid larvae. By 50-mm total length, they become almost exclusively herbivorous.

B. FERTILIZATION AND MANURING

Natural productivity in fish culture ponds will generally support polyculture production levels of 20 to 40 kg/ha. The use of moderate amounts of organic or inorganic fertilizers can increase production 200 to 400 kg/ha. This increase is accomplished by a concomitant increase in primary and detrital production which, in turn, increases the amount of natural food available to the carp. Fertilizers may be the only nutrient source applied to ponds in Europe (Brown 1977). They are also used as relatively inexpensive food-producing materials in conjunction with intensive monoculture or polyculture systems when combined with the use of pelleted feeds (Hepher and Pruginin 1981). Heavy manuring has also been employed as the only source of nutrients in intensive carp polyculture (Tapidor et al. 1977; FAO 1979). There is ample literature available on the experimental use of inorganic and organic fertilizers in carp culture (Dimitrov 1974; Klekot 1979; Szlauer and Szlauer 1979; Debeljak et al. 1979; Hepher and Pruginin 1981; Malacha et al. 1981; Barash et al. 1982; Kafuku and Ikenoue 1983).

C. FEEDS AND FEEDING

The composition of pelleted carp diets has been widely reported (Brown 1977; Atack et al. 1979; Viola and Rappaport 1979; Viola et al. 1981a,b, 1982; Christensen 1981; Hepher and Pruginin 1981; Anwar et al. 1982; Capper et al. 1982; Kafuku and Ikenoue 1983; National Research Council 1983; Viola and Arieli 1983; Ufodike and Matty 1983; Jhingran and Pullin 1985). Common ingredients in carp feed include fish meal, soybean meal, wheat, barley, rye, oats, corn, sorghum, peas, lupine, cottonseed meal, wheat bran, rice bran, salt, inorganic phosphate, and vitamin and mineral premixes. Feed generally contain from 25 to 40% crude protein and 3 to 6% crude fat. Now that the ingredients that perform well in common carp feed have been determined, research continues on the digestibility and efficiency of various feed components, and attempts are being made to find less expensive alternative ingredients (Grabner and Hofer 1985; Sturmbauer and Hofer 1986; Kirchgessner et al. 1986; Schwarz et al. 1988; Hossain and Jauncey 1989).

Research has been conducted on the use of steroid hormones as growth-promoting feed additives (Matty and Lone 1979; Lone and Matty 1981, 1982, 1983). Research on the use of subtherapeutic doses of antibiotics in carp feeds as growth-rate enhancers has shown positive results for terramycin and virginiamycin (Viola and Arieli 1987; Ahmad and Matty 1989). Studies on the development of artificial feeds for young carp have centered on powdered yeast, organ meats (e.g., liver and spleen), egg, and fish (as supplements or replacements for live food) (Appelbaum and Dor 1978; Dabrowski et al. 1978; Bryant and Matty 1981; Goolish and Adelman 1984; Charlon et al. 1986; Szlaminska and Przybyl 1986; Csengeri and Petitjean 1987; Dabrowski and Poczyzynski 1988; Alami-Durante et al. 1991).

Optimum feeding rates for carp vary depending on fish size and age, water temperature, relative amount of natural food available, and the quality of the feed being offered. Feeding rates for production ponds in Israel are generally 4 to 5% of body weight daily (Hepher and Pruginin 1981). Carp are fed either once in the morning or continuously with automatic feeders. Some attempts have been made to feed grass carp pelleted diets, but that practice is not common (Dabrowski and Kozak 1979; Huisman and Valentijn 1981).

D. NUTRITIONAL REQUIREMENTS

The nutritional requirements of carp have been widely researched and are well documented (Ogino and Saito 1970; Ogino et al. 1970, 1976, 1978, 1979; Watanabi et al. 1970, 1975a, 1975b; Ogino and Kamizono 1975; Ogino and Chiou 1976; Ogino and Takeda 1976; Viola 1977; Takeuchi and Watanabi 1977; Farkas et al. 1977, 1978, 1980; Sakamoto and Yone 1978; Viola and Amidan 1978; Sato et al. 1978; Sen et al. 1978; Takeuchi et al. 1978, 1979a, 1979b, 1979c, 1980; Ogino and Yang 1979, 1980; Yone and Toshima 1979; Dabrowski et al. 1979; Ogino 1980; Murata and Higashi 1980; Plakas et al. 1980; Steffens 1981; Huisman and Valentijn 1981; Plakas and Katayama 1981; Shimemo et al. 1981; Viola and Arieli 1982; Dabrowski 1983; Jonas et al. 1983; Zeitler et al. 1984). Those requirements have been the subject of two review publications (Jauncey 1982; National Research Council 1983), and the interested reader is directed to those sources for further information.

VII. DISEASES

One of the most common and important diseases of the common carp is infectious carp dropsy (Hepher and Pruginin 1981), also known as bacterial hemorrhagic septicemia and infectious abdominal dropsy (Amlacher 1970). It is primarily caused by the bacterium *Aeromonas punctata* (synonymous with *A. liquefaciens*), though it is often found to occur in conjunction with *Pseudomonas fluorescens*. The most obvious sign of the disease is accumulation of clear fluid in the abdominal cavity which leads to distension of the abdomen. Other signs of pathology include ulcers in the liver, decreased erythrocyte numbers and consequent lower hemoglobins, increased leucocytes, and lesions in the intestinal tract

(Amlacher 1970). The disease occurs after the carp have been stressed. It is most common in the early spring after a hard winter when the stored energy reserves of the fish have been depleted. Carp with high levels of blood and serum proteins resist exposure to the disease (Amlacher 1970).

Infectious carp dropsy can be prevented through employment of the following prophylactic measures:

1. Thorough drying of rearing ponds during the winter to remove carrier wild fish and parasites
2. Fertilization of rearing ponds to establish a good natural food supply
3. Stocking of only healthy fish into rearing ponds
4. Adequate provision of feed in the fall to allow the accumulation of large nutrient reserves (primarily fat)
5. Injection of fish during the spring with antibiotics such as chloromycetin or streptomycin

Columnaris disease, caused by the bacterium *Flexibacter columnaris* is a common fish pathogen which has only since the mid-1970s been reported from carp in Europe (Bootsma and Clerx 1976). Virulent strains produce external ulcers surrounded by red areas while other strains produce different signs. Successful treatment can be effected with penicillin and ampicillin.

When *F. columnaris* attacks the gills it produces "gill necrosis", which is a widespread disease of carp in Europe that leads to a high mortality rate (Farkas and Olah 1986). After some stressfull incident, usually when water temperature is above 20°C, the bacteria attack the gills producing a gray-white coating. If fish survive initial infection, large necrotic areas develop on their gills.

Research in Israel (Hines et al. 1974) has shown that the susceptibility of carp to certain diseases is related to recessive genes. Fish pox, epidermal epithelioma, and swim-bladder inflammation have all been shown to occur only in fish which were homozygous for certain recessive traits and were members of highly inbred lines. The use of crossbreeding in Israel has greatly decreased the incidence of the diseases mentioned. Susceptibility to carp erythrodermatitis, an inflamatory and ulcerous skin infection caused by the bacterium *Aeromonas salmonicida*, in an inbred strain of carp in Hungary was greatly decreased by crossing fish from that strain with a Japanese colored carp strain (Sovenyi et al. 1988).

External parasites are another common problem (Hepher and Pruginin 1981, Jhingran and Pullin 1985, Michaels 1988). The monogenetic trematodes *Dactylogyrus* sp. (gill flukes) and *Gyrodactylus* sp. (skin flukes) are problems in fry but can be controlled with 0.2 ppm of Bromex® in ponds. The parasitic copepods known as fish lice (*Argulus* sp.) and anchor worms (*Lernaea* sp.), while mainly problems of esthetics in adult carp, can be fatal when present in high concentrations on young fish. Fish lice can be controlled with low levels of Lindane® (0.02 ppm), Malathion®, or Bromex® (0.1 to 0.2 ppm) in ponds or as a 10-min bath in 100 ppm potassium permanganate. Adult anchor worms attached to carp are resistant to pesticide treatments, but the free-swimming larvae can be killed with Bromex at levels of 0.12 to 0.15 ppm active ingredient in ponds. The chemical should be applied 3 times at 1-week intervals during summer. Anchor worms and fish lice can also be controlled with 0.2 to 0.3 ppm Dipterex applied to ponds at 2- to 3-week intervals (Kafuku and Ikenoue 1983).

The protozoan *Costia necatrix*, which also attacks the gills and skin of young carp, can be removed with a 10-min bath in 20 ppt salt, a 30-min bath in 300 ppm formalin, or in ponds with 0.667 ppm malachite green for 10 h (Michaels 1988). The protozoan *Ichthyophthirius multifiliis* can also be a problem in carp culture (Amlacher 1970; Kafuku and Ikenoue 1983; Jhingran and Pullin 1985; Michaels 1988). It can be eliminated by raising the water temperature above 25°C, by treating ponds with either 0.10 to 0.15 ppm malachite green or 20 ppm formalin, or with a 1-h bath in 1 to 2 ppm malachite green and 167 to 250 ppm formalin.

The fungus *Saprolegnia* sp. is commonly seen as a secondary infection during the winter when carp are weakened following injury or handling stress. The fungus can be controlled prophylactically in rearing ponds by the application of 25 ppm formalin.

VIII. BUFFALO CULTURE

There are three species of buffalo native to North America: bigmouth buffalo (*Ictiobus cyprinellus*), smallmouth buffalo (*I. bubalus*), and black buffalo (*I. niger*). Of these, the bigmouth buffalo is the primary species considered for culture because it grows most rapidly, is best adapted to pond and lake environments, matures earliest, and has a higher fecundity than the other species (Brady and Hulsey 1959). It can be distinguished from the smallmouth and black buffalo by the location of its mouth, which is terminal, while the mouths of the other two are inferior. The bigmouth buffalo feeds primarily on zooplankton.

Buffalo are raised primarily in Arkansas and nearby states either in monoculture in fertilized rice field ponds undergoing a fish cycle of crop rotation or in polyculture with channel catfish (FAO 1979). The Payment In Kind (PIK) program initiated by the federal government produced some shift in acreage from grain to bigmouth buffalo. In the last decade, production has averaged about 11,000 tons per year in the U.S. (FAO 1987, 1990).

The bigmouth buffalo is spawned and the fry reared in a manner similar to that described above for common carp (Walker and Frank 1952; Swingle 1956; Brady and Hulsey 1959). The fish spawn in the early spring when water temperatures rise to from 16 to 21°C. They spawn during flooding periods and lay their adhesive eggs on vegetation.

Broodfish of 1.5 to 4.0 kg are selected in the late fall or early winter from the most rapidly growing stock. They are treated prophylactically to remove external parasites and are placed in overwintering ponds at high density (500 to 1,000 kg/ha) to inhibit premature spawning in the spring. Spawning ponds are prepared by draining to kill disease organisms and predatory wild fish. Grass is allowed to grow in the ponds to provide spawning substrate. Spawning ponds are filled immediately before the broodfish are stocked.

Spawning generally occurs within a few days. If the fry are to be transferred to a nursery pond, 18 to 22 pairs of broodfish are stocked per hectare of spawning pond. If the spawning pond is also to be used as a nursery pond, only three to five pairs of brood fish are stocked per hectare. Males can be differentiated from females by the rough sandpaper-feel of their scales during breeding season.

If spawning has not occurred within 2 weeks of stocking, the ponds can be drained and refilled. This will sometimes induce the fish to spawn. Broodfish can be stripped in the hatchery and the wet method of fertilization used with stirring for 1 h in a cornstarch solution to prevent adhesion until the eggs water harden.

The fish can then be hatched in jars (Walker and Frank 1952). Within the range of 16 to 21°C, eggs will hatch in 5 to 10 d (Walker and Frank 1952; Swingle 1956; Brady and Hulsey 1959).

Ponds are fertilized at the time of spawning to provide food for the larvae following yolk sac absorption. Ponds can be periodically treated with oil to kill predatory insects. At 10 to 15 d, supplemental feeding with finely ground fish feed can be implemented, particularly in instances where the fry are observed to school along the edge of the pond in search of food.

At 3 to 5 cm long the fry can be seined and moved to secondary nursery ponds. Such ponds are commonly stocked at 4500 to 22,000 fish per hectare and are fertilized at frequent intervals to provide natural zooplankton. Fingerlings for subsequent stocking can be produced by the end of the first 8-month growing season.

Buffalo are usually grown out in extensive ponds ranging in size from 5 to 100 ha. The final market size of fish desired controls stocking density; the larger the fish needed, the lower the stocking density. The range is generally from 50 to 150 fish per hectare (Swingle 1956; Brady and Hulsey 1959). Monoculture ponds are either fertilized only or are fertilized and provided with supplemental sinking feed.

In polyculture with channel catfish, buffalo are expected to feed off natural foods and excess catfish ration. Polyculture research has been conducted with combinations of bigmouth buffalo, channel catfish, crayfish, and paddlefish (Tuten and Avault 1981), as well as with hybrid bigmouth × smallmouth buffalo, common carp, tilapia, and channel catfish (Williamson and Smitherman 1975). Polyculture research in Bulgaria has shown that a combination of 2000 black buffalo per hectare, 2500 common carp per hectare, and 1200 to 1500 silver carp per hectare gives the highest economic return (Dimitrov 1987).

Buffalo are marketed at 1 to 3 kg. The larger fish sometimes bring a higher price per kilogram than the smaller ones.

REFERENCES

Ahmad, T.S. and Matty, A.J., The effect of feeding antibiotics on growth and body composition of carp (*Cyprinus carpio*), *Aquaculture*, 77, 211, 1989.

Alami-Durante, H., Charlon, N., Escaffre, A.M., and Bergot, P., Supplementation of artificial diets for common carp (*Cyprinus carpio* L.) larvae, *Aquaculture*, 93, 167, 1991.

Amlacher, E., *Textbook of Fish Diseases*, Conroy, D. A. and Herman, R. L., (transl.,) T.F.H. Publ., Hong Kong, 1970, 1.

Anon., Transgenic carp: pond-ready?, *Science*, 247, 1298, 1990.

Anon., Transgenic carp now in research pond at Auburn experiment station, *Catfish Aquaculture News*, 5(12), 13, 1991.

Anwar, A., Ishak, M.M., El-Zeiny, M., and Hassanen, G.D., Activated sewage sludge as a replacement for bran-cotton seed meal mixture for carp, *Cyprinus carpio* L., *Aquaculture*, 28, 321, 1982.

Appelbaum, S. and Dor, U., Ten day experimental nursing of carp (*Cyprinus carpio* L.) larvae with dry feed, *Bamidgeh*, 30, 85, 1978.

Atack, T.H., Jauncey, K., and Matty, A.J., The utilization of some single cell proteins by fingerling mirror carp (*Cyprinus carpio*), *Aquaculture*, 18, 337, 1979.

Bailey, W.M., Meyer, F.P., Martin, J.M., and Gray, D.L., Fish farm production in Arkansas during 1972, *Proc. S. E. Assoc. Game Fish Comm.*, 27, 750, 1973.

Balon, E.K., Domestication of the Carp *Cyprinus carpio* L., Royal Ontario Museum Life Sciences Misc. Publ., Toronto, Canada, 1974, 1.

Barash, H., Plavnik, I., and Moav, R., Integration of duck and fish farming: experimental results, *Aquaculture*, 27, 129, 1982.

Beck, M.L. and Biggers, C.J., Chromosomal investigation of *Ctenopharyngodon idella × Aristichthys nobilis* hybrids, *Experimentia*, 38, 39, 1982.

Beck, M.L. and Biggers, C.J., Erythrocyte measurements of diploid and triploid *Ctenopharyngodon idella × Hypophthalmichthys nobilis* hybrids, *J. Fish Biol.*, 22, 497, 1983a.

Beck, M.L. and Biggers, C.J., Ploidy of hybrids between grass carp and bighead carp determined by morphological analysis, *Trans. Amer. Fish. Soc.*, 112, 808, 1983b.

Bieniarz, K., Epler, P., Thuy, L.-N., and Kogut, E., Changes in the ovaries of adult carp, *Aquaculture*, 17, 45, 1979.

Bisshai, H.M., Ishak, M.M., and Labib, W., Fecundity of the mirror carp *Cyprinus carpio* L. at the Serow Fish Farm (Egypt), *Aquaculture*, 4, 257, 1974.

Bootsma, R. and Clerx, J.P.M., Columnaris disease of cultured carp *Cyprinus carpio* L. Characterization of the causitive agent, *Aquaculture*, 7, 317, 1976.

Bowen, J.T., A history of fish culture as related to the development of fisheries programs, in *A Century of Fisheries in North America*, Benson, N. G., Ed., American Fisheries Society, Washington, D.C., 1970, 71.

Brady, L. and Hulsey, A., Propagation of buffalo fishes, *Proc. Annu. Conf. S.E. Assoc. Fish Wildl. Agen.*, 13, 80, 1959.

Brody, T., Wohlfarth, G., Hulata, G., and Moav, R., Application of electrophoretic genetic markers to fish breeding. IV. Assessment of breeding value of full-sib families, *Aquaculture*, 24, 175, 1981.

Brody, T., Moav, R., Abramson, Z.V., Hulata, G., and Wohlfarth, G., Applications of electrophoretic genetic markers to fish breeding. II. Genetic variation within maternal half-sibs in carp, *Aquaculture*, 9, 351, 1976.

Brody, T., Storch, N., Kirsht, D., Hulata, G., Wohlfarth, G. and Moav, R., Application of electrophoretic genetic markers to fish breeding. III. Diallel analysis of growth rate in carp, *Aquaculture*, 20, 371, 1980.

Brown, E.E., *World Fish Farming: Cultivation and Economics*, AVI Publishing, Westport, CT, 1977, 1.

Bryant, P.L. and Matty, A.J., Optimization of *Artemia* feeding rate for carp larvae (*Cyprinus carpio* L.), *Aquaculture*, 21, 203, 1980.

Bryant, P.L. and Matty, A J., Adaptation of carp (*Cyprinus carpio*) larvae to artificial diets. I. Optimum feeding rate and adaptation age for a commercial diet, *Aquaculture*, 23, 275, 1981.

Capper, B.S., Wood, J.F., and Jackson, A.J., The feeding value for carp of two types of mustard seed cake from Nepal, *Aquaculture*, 29, 373, 1982.

Cassani, J.R. and Caton, W.E., Induced triploidy in grass carp, *Ctenophyaryngodon idella* Val., *Aquaculture*, 46, 37, 1985.

Cassani, J.R. and Caton, W.E., Efficient production of triploid grass carp (*Ctenopharyngodon idella*) utilizing hydrostatic pressure, *Aquaculture*, 55, 43, 1986.

Chadhuri, H., Chakrabarty, R.D., Sen, P. R., Rao, N.G., and Jena, S., A new high in fish production in India with record yields by composite fish culture in freshwater ponds, *Aquaculture*, 6, 343, 1975.

Chapman, D.C., Hubert, W.A., and Jackson, U.T., Phosphorus retention by grass carp (*Ctenopharyngodon idella*) fed Sago pondweed (*Potamogeton pectinatus*), *Aquaculture*, 65, 221, 1987.

Charlon, N., Durante, H., Escaffre, A.M., and Bergot, P., Alimentation artificielle des carpe (*Cyprinus carpio* L.), *Aquaculture*, 54, 83, 1986.

Chervinski, J., personal communication, 1979.

Christensen, M.S., Preliminary tests on the suitability of coffee pulp in the diets of common carp (*Cyprinus carpio* L.) and catfish (*Clarias mossambicus* Peters), *Aquaculture*, 25, 235, 1981.

Christiansen, B., Lomholt, J.P., and Johansen, K., Oxygen uptake of carp, *Cyprinus carpio*, swimming in normoxic and hypoxic water, *Environ. Biol. Fish.*, 7, 291, 1982.

Csengeri, I. and Petitjean, M., Fresh liver powder: a new starter diet for the larvae of a cyprinid fish, *Aquaculture*, 65, 189, 1987.

Dabrowski, K., Comparative aspects of protein digestion and amino acid absorption in fish and other animals, *Comp. Biochem. Physiol.*, 74A, 417, 1983.

Dabrowski, K. and Kozak, B., The use of fish meal and soybean meal as a protein source in the diet of grass carp fry, *Aquaculture*, 18, 107, 1979.

Dabrowski, K. and Poczyczynski, P., Comparative experiments of starter diets for grass carp and common carp, *Aquaculture*, 69, 317, 1988.

Dabrowski, K., Dabrowski, H., and Grudniewski, C., A study of the feeding of common carp larvae with artificial food, *Aquaculture*, 13, 257, 1978.

Dabrowski, H., Grudniewski, C., and Dabrowski, K., Artificial diets for the common carp: effect of the addition of enzymes extracts, *Prog. Fish-Cult.*, 41, 196, 1979.

Das, K.M. and Tripathi, S.D., Studies on the digestive enzymes of grass carp, *Ctenopharyngodon idella* (Val.), *Aquaculture*, 92, 21, 1991.

Debeljak, L., Pleic, D., Turk, M., Rearing of cyprinid fish fry and zooplankton in Croatia, in *Cultivation of Fish Fry and Its Live Food*, Styczynska-Jurewicz, E., Backiel, T., Jaspers, E., and Persoone, G., Eds., European Mariculture Soc., Bredene, Belgium, 1979, 179.

DeSilva, S.S. and Weerakoon, D.E.M., Growth, food intake and evacuation rates of grass carp, *Ctenopharyngodon idella*, fry, *Aquaculture*, 25, 67, 1981.

Dimitrov, M., Mineral fertilization of carp ponds in polycultural rearing, *Aquaculture*, 3, 273, 1974.

Dimitrov, M., Intensive polyculture of common carp, *Cyprinus carpio* L., silver carp, *Hypophthalmichthys molitrix* (Val.) and black buffalo, *Ictiobus niger* (Raf.), *Aquaculture*, 65, 119, 1987.

Doughty, R.W., Wildlife conservation in late nineteenth-century Texas: the carp experiment, *Southwest. Hist. Q.*, 84, 169, 1980.

Epler, P., Bieniarz, K., and Marosz, E., The effect of low doses of carp hypophysal homogenate and some steroid hormones on carp (*Cyprinus carpio* L.) oocyte maturation *in vitro*, *Aquaculture*, 26, 245, 1982.

FAO, Aquaculture Development in China, Aquaculture Development and Coordination Program, ADCP/REP/79/10, Food and Agriculture Organization, Rome, 1979, 1.

FAO, Yearbook of Fishery Statistics Catches and Landings 1979, Food and Agriculture Organization, Rome, 1981, 1.

FAO, Yearbook of Fishery Statistics Catches and Landings 1985, Food and Agriculture Organization, Rome, 1987, 1.

FAO, Yearbook of Fishery Statistics Catches and Landings 1988, Food and Agriculture Organization, Rome, 1990, 1.

Farkas, J. and Olah, J., Gill necrosis — a complex disease of carp, *Aquaculture,* 58, 17, 1986.

Farkas, T., Csengeri, I., Majoros, F., and Olah, J., Metabolism of fatty acids in fish. I. Development of essential fatty acid deficiency in the carp, *Cyprinus carpio* Linnaeus 1758, *Aquaculture,* 11, 147, 1977.

Farkas, T., Csengeri, I., Majoros, F., and Olah, J., Metabolism of fatty acids in fish. II. Biosynthesis of fatty acids in relation to diet in the carp, *Cyprinus carpio* Linnaeus 1758, *Aquaculture,* 14, 57, 1978.

Farkas, T., Csengeri, I., Majoros, F., and Olah, J., Metabolism of fatty acids in fish. III. Combined effect of environmental temperature and diet formulation and deposition of fatty acids in the carp, *Cyprinus carpio* Linnaeus 1758, *Aquaculture,* 20, 29, 1980.

George, T.T., The Chinese grass carp, *Ctenopharyngodon idella,* its biology, introduction, control of aquatic macrophytes and breeding in the Sudan, *Aquaculture,* 27, 317, 1982.

Gervai, J., Peter, S., Nagy, A., Horvath, L., and Csanyi, V., Induced triploidy in carp, *Cyprinus carpio* L., *J. Fish Biol.,* 17, 667, 1980.

Goolish, E.M. and Adelman, I.R., Effects of ration size and temperature on the growth of juvenile common carp (*Cyprinus carpio* L.), *Aquaculture,* 36, 27, 1984.

Grabner, M. and Hofer, R., The digestibility of the proteins of broad bean (*Vica faba*) and soya bean (*Glycine max*) under in vitro conditions simulating the alimentary tracts of rainbow trout (Salmo gardneri) and carp (*Cyprinus carpio*), *Aquaculture,* 48, 111, 1985.

Grygierek, E., The influence of phytophagous fish on pond zooplankton, *Aquaculture,* 2, 197, 1973.

Hajra, A., Biochemical evaluation of common land grass as feed for grass carp, *Ctenopharyngodon idella* (Val.), in the tropics, *Aquaculture,* 47, 293, 1985.

Hajra, A., Biochemical investigations on the protein-calorie availability in grass carp (*Ctenopharyngodon idella* Val.) from an aquatic weed (*Ceratophyllum demersum* Linn.) in the tropics, *Aquaculture,* 61, 113, 1987.

Henderson, S. and Freeze, M., The Aquaculture Industry of Arkansas in 1978, Arkansas Game and Fish Commission, Little Rock, AR, 1979.

Hepher, B. and Pruginin, Y., *Commercial Fish Farming,* John Wiley & Sons, New York, 1981, 1.

Hines, R.S., Wohlfarth, G., Moav, R., and Hulata, G., Genetic differences in susceptibility to two diseases among strains of the common carp, *Aquaculture,* 3, 187, 1974.

Hollebecq, M.G., Chourrout, D., Wohlfarth, G., and Ballard, R., Diploid gynogenesis induced by heat shocks after activation with UV-irradiated sperm in common carp, *Aquaculture,* 54, 69, 1986.

Hossain, M.A. and Jauncey, K., Studies on the protein, energy and amino acid digestibility of fish meal, mustard oilcake, linseed and sesame meal for common carp (*Cyprinus carpio*), *Aquaculture,* 83, 59, 1989.

Huisman, E.A. and Valentijn, P., Conversion efficiencies in grass carp (*Ctenopharyngodon idella,* Val.) using a feed for commercial production, *Aquaculture,* 22, 279, 1981.

Hulata, G. and Rothbard, S., Cold storage of carp semen for short periods, *Aquaculture,* 16, 267, 1979.

Hulata, G., Moav, R., and Wohlfarth, G., Genetic differences between the Chinese and European races of the common carp. III. Gonad abnormalities in hybrids, *J. Fish Biol.,* 16, 369, 1980.

Hulata, G., Moav, R., and Wohlfarth, G., Effects of crowdings and availability of food on growth rate of fry in the European and Chinese races of the common carp, *J. Fish Biol.,* 20, 323, 1982.

Jalbert, J., Breton, B., Brzuska, E., Fostier, A., and Wieniawski, J., A new tool for induced spawning: the use of 17-α-hydroxy-20-β-dihydroprogesterone to spawn carp at low temperature, *Aquaculture,* 10, 353, 1977.

Jauncey, K., Carp (*Cyprinus carpio* L.) nutrition — a review, in, *Recent Advances in Aquaculture*, Miur, J. F., and Roberts, R. J., Eds., Westview Press, Boulder, CO, 1982, 215.

Jhingran, V.G. and Pullin, R.S.V., *A Hatchery Manual for the Common, Chinese and Indian Major Carps*, ICLARM Studies and Reviews 11, Asian Development Bank and International Center for Living Aquatic Resources Management, Manila, 1985, 1.

Johnston, I.A., A comparative study of glycolysis in red and white muscles of the trout (*Salmo gairdneri*) and mirror carp (*Cyprinus carpio*), *J. Fish Biol.*, 11, 575, 1977.

Jonas, E., Ragyanszki, M., Olah, J., and Boross, L., Proteolytic digestive enzymes of carniverous (*Silurus glanis* L.), herbiverous (*Hypophthalmichthys molitrix* Val.) and omniverous (*Cyprinus carpio* L.) fishes, *Aquaculture*, 30, 145, 1983.

Kafuku, T. and Ikenoue, H., *Modern Methods of Aquaculture in Japan*, Elsevier Science, New York, 1983, 1.

Kirchgessner, M., Kurzinger, H., and Schwarz, F.J., Digestibility of crude nutrients in different feeds and estimation of their energy content for carp (*Cyprinus carpio* L.), *Aquaculture*, 58, 185, 1986.

Klekot, L., Pilot-scale field experiment on the utilization of biologically purified municipal-industrial waste waters for fish farming, in *Cultivation of Fish Fry and Its Live Food*, Styczynska-Jurewicz, E., Backiel, T., Jaspers, E., and Persoone, G., Eds., European Mariculture Soc., Bredene, Belgium 1979, 475.

Komen, J., Duynhouwer, J., Richter, C.J.J., and Huisman, E A., Gynogenesis in common carp, (*Cyprinus carpio* L.). I. Effects of genetic manipulation of sexual products and incubation conditions of eggs, *Aquaculture*, 69, 227, 1988.

Komen, J., Bongers, A.B.J., Richter, C.J.J., Van Muiswinkel, W.B., and Huisman, E.A., Gynogenesis in common carp (*Cyprinus carpio* L.). II. The production of homozygous gynogenetic clones hybrids, *Aquaculture*, 92, 127, 1991.

Komen, J., Lodder, P., Huskens, F., Richter, C., and Huisman, E., Effects of oral administration of 17 alpha-methyltestosterone and 17 beta-estradiol on gonadal development in common carp, *Cyprinus carpio* L., *Aquaculture*, 78, 349, 1989.

Krupauer, V., Pond Fish Culture in Czechoslovakia, EIFAC Occasional Paper No. 8, Food and Agricultural Organization, Rome, 1973, 1.

Law, A.T., Digestibility of low-cost ingredients in pelleted feed by grass carp (*Ctenopharyngodon idella* C. et V.), *Aquaculture*, 51, 97, 1986.

Li, K., Rice-fish culture in China: a review, *Aquaculture*, 71, 173, 1988.

Ling, S.W., *Aquaculture in Southeast Asia*, Washington Sea Grant Publication, College of Fisheries, University of Washington, Seattle, 1977, 103.

Linhart, O., Kvasnicka, P., Slechtová, V., and Pokorny, J., Induced gynogenesis by retention of the second polar body in common carp, *Cyprinus carpio* L., and heterozygosity of gynogenic progeny in transferrin and Ldh-B[1], *Aquaculture*, 54, 63, 1986.

Lone, K.P. and Matty, A.J., Uptake and disappearance of radioactivity in blood and tissue of carp (*Cyprinus carpio*) after feeding ^{3}H-testosterone, *Aquaculture*, 24, 315, 1981.

Lone, K.P. and Matty, A.J., Cellular effects of adrenosterone feeding to juvenile carp, *Cyprinus carpio* L., effect on liver, kidney, brain and muscle protein and nucleic acids, *J. Fish Biol.*, 21, 33, 1982.

Lone, K.P. and Matty, A.J., The effect of Ethylestrenol on growth, food conversion and tissue chemistry of the carp, *Cyprinus carpio*, *Aquaculture*, 32, 39, 1983.

Lovell, R.T., Fish Culture in Poland, International Center for Aquaculture, Res. Dev. Ser. No. 16, Agricultural Experiment Station, Auburn University, Auburn, AL, 1977.

Malacha, S.R., Buck, H.D., Baur, R.J., and Onizuka, D.R., Polyculture of the freshwater prawn, *Macrobrachium rosenbergii*, Chinese and common carps in ponds enriched with swine manure. I. Initial trials, *Aquaculture*, 25, 101, 1981.

Manzoor Ali, P. and Satyanarayana Rao, G., Growth improvement in carp, *Cyprinus carpio* (Linnaeus), sterilized with 17 α-methyltestosterone, *Aquaculture*, 76, 157, 1989.

Martin, J.M., personal communication, 1984.

Matty, A.J. and Lone, K.P., The effect of androgenic steroids as dietary additives on the growth of carp (*Cyprinus carpio*), *Proc. World Maricult. Soc.*, 10, 735, 1979.

Michaels, V.K., *Carp Farming*, Fishing News Books, Farnham, England, 1988, 1.

Murata, H. and Higashi, T., Selective utilization of fatty acid as energy source in carp, *Bull. Jpn. Soc. Sci. Fish.*, 46, 1333, 1980.

Nagy, A., Rajki, K., Horvath, L., and Csanyi, V., Investigation on carp, *Cyprinus carpio* L. gynogenesis, *J. Fish Biol.*, 13, 215, 1978.

Naskar, K.R., Banerjee, A.C., Chakraborty, N.M., and Gosh, A., Yield of *Wolffia arrhiza* (L.) Horkel ex Wimmer from cement cistern with different sewage concertrations, and its efficacy as a carp feed, *Aquaculture*, 51, 211, 1986.

National Research Council, *Nutrient Requirements of Warmwater Fishes and Shellfishes*, rev. ed., National Academy Press, Washington, D.C., 1983, 1.

Ogino, C., Protein requirements of carp and rainbow trout, *Bull. Jpn. Soc. Sci. Fish.*, 46, 385, 1980.

Ogino, C. and Chiou, J.Y., Mineral requirements in fish. II. Magnesium requirements in carp, *Bull. Jpn. Soc. Sci. Fish.*, 42, 71, 1976.

Ogino, C. and Kamizono, M., Mineral requirements in fish. I. Effects of dietary salt-mixture levels on growth, mortality, and body composition in rainbow trout and carp, *Bull. Jpn. Soc. Sci. Fish.*, 41, 429, 1975.

Ogino, C. and Saito, K., Protein nutrition in fish. I. The utilization of dietary protein by young carp, *Bull. Jpn. Soc. Sci. Fish.*, 36, 250, 1970.

Ogino, C. and Takeda, H., Mineral requirements in fish. III. Calcium and phosphorus requirements in carp, *Bull. Jpn. Soc. Sci. Fish.*, 42, 793, 1976.

Ogino, C. and Yang, G.-Y., Requirement of carp for dietary zinc, *Bull. Jpn. Soc. Sci. Fish.*, 45, 967, 1979.

Ogino, C. and Yang, G.-Y., Requirements of carp and rainbow trout for dietary manganese and copper, *Bull. Jpn. Soc. Sci. Fish.*, 46, 455, 1980.

Ogino, C., Chiou, J.-Y., and Takeuchi, T., Protein nutrition in fish. VI. Effects of dietary energy sources on the utilization of proteins by rainbow trout and carp, *Bull. Jpn. Soc. Sci. Fish.*, 42, 213, 1976.

Ogino, C., Cowey, C.B., and Chiou, J.-Y., Leaf protein concentrate as a protein source in diets for carp and rainbow trout, *Bull. Jpn. Soc. Sci. Fish.*, 44, 49, 1978.

Ogino, C., Takeuchi, L., Takeda, H., and Watanabi, T., Availability of dietary phosphorus in carp and rainbow trout, *Bull. Jpn. Soc. Sci. Fish.*, 45, 1527, 1979.

Ogino, C., Watanabi, T., Kakino, J., Iwanaga, N., and Mizunox, M., B vitamin requirements of carp. III. Requirements for biotin, *Bull. Jpn. Soc. Sci. Fish.*, 36, 734, 1970.

Peter, R. E., Lin, H.R., and Van der Kraak, G., Induced ovulation and spawning of cultured freshwater fish in China: advances in application of GnRH analogues and dopamine antagonists, *Aquaculture*, 74, 1, 1988.

Plakas, S.M. and Katayama, T., Apparent digestibilities of amino acids from three regions of the gastrointestinal tract of carp (*Cyprinus carpio*) after ingestion of a protein and a corresponding free amino acid diet, *Aquaculture*, 24, 309, 1981.

Plakas, S.M., Katayama, T., Tanaka, Y., and Deshimaru, O., Changes in the level of circulating plasma free amino acids of carp (*Cyprinus carpio*) after feeding a protein and an amino acid diet of similar composition, *Aquaculture*, 21, 307, 1980.

Rabelahatra, A., Improvement of techniques for carp fry production in the laboratory, *Aquaculture*, 27, 307, 1982.

Rappaport, U. and Sarig, S., The effect of population density of carp in monoculture under conditions of intensive growth, *Bamidgeh*, 31, 26, 1979.

Rothbard, S., Induced reproduction in cultivated cyprinids — the common carp and the group of Chinese carps. I. The technique of induction, spawning and hatching, *Bamidgeh*, 33, 103, 1981.

Rothbard, S., Induced reproduction in cultivated cyprinids — the common carp and the group of Chinese carps: II. The rearing of larvae and the primary nursing of fry, *Bamidgeh*, 34, 20, 1982.

Sakamoto, S. and Yone, Y., Iron deficiency symptoms of carp, *Bull. Jpn. Soc. Sci. Fish.*, 44, 1157, 1978.

Sato, M., Yoshinaka, R., Yamamoto, Y., and Ikeda, S., Nonessentiality of ascorbic acid in the diet of carp, *Bull. Jpn. Soc. Sci. Fish.*, 44, 1151, 1978.

Schwarz, F.J., Kirchgessner, M., Steinhart, H., and Runge, G., Influence of different fats with varying additions of α-tocopheryl acetate on growth and body composition of carp (*Cyprinus carpio* L.), *Aquaculture*, 69, 57, 1988.

Sen, P.R., Rao, N.G.S., Gosh, S.R., and Rout, M., Observations on the protein and carbohydrate requirements of carps, *Aquaculture*, 13, 245, 1978.

Shelton, W., Broodstock development for monosex production of grass carp, *Aquaculture*, 57, 311, 1986.

Shen, P. and Xu, B., Pen culture in Wuli Lake, Jaingsu, China, *Aquaculture*, 71, 301, 1988.

Shimemo, S., Takeda, M., Takayama, S., Fukui, A., Sasaki, H. and Kajiyama, H., Adaptation of hepatopancreatic enzymes to dietary carbohydrate in carp, *Bull. Jpn. Soc. Sci. Fish.*, 47, 71, 1981.

Sin, A. W.-C., Stock improvement of the common carp in Hong Kong through hybridization with the introduced Israeli race "Dor-70", *Aquaculture*, 29, 299, 1982.

Singh, S.B., Dey, R.K., and Reddy, R.V.G.K., Observations on feeding of young grass carp on mosquito larvae, *Aquaculture*, 12, 361, 1977.

Sinha, V.R.P. and Vijaya, G., On the growth of grass carp, *Ctenopharyngodon idella* Val., in composite fish culture at Kalyani, West Bengal (India), *Aquaculture*, 5, 283, 1975.

Soin, S.G., Some features of the development of the carp, *Cyprinus carpiof*, under hatchery conditions, *Vopr. Ikhtiol.*, 17, 759, 1977.

Sovenyi, J.F., Bercsenyi, M., and Bakos, J., Comparative examination of susceptibility of two genotypes of carp (*Cyprinus carpio* L.) to infection with *Aeromonas salmonicida*, *Aquaculture*, 70, 301, 1988.

Spataru, P. and Hepher, B., Common carp predating on tilapia fry in a high density polyculture fish pond system, *Bamidgeh*, 29, 25, 1977.

Stanley, J. and Jones, J.B., Morphology of androgenic and gynogenic grass carp, *Ctenopharyngodon idella* (Valencienes), *J. Fish Biol.*, 9, 523, 1976.

Stanley, J.G., Martin, J. M., and Jones, J.B., Gynogenesis as a possible method for producing monosex grass carp (*Ctenopharyngodon idella*), *Prog. Fish-Cult.*, 37, 25, 1975.

Steffens, W., Protein utilization by rainbow trout (*Salmo gairdneri*) and carp (*Cyprinus carpio*): a brief review, *Aquaculture*, 23, 337, 1981.

Stevenson, J.M., Observations on grass carp in Arkansas, *Prog. Fish-Cult.*, 27, 203, 1965.

Sturmbauer, C. and Hofer, R., Compensation for amalase inhibitors in the intestine of the carp (*Cyprinus carpio*), *Aquaculture*, 52, 31, 1986.

Suzuki, R. and Yamaguchi, M., Improvement of quality in the common carp by crossbreeding, *Bull. Jpn. Soc. Sci. Fish.*, 46, 1427, 1980.

Suzuki, R., Yamaguchi, M., and Ishikawa, K., Differences in the growth rate of two races of the common carp at various water temperatures, *Bull. Freshwater Fish. Res. Lab., Tokyo*, 27, 21, 1977.

Swingle, H.S., Revised procedures for commercial production of bigmouth buffalo fish in ponds in the southeast, *Proc. Annu. Conf. S. E. Assoc. Fish Wildl. Agen.*, 10, 162, 1956.

Szlaminska, M. and Przybyl, A., Feeding of carp (*Cyprinus carpio* L.) larvae with an artificial dry food, living zooplankton and mixed food, *Aquaculture*, 54, 77, 1986.

Szlauer, L. and Szlauer, B., Utilization of waste effluent of a fertilizer plant for rearing of zooplankton and carp fry, in *Cultivation of Fish Fry and Its Live Food*, Styczynska-Jurewicz, E., Backiel, T., Jaspers, E., and Persoone, G., Eds., European Mariculture Soc., Bredene, Belgium, 1979, 501.

Takeuchi, L., Takeuchi, T., and Ogino, C., Riboflavin requirement in carp and rainbow trout, *Bull. Jpn. Soc. Sci. Fish.*, 46, 733, 1980.

Takeuchi, T. and Watanabi, T., Requirement of carp for essential fatty acids, *Bull. Jpn. Soc. Sci. Fish.*, 43, 541, 1977.

Takeuchi, T., Watanabi, T., and Ogino, C., Use of hydrogenated fish oil and beef tallow as a dietary source for carp and rainbow trout, *Bull. Jpn. Soc. Sci. Fish.*, 44, 875, 1978.

Takeuchi, T., Watanabi, T., and Ogino, C., Availability of carbohydrate and lipid as dietary energy sources for carp, *Bull. Jpn. Soc. Sci. Fish.*, 45, 977, 1979a.

Takeuchi, T., Watanabi, T., and Ogino, C., Digestibility of hydrogenated fish oils in carp and rainbow trout, *Bull. Jpn. Soc. Sci. Fish.*, 45, 1521, 1979b.

Takeuchi, T., Watanabi, T., and Ogino, C., Optimum ratio of dietary energy to protein for carp, *Bull. Jpn. Soc. Sci. Fish.*, 45, 983, 1979c.

Tamas, G., Rearing of common carp fry and mass cultivation of its food organisms in ponds, in *Cultivation of Fish Fry and Its Live Food*, Styczynska-Jurewicz, E., Backiel, T., Jaspers, E., and Persoone, G., Eds., European Mariculture Soc., Bredene, Belgium, 1979, 281.

Tamas, G. and Horvath, L., Growth of cyprinids under optimal zooplankton conditions, *Bamidgeh*, 28, 50, 1976.

Tapidor, D.D., Henderson, H.F., Delmendo, M.N., and Tsutsui, H., Freshwater fisheries and aquaculture in China, FAO Fisheries Technical Paper No. 168, FIR/T168, Food and Agriculture Organization, Rome, 1977.

Tomita, M., Twahashi, M., and Suzuki, R., Number of spawned eggs and ovarian eggs and egg diameter and percent eyed eggs with reference to the size of female carp, *Bull. Jpn. Soc. Sci. Fish.*, 46, 1077, 1980.

Trzebiatowski, R., Rearing carp juveniles in heated waste-water of a power station, in *Cultivation of Fish Fry and Its Live Food*, Styczynska-Jurewicz, E., Backiel, T., Jaspers, E., and Persoone, G., Eds., European Mariculture Soc., Bredene, Belgium, 1979, 507.

Tuten, J.S. and Avault, J.W., Growing red swamp crayfish (*Procambarus clarkii*) and several North American fish species together, *Prog. Fish-Cult.*, 43, 97, 1981.

Ufodike, E.B.C. and Matty, A.J., Growth responses and nutrient digestibility in mirror carp (*Cyprinus carpio*) fed different levels of cassava and rice, *Aquaculture*, 31, 41, 1983.

Venkatesh, B. and Shetty, H.P.C., Studies on the growth rate of the grass carp *Ctenopharyngodon idella* (Valenciennes) fed on two aquatic weeds and a terrestrial grass, *Aquaculture*, 13, 45, 1978.

Viola, S., Energy values of feedstuffs for carp, *Bamidgeh*, 29, 29, 1977.

Viola, S. and Amidan, G., The effects of different dietary oil supplements on the composition of carp's body fat, *Bamidgeh*, 30, 104, 1978.

Viola, S. and Arieli, Y., Nutrition studies with a high-protein pellet for carp and *Sarotherodon* spp. (tilapia), *Bamidgeh*, 34, 39, 1982.

Viola, S. and Arieli, Y., Evaluation of different grains as basic ingredients in complete feeds for carp and tilapia in intensive culture, *Bamidgeh*, 35, 38, 1983.

Viola, S. and Arieli, Y., Nonhormonal growth promoters for tilapia and carp. I. Screening tests in cages, *Bamidgeh*, 39, 31, 1987.

Viola, S. and Rappaport, U., The "extra-caloric effect" of oil in the nutrition of carp, *Bamidgeh*, 31, 51, 1979.

Viola, S., Arieli, Y., Rappaport, U., and Mokady, S., Experiments in the nutrition of carp replacement of fishmeal by soybean meal, *Bamidgeh*, 33, 35, 1981a.

Viola, S., Rappaport, U., Arieli, Y., Amidan, G., and Mokady, S., The effects of oil-coated pellets on carp (*Cyprinus carpio*) in intensive culture, *Aquaculture*, 26, 49, 1981b.

Viola, S., Mokady, S., Rappaport, U., and Arieli, Y., Partial and complete replacement of fish meal by soybean meal in feed for intensive culture of carp, *Aquaculture*, 26, 223, 1982.

Walker, M. C. and Frank, P.T., The propagation of buffalo, *Prog. Fish-Cult.*, 14, 129, 1952.

Watanabi, T., Takeuchi, T., and Ogino, C., Effect of dietary methyl lineolate and linolenate on growth of carp, *Bull. Jpn. Soc. Sci. Fish.*, 41, 263, 1975a.

Watanabi, T., Utsue, O., Kobayshi, I., and Ogino, C., Effect of dietary methyl linoleate and linolenate on growth of carp, *Bull. Jpn. Soc. Sci. Fish.*, 41, 257, 1975b.

Watanabi, T., Takashima, F., Ogino, C., and Hibiya, T., Requirement of young carp for alpha-tocopherol, *Bull. Jpn. Soc. Sci. Fish.*, 36, 972, 1970.

Watkins, C.E., Shireman, J.V., and Rottman, R.W., Food habits of fingerling grass carp, *Prog. Fish-Cult.,* 43, 95, 1981.

Williamson, J. and Smitherman, R.O., Food habits of hybrid buffalofish, tilapia, Israeli carp and channel catfish in polyculture, *Proc. Annu. Conf. S. E. Assoc. Fish Wildl. Agen.,* 29, 86, 1975.

Wohlfarth, G., Moav, R., and Hulata, G., A genotype-environment interaction for growth rate in the common carp, growing in intensively manured ponds, *Aquaculture,* 33, 187, 1983.

Wohlfarth, G., Moav, R., Hulata, G., and Beiles, A., Genetic variations in seine escapability of the common carp, *Aquaculture,* 5, 375, 1975.

Wohlfarth, G., Lahman, M., Hulata, G., and Moav, R., The story of "Dor-70," a selected strain of the Israeli common carp, *Bamidgeh,* 32, 3, 1980.

Yone, Y. and Toshima, N., The utilization of phosphorus in fish meal by carp and black sea bream, *Bull. Jpn. Soc. Sci. Fish.,* 45, 753, 1979.

Zeitler, M., Kirchghessner, M., and Schwarz, F. J., Effects of different protein and energy supplies on carcass composition of carp (*Cyprinus carpio* L.), *Aquaculture,* 36, 37, 1984.

Zur, O., The appearance of chironomid larvae in ponds containing common carp (*Cyprinus carpio*), *Bamidgeh,* 31, 105, 1979.

Zur, O., The importance of chironomid larvae as natural feed and as a biological indicator of soil condition in ponds containing common carp (*Cyprinus carpio*) and tilapia (*Sarotherodon aurea*), *Bamidgeh,* 32, 66, 1980.

Zur, O. and Sarig, S., Observation on the feeding of common carp (*Cyprinus carpio*) on chironomid larvae, *Bamidgeh,* 32, 25, 1980.

Chapter 5

CENTRARCHIDS

J. Holt Williamson, Gary J. Carmichael, Kerry G. Graves, B. A. Simco, and J. R. Tomasso

TABLE OF CONTENTS

I. INTRODUCTION[*],[**]

Historical demands for cultured black basses, crappies, and sunfish (family Centrarchidae) have been for stocking into natural waters as a means of establishing or supplementing existing sport fisheries. Federal hatcheries within the U. S. Fish Commission were established at the end of the last century and the beginning of the present century to meet public demands for these sportfishes. Initiation of the Federal Farm Pond Program in the late 1930s, and its expansion through the 1940s and into the 1960s added greater emphasis to sportfish aquaculture. Over the last century, both smallmouth bass *Micropterus dolomieui* (SMB) and largemouth bass *M. salmoides* (LMB) have been broadly introduced throughout North, Central, and

---

[*]   Use of trade names in this chapter does not imply endorsement by the U. S. Fish and Wildlife Service nor the individual authors of the chapter.

[**]   Current policy, as of 1992, on use of chemicals in aquaculture is uncertain. Fish culturists should be knowledgable of prevailing regulations. For information on chemical status for use in aquaculture contact the U. S. Fish and Wildlife Service, National Fisheries Research Center, P. O. Box 818, LaCrosse, Wisconsin 54601-0818.

South America and around the temperate areas of the world (Robbins and MacCrimmon 1974; MacCrimmon and Robbins 1975). Today, sportfishing for LMB alone is the foundation of a multibillion dollar industry in the U.S.

Under the Federal Farm Pond Program, LMB were extensively stocked, usually at sizes of 25 to 50 mm. Producing fish in that size range was economical, and stocking still is successful in new or recently renovated reservoirs where predation and competition from existing fishes are absent or greatly limited. In some cases, small fingerlings were stocked regardless of conditions in receiving waters, and survival was highly variable. Nevertheless, the Farm Pond Program had a positive effect on the development of improved management methods.

Research at Auburn University pioneered fisheries research related to farm pond dynamics. That work clarified predator-prey relationships for stocking combinations of LMB and bluegills (*Lepomis macrochirus*) which are the species recommended for southeastern farm ponds (Swingle and Smith 1938, 1942; Swingle 1945, 1950). Success derived from broad implementation of their stocking recommendations contributed to increased demand for sunfish, (*Lepomis* spp.) leading to the annual production of millions of centrarchid fishes by federal, state, and private hatcheries.

Improved management programs stressed integrated approaches to restore pond balance and promote rapid growth of both predator and prey species with the goal of improving angling success (Davies et al. 1982). Although LMB have dominated statistics, SMB, bluegill, and various hybrid sunfish also have been key elements of many stocking programs. Since the early 1970s, Federal production declined while production by states increased. In 1989, Federal hatcheries produced 3.2 million LMB (3800 kg) and 630,000 SMB (1100 kg) for stocking into public waters (Anonymous 1989).

Many basic culture techniques are surprisingly similar among centrarchid fishes; therefore, our emphasis in reporting methodology has been placed on LMB production. Techniques developed for LMB have been used with success on SMB, Guadalupe bass (*M. treculi*), and spotted bass (*M. punctulatus*). Some differences occur and are mentioned, as appropriate, later in the chapter.

II. LARGEMOUTH BASS

Hatchery production of 38 to 50 mm LMB fingerlings has primarily been to establish sport fisheries. Larger fish have been successfully used for corrective restocking, research studies, "put and take" fisheries, and trophy fisheries and have been maintained on hatcheries as broodstock. Largemouth bass have been reared in polyculture with other species including channel catfish, *Ictalurus punctatus* (Buck et al. 1973), and grass carp, *Ctenopharyngedon idella* (Buck et al. 1978). They have been reared commercially as a food fish to supply a live fish market on the west coast of the U.S. and in the Orient. Production and stocking of advanced fingerlings (>100 mm) is an effective management technique to control established populations of forage fish, including sunfish. Lawrence (1958) determined the size of LMB required to prey upon various sizes of bluegills.

In the 1960s, Snow and associates at the U.S. Fish and Wildlife Service laboratory near Marion, AL, developed the "Marion Program" which established principles and basic culture technology that are currently used in modified form for the culture of advanced LMB fingerlings. Refinement of spawning and rearing techniques now permits intensive production of LMB. Most production of LMB today is through extensive and intensive methods or combinations of the two approaches. Ponds are still the central units of production. Snow (1975) provided an excellent, detailed account of the various production methods, as well as hatchery design. Improved identification techniques, including the use of biochemical genetic markers, have encouraged development of controlled breeding and broodstock programs. Continued improvement of culture techniques and strategies will depend on increased demand and new objectives, as well as the originality and determination of managers.

A. BROODSTOCK
1. Origins
The natural distribution of LMB in North America was briefly described by Lee (1980). Wide geographic distribution and partial isolation of LMB over a period of time have resulted in genetic differentiation among populations (Philipp et al. 1981, 1983). Historically, LMB broodfish were collected by culturists within their natural range. Specimens from local populations near the hatchery were removed by seining, electrofishing, or other means. Hatchery broodstocks were derived from the wild populations. The hatchery-reared offspring were then stocked back into the same or surrounding waters. Over time, hatchery broodstocks adapted to culture conditions and became domesticated due to intentional or inadvertent selection (Williamson 1983). There were occasional exchanges of broodfish among hatcheries, particularly in the Federal hatchery system. Eventually, hatchery broodstocks bore little relation to original donor populations (Harvey et al. 1980).

In recent times, hatchery managers have become more aware of the role that genetics and continuity play in hatchery performance of LMB (Williamson 1983, 1986a; Carmichael et al. 1988; Williamson and Carmichael 1989, 1990; Kliensasser et al. 1991). Now, initial choices of LMB stocks by hatchery managers may be strain dependent, based on production or management needs. Awareness of LMB genetics in hatchery and fisheries management programs is most obvious in the proliferation of Florida LMB production and stocking programs across the country. In a few cases, recognition of the value of local adaptations inherent in wild populations has resulted in a resurgence in using fish from such populations in new broodstock development efforts. The geographic or genetic origin of a particular broodstock is now considered a primary concern in production and management programs for LMB (Philipp et al. 1981; Williamson 1986a). What was once done by hatchery managers for convenience, is now done for biological reasons and is considered sound management strategy.

2. Marking and Identifying

Identification and marking of broodfish and keeping accurate performance records are positive steps toward establishing biologically proper breeding programs (Williamson 1986a). Broodfish should be marked by strain, sex, and year class, at least. New high-technology tags, such as the internal passive integrated transponders or PIT-tags can be used to mark individual broodfish (Harvey and Campbell 1989). Tags that mark individuals are injected into the body cavity or musculature of the fish. The tag is permanently retained, based on available information, and can be read throughout the life of the fish. Pedigree, breeding, production, and performance data on individual fish can, therefore, be electronically collected, stored, and retrieved. This capability permits implementation of sophisticated broodstock development, pedigree analysis, and breeding programs.

Biochemical genetic markers (specific protein and enzyme characters) permit genetic identification of LMB stocks, strains, or individuals. These markers allow for the generation of genetic profiles and also provide evidence of strain contamination and inbreeding. Secondary physical marks, such as fin clips, acrylic dyes, dangler-type tags, or other types of external tags, supplement the PIT-tag at the convenience of the fish-culturist (Williamson 1986a). Evaluation of liver, fin, and other nonlethally sampled tissues by electrophoretic genetic techniques now permits identification of Florida and northern strains of LMB and intergrades (Harvey et al. 1984; Carmichael et al. 1986; Williamson et al. 1986; Morizot et al. 1991a), as well as other closely related centrarchid basses (Morizot et al. 1991b).

Effective methods for marking individual fish include routine methods of genetic analysis, nonlethal tissue sampling techniques for genetic identification, biochemical or genetic markers, and computerized data collection, storage, and retrieval. All are valuable tools in advancing LMB breeding, and most probably, production programs in the future.

3. Sexual Maturity

The sexual maturity of LMB depends on body size rather than age (Moorman 1957). In the southern U.S., a LMB may mature at 8 months (22- to 24-cm long and 0.18 kg) according to Swingle (1950) and Crumpton et al. (1977). Since growth is temperature dependent, the time required to reach spawning size increases with increasing latitude and altitude except under unique conditions. Fish living in water receiving thermal effluent from power plants or geothermal spring water may mature earlier than fish at the same latitude which are maintained under prevailing ambient temperatures. Males mature more rapidly than females under similar conditions (Williamson 1986a).

According to Snow et al. (1964), older broodfish should be replaced as spawning is less certain in them and they are larger and more difficult to handle than young adults. However, as more money is invested in broodstock development, particularly in genetic marking and tagging of individual fish, managers may be more reluctant to replace older fish unless the productivity of those animals is substantially reduced. At the San Marcos National Fish Hatchery and Technology Center (SM NFH & TC), San Marcos, TX, 8-year-old LMB broodfish have been successfully used to produce offspring.

4. Identification of Sexes

Random sampling from wild donor LMB populations should provide broodstock of both sexes in approximately equal numbers, since sex ratios tend to be equal in natural LMB populations. However, male fish apparently do not live as long as females, and sex ratios may be altered in older populations (Padfield 1951; Heidinger 1976).

The sexing of broodfish is easiest in early spring prior to spawning. Females can often be distinguished by the presence of distended abdomens. Usually, LMB longer than 35 cm can be sexed by examining the scaleless area adjacent to the urogenital opening. That area is nearly circular in males, but is elliptical or pear-shaped in females (Parker 1971). However, Manns and Whiteside (1980) verified the potential for inaccurate sexual identification using external appearance of the sexual characters. In general, however, ripe females are selected on the basis of an obviously distended and soft abdomen and a red, swollen, and protruding vent. Benz and Jacobs (1986) reported using a "probe" to sexually identify LMB. Presently, the most accurate way to sex LMB using physical characters is by insertion of an otoscope or capillary tube as a catheter into the vent (oviduct or urethra) to detect the presence of eggs or milt (Driscoll 1969, Heidinger 1976, Williamson 1986a). Ripe males will emit sperm when stripped. Accuracy of sexual identification is particularly important in breeding programs and for calculating the effective breeding population size.

5. Maintenance

Maintenance of broodfish in good physical condition is extremely important; healthy fish are more resistant to environmental stresses, have higher survival rates over winter, and are more productive during spawning, as compared with less healthy fish (Figure 1). Broodstock of LMB should be fed at a reduced rate during the warmer parts of the winter to maintain reproductive condition for spring spawning. Feeding should be commensurate with fish metabolism, which is temperature dependent.

Live forage fishes have been the customary diet of LMB broodfish. There are advantages, however, to feeding pelleted rations, as the actual amount of food fed can be more accurately measured. Also, parasites and diseases introduced with live forage fishes are avoided. If needed, drugs and other types of supplements can be incorporated into the feed and fed directly to the fish. Finally, when pelleted feed is provided, there is no cost associated in locating, rearing, or transporting forage fishes. More accurate estimates of pond biomass or standing crops are possible when pelleted feed is used.

Rosenblum et al. (1991) compared female gonad development and hormonal cycles associated with oogenesis among pellet- and forage-fed LMB. They

FIGURE 1.   Largemouth bass broodfish.

emphasized that egg development is a continuous process carried on throughout the year. Egg development processes can be characterized by specific hormonal events triggered by, or associated in, a predictable fashion with environmental cues or circumstances, usually of a seasonal nature. They alluded to the relationship of reproduction and the general physiological condition of the fish, particularly nutritional state. Overall, they concluded that pelleted rations may enhance reproductive performance in female fish when compared to LMB reared on forage fish. Finally, they emphasized the substantial role that nutrition management plays in broodstock maintenance.

Snow and Maxwell (1970) fed broodstock with the Oregon Moist Pellet diet through seven seasons without causing discernible differences in fecundity. Broodfish at the SM NFH&TC have been maintained in reproductive condition for 7 years exclusively on formulated feed. Biodiet® (Bioproducts, Warrenton, OR), a semi-moist pellet, has been a good broodstock ration, especially in the larger pellet sizes (15 to 25 mm in diameter, see Table 1). Commercial semi-moist diets are now available from other major feed manufacturers as well.

In northern climates with severe winters, formulated diets have been associated with a "winterkill" phenomenon (Roem and Stickney 1989). Although the authors

TABLE 1
Pellet Sizes and Approximate Percentage of Body Weight to Feed Largemouth Bass Fingerlings under Summer Conditions[a]

Fish length (cm)	Pellet diameter (mm)	Body weight (%)
<2.5	0.8–1.2	15.0
2.5–3.2	1.0–1.3	15.0
3.2–4.4	1.5	15.0
4.4–8.9	2.5	10.0
8.9–11.4	3.0	10.0
11.4–12.7	4.0	7.5
12.7–15.2	5.0	5.0
15.2–20.3	6.0	3.0
20.3–25.4	9.0	2.0
25.4–30.5	12.0	2.0
>30.5	19.0	2.0

[a] This practical feeding guideline was developed from data collected at the SM NFH&TC; the feed used was Biodiet®.

were uncertain of the cause of death, they associated the problem with fatty livers and potentially impaired lipid metabolism associated with the prepared diets. Salmon diets, high in choline, seemed to ameliorate the problem. At the SM NFH&TC, LMB broodfish are typically fed at 3% of their body weight daily. During the winter the feeding rate may be reduced to 1% daily.

Forage fish are often kept in the ponds to supplement pelleted rations. Where ponds typically ice over during winter, forage should be placed in broodfish ponds. Care of broodfish should continue diligently following spawning events. Broodfish may lose 10 to 30% in body weight due to spawning activity (White 1985). This weight loss may be affected by number of spawns and the length of time broodfish are held in ponds without feed. Following removal from spawning ponds, broodfish should be fed sufficiently to replace the weight lost, as well as to provide energy required for growth and developing future sex products. Gonads begin to redevelop immediately following spawning, so a good nutritional program should begin as soon as broodfish are removed from spawning ponds.

Weight gain in broodfish is a positive indication that egg development is occurring, but the minimal amount of weight gain required to maximize egg production has not been determined. Maximizing broodfish weight gain, therefore, may not maximize production in terms of offspring generated.

When feeding forage fishes to broodfish, approximately 3 kg of forage fish per kilogram of LMB are required for maintenance, and 5 kg of forage fish will be required for each kilogram of LMB for weight gain during overwintering (Snow et al. 1964). If live forage is fed, it is important that it be of a size available to the broodfish (Lawrence 1958). The ration for the year should never be placed in the pond at one time as some of the forage will become too large for the LMB to eat.

Forage of the appropriate size should be stocked at intervals into the holding or maintenance ponds, with the heaviest introductions following spawning.

Common forage fish used to maintain LMB broodstocks are goldfish (*Carassius auratus*), fathead minnows (*Pimephales promelas*), golden shiners (*Notemigonas crysoleucas*), bluegill sunfish, threadfin shad (*Dorsoma petenense*), gizzard shad (*D. cepedianum*), and tilapia (*Tilapia* spp.). Culture techniques for some of these species are reported in Chapters 3 and 10 in this book and also in Higginbotham et al. (1983). Suggested stocking rates for broodfish are 336 to 448 kg/ha (White 1985), with higher rates possible if there are no forage fish already in the pond.

6. Diseases

Disease problems associated with raising LMB, and for fish in general, are usually due to poor management practices which impose stress on the fish. Fish behavior associated with stress should be recognizable. Culturists should become aware of environmental conditions and culture activities which stress fish (such as harvesting, grading, spawning, and transporting). All practices which result in scale loss and skin abrasion are traumatizing. Breaking the surface of the skin of fishes permits invasion of pathogens. When water quality is poor and fish are confined at high densities, sublethal stress is chronic, and pathogenic infections and infestations can quickly evolve into substantial losses. Stress and its alleviation or prevention are discussed in more detail later in this chapter.

Hatchery managers should use anesthetics and a saline/antibacterial solution at times when broodfish are handled or hauled to reduce the effects associated with stress and trauma. Handling fish in a 1% salt solution also reduces the number of external parasites typically carried by broodfish held in ponds. A 3% salt-dip is frequently employed to reduce external parasites whenever broodfish are handled.

Campbell and Johnson (1989) compared three methods of administering antibiotics to LMB and found intraperitoneal injection of tetracycline to be superior to hyperosmotic infiltration in treating systemic bacterial (*Aeromonas hydrophila*) infections. However, the authors suggested alternative methods by which hyperosmotic infiltration may be more effectively used to administer drugs and chemicals.

Bacterial pathogens and fungal infections are often seen on broodfish in late winter and early spring, about the time fish are preparing to spawn. Feeding medicated feed in anticipation of this stressful period may prevent some broodfish loss.

The anchorworms, *Lernaea* spp., are common external parasites, often transmitted to LMB broodfish by forage fish. A common pond treatment, dependent on water hardness and temperature, is an application of Dylox® at 0.25 to 0.50 mg/l once a week for four weeks. While not usually a significant problem, LMB are frequently infected by digenetic trematodes which encyst as metacercaria in the fish's eyes, internal organs, mesenteries, and musculature. In most instances, snails are the first intermediate host, fish the second, and a bird is the definitive or terminal host. Interrupting the life cycle of the parasite provides the most effective treatment. Monogenetic trematodes, such as *Dactylogyrus* spp. (gill flukes), occasionally infect LMB but can be treated with 1:2000 parts 40% formalin (or 0.25 mg/l) for

30 min to 1 h. A 3% salt dip is also beneficial. Common ciliated protozoans, such as *Trichodina* and *Costia*, can also be treated with formalin and salt. The largemouth bass tapeworm, *Protepcephalus ambloplitis*, has been treated with dibutyl tin oxide mixed into the food and fed such that each fish received 250 mg/kg body weight for 3 d.

A potential cure for LMB tapeworm does exist. Razorback suckers (*Xyrauchen texanus*) and Colorado squawfish (*Ptychocheilus lucius*) have been successfully treated for the Asian tapeworm (*Bothriocephalus acheilognath*) by holding broodfish for 24 h in a 1 mg/l Praziquantel® solution.

Other maladies associated with fish cultural stress encountered by LMB during their production are acute corneal cloudiness (Brandt et al. 1986) and swim bladder stress syndrome (Carmichael and Tomasso 1984b). Both of these maladies have been observed following handling and transport.

7. Predation

Predation on LMB broodfish is usually highly restricted due to their size. However, mammals such as raccoons, otters, and minks certainly have the inclination and ability to take LMB or other fish when available. Reptiles such as water snakes (*Nerodia* spp.) and turtles can take smaller LMB.

Birds are major predators, particularly on broodfish when they are spawning in shallow water or are concentrated near the surface, which usually occurs when fish are feeding or water quality is poor. The great blue heron (*Ardea herodius*) kills many adult LMB as the fish occupy nests during spawning. The osprey (*Pandion heliaetus*) dives from great heights to take bass from ponds being drained during harvest. Bird predation and control has been discussed by Salmon and Conte (1981), Stickley (1990), and Littauer (1990).

Man, however, is the most devastating predator of LMB on hatcheries. Poachers frequent hatcheries and will often take as many fish as they can carry, including the largest fish on the facility and broodfish. Security measures are required to protect broodfish.

8. Condition

Variations on measurements of body weight and length, organ weight, and physiological variables have been used or proposed for use as descriptors of the health or well being of LMB. However, no descriptions of quantitative individual or population variation have been reported between the indices and the changes in them within a growing, healthy LMB population.

In order to test such variation, LMB were maintained in a 0.04-ha earthen pond at the SM NFH&TC from December 1983 through December 1984 (age 2.5 years at stocking). They were fed commercial pelleted food (Biodiet®). Nine to twenty fish were sampled approximately monthly by angling. A total of 160 LMB were sampled throughout the annual cycle (Carmichael 1987).

The mean weight ± SE of all fish was 625 g ± 15.8, and standard length was 340 mm ± 2.3. The mean weight and length of males was 630 grams ± 23.8 and 340 mm ± 3.5, respectively; that of females was 622 grams ± 21.2 and 338 mm ± 3.2.

Variation in indices of health or well-being was examined, and levels of significant differences (*p* values) are presented in Table 2. The condition indices of Fultons condition factor (K), relative weight (Wr), gonadosomatic index (GSI), relative gonadal index (RGI), liver-somatic index (LSI), relative liver index (RLI), visceral somatic index (VSI), and relative viscera index (RVI) varied significantly ($p < 0.0016$) throughout the year in this hatchery population of growing and healthy LMB. Standard length, body weight, and organ weights also changed significantly throughout the year, ($p < 0.0001$). Levels of plasma chloride, osmolality, protein, and hematocrit varied seasonally (see later discussion of handling and transport) in both males and females ($p < 0.0062$). During the year, significant diffences existed between males and females for gonad weight, GSI, RGI, LSI, RLI, hematocrit, and plasma protein.

Comparisons of condition indices derived from individual fish (K, GSI, LSI, and VSI) with "relative" condition indices, based on standardized "optimal" fish-size

TABLE 3
Coefficients of Correlation (r) Between
Relative and Conventional Condition Indices[a]
in Pond-Reared Largemouth Bass[b]

Relative condition index	Sex	Conventional condition index			
		K	GSI	LSI	VSI
Wr	Male	0.985			
	Female	0.962			
	M and F	0.972			
RGI	Male		0.969		
	Female		0.992		
	M and F		0.994		
RLI	Male			0.941	
	Female			0.974	
	M and F			0.959	
RVI	Male				0.832
	Female				0.915
	M and F				0.878

Note: The condition indices are Fultons condition factor (K), relative weight (Wr), gonadosomatic index (GSI), relative gonadal index (RGI), liver-somatic index (LSI), relative liver index (RLI), visceral somatic index (VSI), and relative viscera index (RVI).

[a] K = [weight (g)/length (mm)^3] x 100,000; GSI, LSI, VSI = [organ weight (g)/body weight (g)] x 100. Liver and gonad weight from Nielsen and Johnson (1983).
[b] All indices correlated significantly ($p > 0.0001$).

TABLE 2
One- and Two-Way Analyses of Variance (ANOVA) Probabilities (*p*) for the Physical Factors, Physiological Indicators, and Condition Indices in Pond-Reared Largemouth Bass[a]

Criteria	Male and female (two-way ANOVA)			One way ANOVA		
	Date	Sex	Interaction (date x sex)	Date male	Date female	*p*
Physical Factors						
Weight (g)	0.0001	0.9122	0.7069	0.0001	0.0001	M=F 0.7979
Length (cm)	0.0001	0.4372	0.6692	0.0001	0.0001	M=F 0.5981
Liver (g)	0.0001	0.0845	0.3745	0.0001	0.0001	M=F 0.5289
Viscera (g)	0.0001	0.7321	0.4607	0.0001	0.0001	M=F 0.5243
Gonad (g)	0.0001	0.0001	0.0001	0.0001	0.0001	M=F 0.0001
Condition Indices						
K[b]	0.0001	0.4899	0.6892	0.0158	0.0177	M=F 0.7497
Wr[c]	0.0016	0.3676	0.7452	0.0523	0.0953	M=F 0.5350
GSI[d]	0.0001	0.0001	0.0001	0.0001	0.0001	M<F 0.0001
RGI[c]	0.0001	0.0001	0.0001	0.0001	0.0001	M<F 0.0001
LSI[d]	0.0001	0.0120	0.1330	0.0001	0.0001	M=F 0.1656
RLI[c]	0.0001	0.0348	0.1463	0.0001	0.0001	M=F 0.2411
VSI[d]	0.0001	0.1844	0.3383	0.0558	0.0012	M=F 0.9245
RVI[c]	0.0001	0.2813	0.0981	0.0394	0.0068	M=F 0.8641

Plasma Indicators

Chloride (Meq/l)	0.0001	0.1881	0.0431	0.0002	0.0001	M=F 0.6239
Osmolality (mosmole/l)	0.0001	0.2516	0.1669	0.0001	0.0001	M=F 0.1025
Hematocrit (%)	0.0001	0.0026	0.8577	0.0001	0.0001	M>F 0.0029
Protein (g/dl)	0.0001	0.0001	0.0121	0.0001	0.0001	M>F 0.0001

Note: The condition indices are Fultoms condition factor (K), relative weight (Wr), gonadosomatic index (GSI), relative gonadal index (RGI), liver-somatic index (LSI), relative liver index (RLI), visceral somatic index (VSI), and rlative viscera index RVI).

[a]The p values for date, sex, and interaction represent analyses on 13 monthly samples; p values represent differences or similarities between males and females in annual means.

[b]$K = $ [weight (g)/length (mm)^3] x 100,000 (Nielsen and Johnson 1983).

[c]$Wr = $ [weight (g)/length-specific standard weight] x 100. Length-specific standard weight $ = -5.316 + 3.191$ log length (Nielsen and Johnson).

[d]GSI, LSI, VSI = [organ weight (g)/body weight (g)] x 100. Viscera - weight - liver and gonad weight from Nielsen and Johnson (1983).

[e]RGI, RLI, RVI = [organ weight (g)/length-specific standard weight] x 100 (Erickson et al. 1985).

FIGURE 2.   Gravel nests in pond.

estimates (Wr, RSI, RLI, and RVI), indicated highly significant correlation
(r > 0.878; $p < 0.0001$) between the two types of indicators (Table 3). As such, no
apparent benefit could be determined for using Wr, RSI, RLI, or RVI vs. K, GSI,
LSI, or VSI for assessing seasonal variation within the population. As significant
seasonal variations in health and condition indices were evident in the hatchery fish,
these variations should be considered when handling, stocking, or monitoring
hatchery LMB.

B. SPAWNING

1. Spawning Pond Preparation

Preparation of spawning ponds depends on the objectives and methods of
culture. For example, if young fish are to be grown out in the spawning pond
(spawning-rearing pond method), fertilization of the pond will be required. Also,
under these conditions, broodfish stocking rates are reduced, as compared to the
method wherein fry are moved to fingerling growout ponds shortly after hatching
(fry transfer method).

Earthen ponds should be drained, disk-harrowed, smoothed, compacted, and
allowed to dry during the fall and winter. Treatment with an approved pre-emergent
herbicide prior to spring pond flooding will help control rooted vegetation and thus
facilitate fry removal (McCraren 1975). Placement of gravel in selected pond sites
(Figure 2) aids in fry collection by inducing nesting and spawning within confined
or desired areas (Huet 1970; Williamson 1986a). Gravel nests should not be placed
in water too deep for wading, as observing young fry around nests will be hampered.

However, spawning LMB are quite vulnerable to blue heron predation if nests are too shallow. Also, light-colored rocks make broodfish easier to observe on nests.

Portable spawning mats (e.g., artificial turf or brown carpeting) have been used successfully in a limited number of instances (Chastain and Snow 1966). Fish culturists at the Sand Ridge State Fish Hatchery, IL, have used lead-centered cord from gill nets to form circular patterns on the underside of such carpets. In raceways, bass seem to prefer that arrangement since it produces a slight circular depression in the carpet while anchoring the nest on the bottom of the raceway.

In ponds, nests should be at least 2 m apart since males exhibit territorial behavior (Moorman 1957). The number of males stocked should not exceed the number of artificial nests available.

No forage fish are placed in spawning ponds because they may compete with the fry and are potential sources of disease and predation. Broodfish trained to consume pelleted feed can be fed in spawning ponds although feeding is not required.

2. Spawning Behavior

The majority of spawning occurs during late afternoon or early morning (White 1985). Heidinger (1976) described the spawning behavior in detail. The male LMB prepares a nest by pivoting in a circle, creating a depression in the sediment with a radius of approximately the length of his body. The substrate used may vary, but nests constructed on firm areas will be shallower than those created on soft substrates.

The male will repeatedly leave the nest in search of a ripe and accommodating female or to chase away intruders. Changes in color patterns and aggressive behavior are used to entice the female to nest. The pair subsequently slowly circle the nest, side by side, with the male nipping and nudging the female. Spawning begins with both fish over the nest and each tilted laterally so their vents are in close proximity. Eggs and sperm are released accompanied by shudders of the body. Spawning takes place over a prolonged period, and multiple spawns of both sexes are common.

The male will guard the nest after the female leaves and may entice additional females to lay eggs within the same nest. In some cases, males have defended a nest even when eggs were dead and covered with fungus. If mats with eggs are removed from underneath a male and replaced with a new mat, he will defend that mat and entice females to spawn on it. A female may spawn more than once and with more than one male.

For the fish culturist to obtain as many eggs as are available in the brood female population, an excess of males is needed (Snow 1970). Bishop (1968) recommended 2:1 to 3:1 male to female ratios. Snow et al. (1964) suggested stocking adults of the same-year class together because they are more likely to spawn at the same time and produce fry of the same size.

3. Spawning Time Manipulation

Largemouth bass normally spawn from spring through early summer at water temperatures of 15 to 24°C (Swingle 1956; Kramer and Smith 1962). The spawning season is generally longer in the southern U.S. than in the north. Fish spawn in central Florida as early as January and may continue until June, though most spawning occurs in March when water temperatures range from 14 to 16°C. Spawning begins as early as February during mild winters at the SM NFH&TC and may continue through June and into July. However, peak spawning is from late March through April when water temperatures range from 17 to 22°C. Young-of-the-year males have been observed constructing nests in hatchery ponds during October and November and produced milt when stripped (Williamson 1986b). Kramer and Smith (1960) reported that LMB spawned in Minnesota as early as mid-April and continued until the first of June.

Variation in spawning time is related to water temperature and other environmental variables, as well as to the physical condition and size of adult fish. Many producers believe that strains vary in spawning times, with the Florida strain spawning earlier than northern populations under the same hatchery conditions. These observations have yet to be verified under controlled conditions.

Premature spawning must be prevented if increased control over production is desired. Mature male LMB in holding ponds will signal the onset of the spawning season by preparing nests (Anonymous 1900; Snow 1975). Various techniques, such as crowding broodfish, manipulating water levels, or running cool water into ponds, are used to prevent premature spawning in holding ponds (Snow 1963, 1968b). However, the most effective method results from segregation of fish by sex.

Holding ponds should be drained and broodfish removed and restocked by sex when pond temperatures warm to 10 to 12°C. The sexes are maintained separately until water temperature stabilizes at around 15 to 16°C. This normally occurs after the last average frost date. Fish can be directly observed for ripeness by removing a few of them periodically by angling or with a seine. When a preponderance of sampled fish are ripe, the broodfish can be placed into spawning ponds. Maturation and spawning are particularly stressful on LMB. Bacterial and fungal infections are most likely to occur at this time, particularly when fish are in water of poor quality or are in poor condition. Broodfish must be handled with care in early spring as waters warm and spawning approaches.

Spawning may be stimulated if the broodfish are stocked while ponds are being filled with water of appropriate temperature (Bishop 1968). In general, the best time to stock spawning ponds corresponds with the date when the main spawning period was observed in previous years. If stocking is delayed until water temperatures rise to 20 to 21°C, males will prepare nests, select females, and spawning will normally occur within 24 to 48 h.

Postponement of the time of spawning may dramatically increase efficiency because the spawning season may be reduced from several weeks to several days. Consequently, fry taken from brood ponds will be more uniform in size and can be distributed or transferred to rearing ponds with minimal losses due to cannibalism and grading stress.

The LMB spawning season has been extended under experimental conditions (Jackson 1979). Adult fish held in cool water (9.5 to 19.5°C) in New York spawned within 1.5 to 11 d after transfer to 23°C water during the period from June 10 to

September 6. In addition, LMB have been induced to spawn out-of-season by temperature and photoperiod manipulation (Brauhn et al. 1972; Carlson and Hale 1972; Carlson 1973). Further experimentation with photoperiod and temperature to manipulate reproductive development and spawning time are necessary.

More precise control of spawning time has been accomplished by means of hormonal injections of ripe fish. Injections with 4000 to 8000 IU/kg human chorionic gonadotropin led to ovulation within 48 h and milt production was maintained in males (Stevens 1970; Wilbur and Langford 1984; Williamson 1986a). Spawns collected on mats may be moved indoors to hatching troughs, or eggs may be treated with a sodium sulfite solution to dissolve the gelatinous matrix that makes them sticky and adhesive (Anonymous 1984). The dissociated eggs may be placed in a Heath incubator tray or a vertical hatching jar for hatching.

Collection is scheduled such that all eggs will hatch within a 24-h time frame. Two days after hatching, fry are transferred to holding troughs until mouth parts develop and fry first begin to feed on exogenous food 5 to 7 d after hatching. At that time, fry are transferred to zooplankton-enriched fingerling rearing ponds. The labor and expense of such intensive methods are not usually warranted for general production purposes, and since large numbers of viable fry are normally produced in spawning ponds, these intensive methods are the exception on most hatcheries. But, where greater control of spawning is required, as in breeding or research programs, such methods must be developed and improved.

The SM NFH&TC concentrated on improving control over LMB reproduction in order to implement breeding programs. Plastic-lined spawning ponds (0.04 ha) were filled with 21°C well water, and artificial nests of two designs were placed in the ponds. One nest design was similar to that described in more detail in Section III of this chapter (Hutson 1983). Essentially, it is a box within a box, each box with a wire-mesh bottom. The smaller, interior box contains spawning gravel (cobbles from 2 to 5 cm in diameter) and has a bottom with wire mesh large enough (10 to 20 mm) for sac-fry to wiggle through easily following hatching. The outside box has mesh too small for fry to move through, thus retaining them. As eggs hatch, fry drift down through the rocks, through the mesh of the interior box, and are trapped in the outer box. Any time prior to swim-up, sac-fry can be collected from the outer box.

The other design consisted of a square (46 cm), metal (round iron [6 mm]) framed mat. Strapped to the frame and forming the interior surface is brown outdoor carpeting with several gravel cobbles glued to the center. Nests are placed from 0.5 to 1.5 m deep and 3.0 meters apart around the circumference of the pond. Adult LMB of the desired strain and in the desired numbers are then placed in the spawning ponds. From the pond bank, spawning LMB can be observed in the clear water over the nests. Periodic wading or swimming through the pond will verify and allow the recording of successful spawns. Nests with eggs (metal-framed nests) or with sac-fry (wooden-framed nests) are removed and replaced with new nests. Males quickly entice other females to the nest, and a new spawn can be produced.

Eggs or fry can be transported in tubs of aerated water to indoor hatching and rearing facilities. Individual egg or fry lots may be reared separately. First-feeding fry are fed zooplankton produced in and collected from intensively managed zooplankton rearing ponds and tanks (Graves and Morrow 1988a and b). Along with zooplankton, fry may be presented with formulated feed dispensed from automatic feeders. As fry learn to eat the prepared diet, zooplankton rations can be reduced. Young LMB can be reared on pelleted food until they can be marked. Once marked, a fish lot can continue to be reared separately or mixed with other marked lots, depending on the study design or available rearing facilities.

Methods to spawn individual pairs of LMB in indoor tanks were explored at SM NFH&TC. Individual pairs were placed into 2000-l indoor tanks. Apparently, incompatibility between individuals prevented successful spawning to a considerable extent. Males would often attack females, but, in some cases, females also attacked males. Nevertheless, enough success was achieved to indicate paired mating in indoor or controlled tanks is possible. Additional studies of spawning behavior and improvement in technique — perhaps temperature and photoperiod manipulation in conjunction with hormone induction — are necessary to further advance controlled spawning of LMB and closely related species.

C. EGG AND FRY DEVELOPMENT
1. Egg Development, Deposition, and Hatch
Spawning female LMB produce from 2000 to 176,000 eggs per kilogram of body weight (Bishop 1968; Snow 1970; Timmons et al. 1980). Fertilized eggs are orange (creamy-white in pellet-fed females), adhesive, and spherical in shape, with a diameter of 1.5 to 2.5 mm. Egg size increases with the size of the producing female. Ovaries contain ova in all stages of development (Stevens 1970, Chew 1974), though most ripe eggs are released with the initial spawn. Each subsequent spawn normally contains approximately half the eggs of the preceding spawn. A range of about 5000 to 43,000 eggs are deposited in a given nest. Some 78% of the eggs hatch, and 58% of the fry survive to swim-up stage, on the average (Snow 1971). Hormone induction and manual stripping of eggs usually results in production of a few eggs at a time.

Egg development and hatching times are primarily functions of temperature. At 10, 18, and 28°C, eggs hatch in 317, 55, and 49 h, respectively (Badenhuizen 1969; Merriner 1971). Snow et al. (1964) found that hatching normally occurs in Alabama 48 to 96 h after spawning. The transparent fry are about 4 mm long and disperse in the substrate of the nest. They become pigmented at between 5 and 7 d of age. By that time, their yolk-sac nutrients are absorbed and their swim bladders are filled. Fry then rise from the nest in a school and become active predators (Johnson 1953). One of the principal differences between LMB and SMB fry is that LMB fry aggregate in schools following hatching and swim-up, while SMB fry scatter as do fry of the Guadalupe and spotted bass.

2. Fry Growth and Development
Fry growth and development are a function of temperature and food availability. Young LMB are strike-feeders, relying on sight to locate and capture prey (Colgan et al. 1986). Swim-up fry consume cladocerans, rotifers, and copepods. In intensive culture studies, Wickstrom and Applegate (1989) showed that

first-feeding LMB (6.5 mm) consume rotifers (*Branchionus* sp.) for the first 4 d. Copepods (*Cyclops vernalis*) were selected during the first 14 d and cladocerans (*Daphnia pulex* and *D. magna*) from day 15 through 25, at which time fry were approximately 30 mm. *D. moina* was consumed in proportion to its relative abundance throughout the study. Copepods and cladocerans were consumed most abundantly by fry larger than 10 mm. Change in food items was related to fry growth and food availability. Larger fry preferred larger food items. Other studies support these data, with some exceptions. Florida LMB fry in Texas ponds did not appear to eat rotifers even when abundant. First-feeding fry selected copepods and cladoceran nauplii, and when larger than 22 mm, fry fed primarily on immature insects (Parmley et al. 1986). Rogers (1968) found that fry greater than 15 mm consumed midge larvae.

Although fry abandon the nest 8 to 10 d after hatching, the male parent continues to guard the fry as long as the school remains intact. School dispersal is usually associated with food availability. Fry can be placed in rearing ponds anytime after they leave the nest, but to avoid injury to the delicate young fish, it is prudent to wait until they are 15 to 20 mm long before moving them. That size is normally reached in 10 d to 4 weeks following hatching, depending upon food and temperature (Snow 1973). When food in spawning ponds becomes scarce, fry must be removed regardless of their size; if not, schools disperse and cannibalism begins; fish numbers diminish rapidly, size variation increases, and the general condition of the young fish deteriorates.

D. FRY REARING

The aggressive, predaceous nature of the LMB makes it a challenging subject for culture. Fingerlings become piscivorous at lengths between 38 and 50 mm. The transitional phase from a diet of microcrustaceans and small insects to one of fish is a critical period in the life cycle of the LMB. During this natural transition period from one food type to another, fish culturists have been successful in modifying fingerling-feeding behavior (i.e., training the fish to accept formulated feed).

Fry less than 50 mm long are usually raised in fertilized ponds. If fingerlings remain in ponds beyond that size or until food becomes scarce, losses from cannibalism will significantly reduce production (Meehan 1939). Depending on production objectives, three culture methods have been outlined (Snow 1975): (1) spawning-rearing pond system, (2) fry transfer system, and (3) intensive system. Snow (1975) presented an excellent comparative analysis of these various methods.

1. Spawning-Rearing Pond System

In this method, young fish are spawned, hatched, and reared in the same pond. With a few exceptions, this method is used when the number of required fingerlings is low, commitment of resources and technical expertise are minimal, or available broodstock are of a quality that would not warrant additional expense. Production is predictably erratic.

Earthen spawning-rearing ponds are usually 0.25 to 2.0 ha. They may be any size or shape. Although smaller ponds are more expensive to build per unit area, they are more manageable, and production per unit area may be greater. Narrow, rectangular ponds with appropriate slopes to the drain are easily managed. They are dried, then disked in the fall and left empty until needed. Pond bottoms should be flat and firm. Ponds may be seeded with winter rye grass to prevent erosion and serve as a "green manure crop" when water is added in the spring. A pre-emergent herbicide may also be applied as directed prior to filling to control rooted aquatic vegetation. Ponds should be filled with groundwater or surface water filtered to keep out wild fish and associated parasites.

Adult fish are stocked into ponds when rising water temperatures stimulate nesting activity. A common stocking rate is 25 to 50 broodfish per hectare with an equal proportion of males and females. The pond is fertilized as necessary to promote development of abundant zooplankton populations (White 1985; Parmley et al. 1986). Organic fertilizers (e.g., cottonseed meal, peanut hay, alfalfa, or animal manure), with or without inorganic (chemical) fertilizers (e.g., phosphoric acid or superphosphate), have been used to establish and maintain zooplankton blooms. Predaceous aquatic insects may be controlled with regular applications of diesel oil or vegetable oil. Effects of diesel oil on fry and zooplankton have not been completely assessed.

Fry should remain in the pond until they reach 25 to 50 mm, when they are harvested. If food is available and fingerlings are in good condition, they may remain in the pond longer for additional growth. Following removal from the pond, fry are usually separated from invertebrate pond organisms and debris (tadpoles, snails, insect larvae, and crayfish), treated for ectoparasites, and loaded on distribution trucks. Depending on temperature, 30 to 50 d are required from spawning to fingerling harvest (White 1985).

Broodfish are returned to holding ponds and are generally maintained on forage fish until the next spawning season. The number of broodfish used in a pond is the principal controlling influence exerted over the number of fry produced.

Predation on fry by adults and larger fry significantly affects the size and number of fingerlings harvested. If fingerlings are kept in fertilized ponds for more than 30 or 40 d, pond vegetation may become dense, making harvest difficult.

Some hatcheries remove fry from spawning ponds in a single operation. Ponds are drained and the fry collected in catch basins. Chicken-wire screens prevent broodfish from entering the basins. As the pond is draining, glass V-traps are used around the catch basin (Figure 3) to collect fry (Anderson 1974). Trapping may be particularly successful when a flow of fresh water is introduced into the catch basin as it attracts the young fish. After fry are harvested for distribution, broodfish are removed by seining from the spawning pond and transferred in a salt solution to holding facilities, usually tanks or raceways. Ponds may be fertilized and refilled within 24 h and broodstock replaced so a second spawn can be produced.

2. Fry Transfer System

The practice of collecting schools of fry from spawning ponds for transfer to fertilized rearing ponds permits increased production through the employment of more efficient management techniques, but it requires more labor, technical expertise, and facilities — particularly ponds. The number of broodfish is increased

FIGURE 3. Glass V-traps.

to 50 to 200 fish per hectare to boost fry production. Since fry are harvested within 2 weeks after hatching, spawning ponds are not usually fertilized because phytoplankton and filamentous algae (*Hydrodiction* spp. and *Spirogyra* spp.) blooms frequently reduce visibility and interfere with harvest.

a. Fry Collection

Fry are most easily captured from the spawning pond while still in schools. They are usually 10 to 20 mm total length (TL), and within a school, there is considerable size uniformity. Vegetation problems can usually be avoided if fry are removed within 2 weeks. Two workers in a boat can maneuver to slide a fine-mesh seine (similar to cheesecloth) under the school. Fry can then be carefully washed from the seine into a tub or bucket of water. More frequently, fry are seined by wading (Figure 4). However, this muddies the water and makes observation of schools difficult. Wading must be done carefully to avoid scattering schools of fry. In lifting the seine with fry, a stone or other weight should be placed in the seine to keep the wind from whipping and inverting the net.

Fry should be carefully dip-netted from the seine into a bucket or tub of water for fry estimation and transfer to fertilized rearing ponds. Several methods have been used to estimate fry numbers, including visual inspection, gravimetric techniques, and volume displacement (Figure 5). Swanson (1982) estimated numbers of fish stocked on the basis of volume displacement. He placed 250 ml of water into a 500-ml beaker and added fry until the water level rose to 500 ml. Numbers of fry in each 250-ml sample were estimated from a sample count taken

FIGURE 4.   Fry seining.

FIGURE 5.   Fry volume displacement.

of 5 ml of fry added to a 50-ml graduated cylinder. High-technology, electronic fry counters are now available, and any of the techniques mentioned should provide an estimate within about 10% of the actual number. Up to 1,250,000 fry per hectare have been harvested from spawning ponds, but Snow (1975) suggested that half that number is more realistic.

Clear calm water, free of debris and vegetation, facilitates fry sightings and removal. Visibility on windy days may be improved by spraying or squirting vegetable oil on the pond surface to create a slick which will retard wind rippling. Schools of fry are difficult to see on cloudy days. As fry are positively phototaxic, night harvesting with lights is sometimes advisable.

Fry must be removed from unfertilized spawning ponds before food supplies are exhausted or the schools will break up. Single fry can sometimes be observed along shorelines searching for food. Trapping along shorelines with such devices as glass V-traps can be an effective method of harvest for fry that have abandoned schooling behavior.

b. Rearing Pond Stocking

Snow (1970) recommended stocking rearing ponds with 120,000 to 180,000 fry per hectare, while McCraren (1974) recommended rates up to 240,000 fry per hectare, if 25 to 50 mm fish are desired. Stocking density is generally inversely proportional to desired production size. If objectives involve production of 100-mm fish, stocking densities as low as 24,000 fry per hectare should be used (White 1985). Stocking rates of 125,000 per hectare in 0.05 ha earthern rearing ponds at SM NFH&TC consistently produced 25 to 35 mm fingerlings. White (1985) provided a table with stocking recommendations used in Texas state hatcheries to produce 100,000 Florida LMB fry of various sizes (Table 4). Obviously, the amount of food in the pond should be considered before deciding on fry stocking rates.

c. Rearing Pond Preparation, Fertilization, and Management

Preparation of fingerling rearing ponds should begin well in advance of stocking. Earthen ponds should be drained following the summer or fall harvest, air-dried, and disked (Bishop 1968). Winter rye grass is often sown to produce a dense growth of vegetation which later provides a source of organic fertilizer upon decay following spring flooding and stimulates prolific zooplankton populations which provide food for fry and fingerlings. Rye grass cover can also reduce erosion of exposed pond bottoms and levees when the ponds are dry.

Application of Casoron® at the rate of 283 kg/ha immediately prior to pond filling controlled submerged and/or rooted vegetation for 5 months at SM NFH&TC. Simazine®, applied by spraying exposed pond bottoms at the rate of 22 kg/ha, has provided protection from aquatic weeds for up to 6 weeks at the A.E. Woods State Fish Hatchery in San Marcos, TX.

TABLE 4
Requirements for the Production of Various Numbers of Florida Bass Fingerlings per Hectare Using the Fry Transfer Method

Size of LMB needed (cm)	Spawning area needed (ha)	Broodfish needed (kg)[a]	Fry stocked (No.)	Fry produced (No.)	Fingerling pond area (ha)	Fingerlings produced	Return (%)
2.5	0.42	60.5	105,600	80,000	0.53	76,000	95
3.2	0.45	63.6	111,300	70,000	0.64	63,000	90
3.8	0.47	67.3	117,600	60,000	0.78	51,000	85
4.4	0.50	70.9	125,000	50,000	1.00	40,000	80
5.1	0.53	75.5	133,200	40,000	1.33	30,000	75
6.4	0.62	87.7	153,750	25,000	2.46	16,250	65
7.6	0.73	103.6	181,800	15,000	4.84	8,250	55
10.2	0.80	113.6	200,000	10,000	8.00	5,000	50

[a]Based on fry production of 250,000 fry per hectare.

Adapted from White, B. L., Proc. Fish Farm Conf. Annu. Conv. Catfish Farmers of Texas, Davis, J., Ed., Texas A&M University, College Station., 1985.

Rearing ponds are fertilized following filling. Snow (1970) suggested fertilization at the rate of 9 kg/ha nitrogen and 8 kg/ha superphosphate at weekly intervals until an abundant zooplankton population is established. Ordinarily, three to four applications are required. Bowling et al. (1984) found that fertilization alone was as effective as rye grass cover in producing zooplankton blooms. Biweekly applications of fertilizer should be made after the zooplankton bloom becomes established and continued until about 10 d prior to anticipated harvest (Snow 1970).

An alternative, successful fertilization scheme employed at SM NFH&TC involved the application of 168 kg/ha of peanut hay to ponds 2 or 3 weeks before stocking. Following the hay application (10 d), the ponds were treated with 100kg/ha of 16-20-0 (N-P-K) inorganic fertilizer, followed by a second application 10 d later at 50 kg/ha.

Application of organic fertilizer (e.g., 1500 kg/ha of alfalfa hay) in the absence of grass cover produces a more immediate zooplankton response than does inorganic fertilizer. Since zooplanktonic organisms feed on detritus, inorganic fertilizer must be incorporated into plants before becoming available to animals in the form of decaying organic matter. The size and quantity of zooplankton available must be appropriate to the size and quantity of LMB fry present; thus, precise timing of fertilization is required. Some fish culturists inoculate ponds with aliquots taken from intensively managed zooplankton cultures. These aliquots are added when the pond is filling, and periodically afterwards in an effort to initiate, accelerate, and maintain desirable zooplankton blooms.

Farquhar (1987) compared liquid and granular inorganic fertilizers supplemented with cottonseed meal and found no differences in production. He recommended using the granular fertilizer as it was less expensive, safer, and more convenient to apply.

Standardization and reliability of production in ponds is problematic. Zooplankton enumeration and water quality analysis have recently been used in studies to measure pond changes that may affect fish production (Farquhar 1987). Difficulties remain with respect to pond studies in the areas of study design, zooplankton and water quality sampling techniques, uncontrolled physical and chemical factors, data analysis and interpretation, and delays in sample analysis. Within and among ponds, variations in environmental and production effects are notorious for confounding the results of pond studies (Williamson 1986a). Greater precision is required either through increased treatment replication or through the greater control of environmental variables, perhaps through the use of plastic-lined ponds.

Traditional and new pond study methods (Geiger 1983) have not proven totally reliable. For instance, an 80-μm mesh Wisconsin plankton net has been used to collect zooplankton; however, many rotifers occur in the range of 50 to 70 μm. Undoubtedly, techniques will be improved for zooplankton and water quality sampling, offering new insight into pond dynamics (Farquhar 1984; Graves and Morrow 1988a and b).

Presently, fingerling condition may be used as a guide for fertilization schedules. If regular samples indicate fish are not growing, adequate food is obviously not available, and fertilization or harvest is necessary. Observations of fish condition coupled with evaluations of zooplankton samples provide the best present indicator of pond or food suitability (Carmichael et al.1984a).

Disease epizootics among fingerlings in rearing ponds are atypical if water quality and pond management practices are adequate. Ectoparasites are somewhat common, but treatment does not usually present a problem if fish are observed carefully in ponds and during sampling. Gill flukes (*Dactylogyrus* spp. and *Trichodina* spp.) are the most common ectoparasites, but can be treated following harvest with salt and/or formalin. The most serious predators are usually wading birds such as the green, great blue, and night herons.

The amount of time required for fingerlings to reach a desired size depends on size at stocking, stocking density, pond productivity, and water quality — particularly water temperature. When fingerlings are in rearing ponds, diligent management is necessary for maximum production. Proper fertilization, vegetation, predator and disease control, and maintenance of good water quality are mandatory. Regular sampling of fingerlings provides important information on fish health and growth rates. Accurate data on fingerling growth rates in ponds permits projection of harvest dates and associated hatchery activities. Fish sampling information, when combined with routine zooplankton sampling, provides instantaneous profiles of fingerling condition and pond productivity.

Fingerlings should be harvested as soon as growth rates decline or size variation becomes obvious. Procedures used to control vegetation and crayfish must begin before the expected harvest date. During harvest, rooted vegetation is usually removed by hand as ponds are being drained to eliminate places for fingerlings to hide. Crayfish control using traps was reported by Rach and Bills (1989). Tadpoles may be controlled following harvest (Carmichael and Tomasso 1983).

Fingerlings reach 38 to 50 cm after 3 to 6 weeks in rearing ponds when water temperatures are in the 21 to 27°C range. White (1985) indicated that 3 to 6 weeks are normally required for the production of 25 to 50 mm fish and 6 to 12 weeks for 50 to 100 mm fish in Texas hatcheries (Table 5). Properly managed rearing ponds may yield 30 to 150 kg/ha of 38 to 50 mm fingerlings. Acceptable survival should be near 90% for 25 mm fish, but may be less than 75% for 50 mm fish and 50% for fish 100 mm or larger.

Plastic-lined ponds are becoming more popular for rearing fingerlings, and new recommendations for pond culture will be necessary for them. Although construction costs are greater, lined ponds may be more economical in the long term. Generally, pond preparation and maintenance costs are lower than those of earthen ponds. Plastic-lined ponds conserve water by preventing seepage loss, and many chemical treatments (e.g., preemergent herbicides) typically applied to unlined ponds may be eliminated. Lined ponds eliminate the cost of labor and herbicide required to control rooted aquatic vegetation; other chemical treatments, including fertilization, may be more effectively applied with greater predictability as the chemicals do not become soilbound or otherwise incorporated into pond sediments. Plastic-lined ponds reduce the loss of fish frequently encountered when respiring or decaying vegetation reduces dissolved oxygen (DO) concentrations. They may

TABLE 5
Largemouth Bass Weight and
Length Relationships

Weight (g)	Length (cm)	Fish (per kg)
0.003	0.6	310,510
0.006	0.8	160,922
0.011	0.9	92,631
0.017	1.1	58,016
0.026	1.3	39,089
0.036	1.4	27,421
0.050	1.6	20,042
0.066	1.7	15,100
0.086	1.9	11,603
0.110	2.1	9,110
0.137	2.2	7,300
0.169	2.4	5,926
0.204	2.5	4,899
0.291	2.9	3,434
0.398	3.2	2,511
0.535	3.5	1,868
0.690	3.8	1,450
0.880	4.1	1,136
1.100	4.5	911
1.351	4.8	740
1.633	5.1	612

Adapted from White, B. L., Proc. Fish Farm
Conf. Annu. Conv. Catfish Farmers of Texas,
Davis, J., Ed., Texas A&M University,
College Station, 1985.

also provide greater stabilization and control of pondwater quality dynamics, thus permitting better management and more predictable production results.

Experimental and production data from plastic-lined ponds should be reported. Some problems with plastic-lined ponds are apparent, such as slick surfaces being hazardous and holes being punched in the lining. Also, the liners may be lifted from their anchors by wind when ponds are left empty. Cleaning lined ponds after draining may present problems.

d. Zooplankton Sampling and Collection

Information about zooplankton densities and species composition is often required for the management of LMB hatchery ponds. Collection of representative samples of zooplankton may be difficult because of the non-random distribution of organisms (Wetzel 1975). Information on zooplankton in hatchery ponds may also be biased by sampling devices and by the limited number of access points. Vertical net tows and pump-type samplers (Farquhar 1984) generally require stations of adequate depth within a reasonable distance of pond levees or roads. Horizontal tows of a net deployed from a boom are generally confined to a single stratum of water, which may not produce representative samples. When heavy plankton blooms are present, tow-nets tend to clog and prevent water from passing through the meshes. Pump samplers mutilate some zooplanktors, making identification impossible.

At the SM NFH&TC, a tube-type sampler was designed to improve sampling accuracy (Graves and Morrow 1988b). The sampler has several advantages: it is light-weight and portable, quick and convenient, and inexpensive (less than $30.00 in 1987). In addition, it samples the entire water column from top to bottom and samples previously inaccessible areas, including areas of dense vegetation, brush, and other submerged structures. The tube sampler is made of a rubber-hinged PVC check valve (3.2 cm diameter opening), located on the bottom of a 2-m section of 2.5-cm PVC pipe. This section of the sampler is attached with a hinge pin to a 4.2-m long (5-cm diameter) aluminum extension tube.

The tube sampler was consistently more efficient and accurate than a Wisconsin plankton net of 80-μm mesh (Graves and Morrow 1988b). It also collected adult calanoid copepods which are able to escape other types of samplers. The tube sampler was found to be efficient for collecting water samples as well as zooplankton.

As intensive production of LMB requires bringing the food to the fish, available food sources are often the limiting factor for rearing success. Ground fish and brine shrimp have been used to feed LMB fry while they are learning to consume feed pellets. Brine shrimp and ground fish are expensive and laborius to obtain and dispense. Live zooplankton have advantages over brine shrimp and ground fish, but until recently, no practical method was available to harvest or collect useful quantities of edible zooplankton.

Research at SM NFH&TC led to development of a practical pump method to harvest hatchery-pond zooplankton (Graves and Morrow 1988a). Modifications were fitted to a propellor-lift pump (model MD, 1/20 horsepower, Fresh-Flo Corporation, Cascade, WI). The intake tube was replaced with a 30-cm length of 7.5 cm schedule 40 PVC having a tee fitting approximately 20 cm above the impellor. This modification permitted directional flow of discharge water. The tube-attachment plate was replaced with a 10×5 cm reducer (with the 5-cm side removed) bolted to the motor housing. The impellor shaft was lengthened to allow pumping during periods of fluctuating water levels. Lengthening the shaft required the addition of a PVC pilot at the base of the intake tube to stabilize the shaft. The outlet pipe was reduced to 5 cm to decrease turbulance in the catch basket. Following modifications, the pumping capacity was 180 l/min. Pumped water was discharged into a $100 \times 45 \times 45$ cm floating Saran® basket with 0.5-mm mesh aperatures.

This collection method allowed harvest of up to 2.7 kg of live zooplankton in one night, limited only, most likely, by the size of the collection basket. From 0.45 to 1.4 kg of live zooplankton were harvested nightly from a single 0.04-ha pond for 90 consecutive days. Cost of the system was less than $110 in 1987. Availability of such quantities of live zooplankton greatly reduced the cost and labor required for training LMB fry to subsequently accept pelleted food.

3. Intensive Culture System

An important contribution to the systematic, intensive culture of LMB was the development of a standard, nutritionally adequate, economical, prepared ration. Although LMB fingerlings have been trained to consume ground fish, beef heart, pet food, and a variety of other meat products, production has been limited and unpredictable (Langlois 1931; Snow 1960). In many cases, diets were nutritionally inadequate, unpalatable, and the supplies undependable. Snow (1968a) successfully trained LMB to accept Oregon Moist Pellet, a salmon ration. The diet led to consistently good yields, food conversion, survival, and rates of gain. Nagle (1976) reported that a larger percentage of LMB accepted feed and grew more rapidly on moist pellets than on commercial dry diets. Minimum nutritional requirements have been established for both LMB and SMB fingerlings (Anderson et al. 1981). Diet improvements continued with development of a stabilized form of soft or semimoist diet which required less stringent storage conditions than the original moist pellets. Less expensive commercial dry diets have been used successfully with larger LMB.

a. Training Conditions

Training fingerlings to convert from natural food organisms to a pelleted diet is a critical step in intensive LMB production (Snow 1960), as fry and small fingerlings do not readily accept prepared feeds. They must be conditioned to accept the diets. Acceptability of such diets may be a matter of physical characteristics or palatability of the diet, or it may be a function of other factors such as initial fish size, developmental stage, condition, or strain (Williamson 1986a; Williamson and Carmichael 1990). Also, feeding regime, fish density, lighting, tank configuration, and color affect LMB feed-training efficiency (Flickinger et al. 1975). Initial fish health is critical as the fish require a considerable amount of energy to sustain them through the training period.

Snow (1960) reported that 30-mm fish are more readily trained on prepared feeds than 13-, 19-, or 70-mm fish. Training fish shorter than 25 mm exclusively on formulated feeds has not been particularly successful in production situations. Brandenburg et al. (1979) demonstrated that fry can be taught to eat carp eggs, later proven to be nutritionally deficient by Willis and Flickinger (1981). LMB fry greater than 10 mm can be trained to eat carp eggs and weaned to Biodiet®, but spinal abnormalities were reported by Willis and Flickinger (1981) and Brandt et al. (1987). The scoliotic condition was transitory, however, as fish appeared normal when fed a nutritionally complete diet for 10 d.

Some success training first-feeding fry on microencapsulated diets such as Biokiowa® has been obtained at SM NFH&TC. Fry accepted food particles that were kept in motion and fed concurrently with zooplankton. If diets are complete and fry accept them, survival may still be a function of nutrient availability associated with development of the larvae's digestive system. Specific enzymes produced by the stomach are required for digestion and nutrient assimilation. Fish less than 25 mm are best trained using zooplankton and ground fish as a transitional diet to formulated feed, preferably a semimoist diet.

Fish attractants or gustatory additives have been used only with limited success to increase acceptance of formulated food (Brandt et al. 1987). A commercial fish attractant was added to moist and dry artificial larval diets and feeding success was improved when added to dry diets but not moist diets (Lovshin and Rushing 1989). Brandt et al. (1987) had limited success feeding 10 mm fry commercial flaked feeds.

Typically, training LMB to consume prepared feeds has been accomplished with fingerlings 38 to 50 mm which have been harvested from rearing ponds where they were consuming zooplankton and other aquatic invertebrates. Following harvest, the fingerlings are cleaned of foreign matter and prophylactically treated for parasites with an approved chemical. Fingerlings are then graded to reduce cannibalism and to establish uniform size classes. Sample numbers, lengths, and weights from each fish lot are recorded, followed by total weights of each fish lot. These values are important to estimate survival, growth rate, and daily feeding rates. Fingerlings are crowded in troughs, circular tanks, or raceways at densities of 4 to 8 kg/m^3 of rearing volume. Crowding the fish provides them with an environment that is so altered from their previous environment that they are forced to modify their feeding behavior to survive. Also, crowding increases competition for food, which in turn encourages the young LMB to learn to accept feed pellets. High training densities can be maintained as long as the fish are healthy and water quality remains acceptable. Outsize fish may be removed to reduce cannibalism.

Fingerlings can be reared in almost any container. Previously, they have been trained in catfish hatching troughs, concrete raceways, cages, and linear and circular fiberglass raceways (Snow 1975, Flickinger and Williamson 1984). Circular tanks have several advantages over linear raceways as they are efficiently self-cleaning with less water flow. Exchange rates can be reduced as long as water quality is maintained. Circular water-flow patterns keep food pellets in suspension longer, thus making them more available to the fish. The environment is more homogeneous, fish are more evenly distributed, and food is more available to them, regardless of how it is dispensed.

Westers and Pratt (1977) developed water quality criteria for salmonids which are generally applicable to LMB. Flow rates of four exchanges per hour were suggested, but fewer exchanges may be permissible if DO is maintained above 5.0 mg/l and un-ionized ammonia remains below 0.025 mg/l in the effluent. At the SM NFH&TC, typical loading rates approximate 5 kg/m^3. Well water (21°C) has been used exclusively during training, and exchange is two to three times hourly. Tanks are cleaned twice daily to maintain water quality.

Ideal temperature ranges during training were suggested to be 25 to 30°C (Nelson et al. 1974; Snow 1975). At higher temperatures, bacterial disease (usually *Chondrococcus columnaris*) can lead to serious problems. Incidence of columnaris disease may be reduced by careful handling, grading, and training in cool water (e.g., less than 25°C), with prophylactic treatments of copper sulfate (CuSO$_4$) and citric acid (5 to 10 mg/l for 1 h daily). LMB fingerlings tolerate those copper concentrations in hard water (250 to 300 mg/l calcium carbonate), but caution should be used in soft water, as copper toxicity varies inversely with water hardness. If columnaris disease occurs, infected and starving fish should be removed as soon as possible and the remaining fish treated on alternate days with potassium permanganate ($KMnO_4$) at 2 mg/l for 2 h) and copper sulfate/citric acid in equal parts (20 mg/l for 1 h) until mortality ceases. Fish should be carefully monitored as the treatment is an additional source of stress. Administration of antibiotics in the training ration may provide some relief from bacterial outbreaks. However, administration through hyperosmotic infiltration may have greater potential for effectively treating fingerlings for columnaris disease (Campbell and Johnson 1989).

b. Training Methods

Feeding regimen is extremely important because it determines to a great extent the number of fish that will accept artificial feed. Fish may be fed immediately after stocking into training units. The more often the fish are fed, the more often they come in contact with pellets and the greater their opportunity to learn to accept the feed. Feeding at least 8 times daily, at 15% of initial body weight per day, 7 d a week is suggested during training.

Broadcasting feed around the culture units increases exposure of fish to the pellets, particularly if they are being fed in linear raceways. Automatic feeders of varying designs (belt, clock, vibrating, solenoid-operated) are being used more frequently during training to replace hand feeding. Such feeders can be adjusted to control the frequency, amount, size of the pellets dispensed, and the duration of each feeding event. Some hand feeding is still recommended on a regular basis to permit the culturist to observe the fish and evaluate the training process. If demand feeders can be designed and built for small fish, important information on fish behavior and learning can be obtained.

The recommended size of pellets is 1.0 to 2.5 mm during the initial phases of training. Pellet size is increased incrementally in accordance with the ability of the fish to consume the larger particles. If fingerlings are reluctant to accept feed, it is advantageous to provide ground fish or zooplankton along with pellets. Bones and skin should be removed from the fish before grinding. By reducing the amount of ground fish or zooplankton over time, while increasing the proportion of pellets, fingerlings can be weaned to a prepared ration. Williamson (1986a) successfully trained 80 to 95% of the fish on trial, depending upon strain, when fingerlings weighing 0.25 to 0.35 g per fish were first hand-fed ground fish for 5 d and subsequently weaned to Biodiet® distributed by automatic feeders.

Feed conversion is not a good indicator of training success during the training period because fish are purposefully fed more than they can consume. The proportion of LMB on feed at the conclusion of training is the only valid measure of success. As the cost of feed is relatively low during training, feeding in excess should be of little concern if the tanks are kept clean and good water quality is maintained.

Grading is necessary during the training phase to maintain uniformity among fingerlings, but caution should be used to limit the amount of stress placed on the fish. Diversity in size encourages cannibalism and reduces training success.

Cannibals are distinguished by their relatively large size or by the presence of tails protruding from their mouths. Emaciated "nonfeeders" should be separated from those that have accepted feed. Grading appears to dissolve any established pecking orders among LMB fingerlings and may lead to as high as 75% of the nonfeeders eventually being trained to accept feed. At the end of the training period, 65 to 95% of the initially stocked fish should be trained and will have doubled their weight (Snow 1975; McCraren 1974). The training period is usually 14 to 21 d. Before trained fish are transferred to growout units, they should be graded into uniform sizes and samples weighed, measured, and counted to determine training success and the number of fish per unit weight.

c. Strain Differences

At the SM NFH&TC, training trials were conducted to evaluate differences in training success among representative populations of the Florida and northern subspecies and their reciprocal crosses (F × N and N × F; Williamson 1986a; Williamson and Carmichael 1990). Differences were found in strain acceptance of pelleted feed measured as the percentage of fish trained at the end of the 19-d training period. Northern LMB (96%) exceeded Florida LMB (80%) and reciprocal crosses were intermediate (N×F = 92%; F×N = 89%). No heterosis was observed for this performance character. The training diet was ground fish, followed by Biodiet® dispensed from electronic feeders.

Another area explored in training LMB to accept pelleted food was the mixing of fish that had already learned to eat pellets with fish that had not yet been trained; the conclusion was that fish do appear to learn from one another. Culturists at the SM NFH&TC used small color-mutant green × orange spotted sunfish hybrids (described later in Section IV.B), which had previously been taught to consume pelleted food, to teach LMB fingerlings. Once the LMB were trained, the lightly pigmented sunfish were easily seen and removed from the tank. Also, wild adult LMB were trained to eat chunks of fresh fish when placed into tanks with pellet-trained adults.

Once trained, fingerling LMB retain the learned behavior indefinitely (McCraren 1974, Nagle 1976). Eight-year-old LMB broodstock have been fed pelleted food since being trained as 25-mm fingerlings at the SM NFH&TC.

Training success depends upon fish density, initial fish size or developmental stage, and initial condition, strain, and availability of a nutritionally adequate and palatable food of an appropriate size. Feeding regimen, fish grading, elimination of cannibalism, disease prevention, and maintenance of good water quality are also important aspects which influence training success.

E. GROWOUT PROCEDURES
1. Growout-Pond Preparation and Fingerling Stocking

Ponds for growing 35- to 50-mm pellet-trained LMB fingerlings to 100 to 200 mm are prepared for stocking much as are ponds for spawning and fingerling rearing. They are not fertilized, and preemergent herbicide treatments are recommended.

Following training, fingerlings of similar size are stocked into growout ponds. Stocking rates in ponds may vary, but 24,000 fish per hectare is typical. LMB may also be reared in raceways at production densities approaching those of trout or channel catfish, as long as water quality is maintained. At SM NFH&TC, LMB do not grow as well in raceways during the typical growing season, likely because densities are higher and temperatures are cooler than in ponds.

2. Feeding

Fingerlings should be tempered or acclimated into new conditions to avoid thermal shock. The fish can be stocked into the catch basin of a pond as it is being filled. Floating mats (1.5 × 1.5 m), as shown in Figure 6, made of wooden frames covered with dark plastic, should be placed in ponds for the first few days. Fingerlings concentrate under such mats for cover or protection, and feeding efficiency is increased by providing feed at the mat locations (Williamson 1986a). Otherwise, fingerlings may scatter throughout the pond, making feed distribution difficult and increasing the number of fish reverting to natural pond food, i.e., zooplankton and other aquatic invertebrates. Growth rates for fish reverting to natural food are low compared to fish feeding on pelleted food (Williamson 1986a).

During the first 10 d following stocking, fingerlings are fed 4 times daily with 2 to 3 mm pellets at 15% of stocked biomass daily. Amounts of food can be varied depending upon food consumption and water quality. Feeding rate, as a percentage of estimated biomass, is gradually reduced during the next week to 10% and is maintained for 30 d after which the fish are fed at the rate of 5% of estimated biomass. When the fish reach 100 mm, the diet may be changed to a dry floating

FIGURE 6.   Floating mat.

or sinking feed of appropriate pellet size and nutritional quality. Nagle (1976) found that dry feed reduced labor and feed costs with no reduction in growth rate. Semimoist growout rations are used at the SM NFH&TC.

Fish should be sampled periodically during growout to determine their general condition and growth rates and to adjust feeding rates. Although advanced fingerlings from 150 to 200 mm are typically harvested during the fall, they can also be fed and maintained during the winter. If LMB are not fed during the winter, they lose weight. A proposed feeding schedule for fish greater than 150 g that was developed by Brandt and Flickinger (1987) is as follows:

- At temperatures above 20°C, feed 2.5 to 3% body weight daily
- At temperatures between 10 and 20°C, feed 2 to 2.5% every other d
- At temperatures between 7.5 and 10°C, feed 1.5 to 2% every 2 to 3 d
- At temperatures between 5 and 7.5 °C, feed 1 to 1.5% every 3 to 4 d
- At temperatures below 5°C, feed 1% every 4 to 7 d.

Carmichael et al. (1988) found that LMB fed little at temperatures under 10°C when the fish were kept indoors in tanks. If temperatures are fluctuating greatly, *ad libitum* feeding is recommended, particularly on warmer days. If fish do not eat the entire ration at one time, splitting and feeding in the morning and evening may be necessary. Experiments at the SM NFH&TC have indicated that demand feeders may be useful in feeding pelleted feed to LMB in ponds and raceways. Feeding rings may be required if floating feed is used.

3. Water Quality

Good water quality is mandatory for the health of the fish and to ensure acceptable production. It should be monitored as frequently as necessary to avoid degradation or catastrophic mortalities. At the SM NFH&TC, DO concentration and temperature are typically monitored and recorded at least twice daily, sunrise and evening. Close observation of pond conditions, particularly as they are associated with management practices, permits prediction of water quality problems allowing sufficient time to react and avoid catastrophic losses. Aerators may have to be operated at night during calm, cloudy, summer conditions.

Aquatic vegetation is closely associated with water quality problems and presents other management difficulties ranging from feeding to harvesting fish. On bright summer days, when aquatic vegetation is abundant, DO concentrations rise rapidly due to photosynthesis, sometimes becoming supersaturated. Before sunrise, DO values may drop to critically low concentrations as a result of plant, animal, and bacterial respiration. After chemicals have been applied to control vegetation (24 to 48 h), DO concentrations tend to decline. They may approach zero for several days as dead and dying vegetation decays and if remedial action is not taken. If DO concentrations are low, feeding should be stopped and aeration applied. Some vegetation may be removed by hand. If possible, ponds should be flushed with water. Considerable expense in the form of labor, chemicals, and lost production is associated with vegetation and its control. Use of lined ponds may help in eliminating or reducing problems associated with aquatic vegetation. Triploid grass carp may have some application where legal.

4. Examples of Growout

At the Marion, AL Fish Hatchery, stocking at 24,000 fish per hectare led to LMB averaging over 20 cm and nearly 115 g in about 100 d at an average water temperature of 27°C (Snow 1968b). Survival of at least 80% was usual, with food conversions of about 1.5 (dry weight fed per weight gain). McCraren (1974) estimated the feed requirements for rearing 20,000 LMB fingerlings to 20 cm at 45 kg of 1.6-mm diameter semimoist pellets, 136 kg of 2.4-mm pellets, and 273 kg of 3.2-mm pellets. An additional 1350 kg of trout feed was required to complete the feeding program.

Pond production of representative populations of the Florida and northern subspecies of LMB and their reciprocal crosses ($N \times F$ and $F \times N$) was compared at the SM NFH&TC (Williamson 1986a; Williamson and Carmichael 1990). In separate pond trials lasting 111 d, northern LMB grew better than other strains of LMB. Mean harvest weight was in grams: northern, $117.3 \pm 2.50 > N \times F$; $90.9 \pm 2.85 > F \times N$; $78.3 \pm 6.21 =$ Florida, 74.5 ± 2.83. Feed conversion ratios were northern, $1.24 \pm 0.05 > N \times F$; $1.55 \pm 0.03 = F \times N$; $1.75 \pm 0.05 =$ Florida, 1.73 ± 0.06. Mean survival ranged from 90 to 97% (Florida, $97.6 \pm 0.8 = F \times N$; $74.5 \pm 2.83 > N \times F$; $95.2 \pm 2.9 >$ northern, 93.9 ± 2.95). Ponds had been stocked with 1.2 to 1.3 g fish at approximately 16,000 fish per hectare.

In communal pond trials, studies lasted 127 d. The northern strain grew better than the Florida strain and both hybrids. Mean harvest weight was in grams: northern, $147.9 \pm 5.91 > N \times F$; $108.5 \pm 2.75 > F \times N$; $90.0 \pm 2.86 >$ Florida, 84.9 ± 3.49. Ponds were stocked with fingerlings weighing 2.0 g at 14,400/ha. No heterosis was expressed for any of these performance characters. Biodiet® was the ration in both communal and separate pond studies.

F. HANDLING AND TRANSPORT

LMB fry may be shipped in plastic bags containing a small amount of water, inflated with oxygen, sealed with rubber bands, and placed in insulated containers. Densities of 2000 to 12,000 fry per liter have been successfully held for 1 to 4 d under those conditions (Snow 1968b). Survival to destinations up to 2500 km from the point of debarkation were excellent (87%), and subsequent survival depended on rearing conditions at receiving hatcheries rather than distance shipped. Larger fish may be transported in hauling tanks equipped with aeration devices. Maximum densities for hauling depend on fish size, temperature, and other water quality characteristics, as well as on the time of the haul (White 1982).

Carmichael and Tomasso (1988) reported that LMB are the third most frequently hauled fish in North America. Respondents to a transportation survey reported that between 6.7 to 2200 LMB per kilogram were hauled for up to 10 h. Most hauled the LMB at temperatures between 10 and 25.5°C. Between 45 g and 1.1 kg of fish were hauled per gallon of water.

Handling, transportation, and changes in water quality are common stresses imposed on LMB during various culture phases. Severe stresses may cause sudden mortalities, but fish often survive temporary stress only to die later of diseases or osmotic dysfunction (Wedemeyer 1970; Mazeud et al. 1977).

Responses of fish to stressors include the general adaptation syndrome described in mammals by Selye (1950) in which corticosteroid and catecholamine hormones are released in response to a variety of nonspecific stimuli as part of the responses of an animal in dealing with a stressor. These primary effects initiate secondary responses associated with energy distribution, such as increased blood glucose, decreased liver glycogen, decreased muscle protein, increased heart rate, increased gill blood flow (Pic et al. 1974; Mazeud et al. 1977; Schreck 1981), as well as diuresis and altered electrolyte and water levels in the blood and tissues (Lewis 1971; Norton and Davis 1977). Ensuing tertiary responses, including decreased disease resistance, decreased inflammatory reaction, and leukocyte migration, can also be attributed largely to the action of corticosteroid hormones (Grant 1967).

Plasma characteristics of resting LMB may provide an important reference for evaluation of the degree of stress imposed by various management practices (Carmichael and Tomasso 1984a; Carmichael et al. 1984b, 1984c). Those studies established normal ranges (mean ± 2 standard deviations) for corticosteroids (0.0 to 3.4 µg/100 ml), glucose (13 to 93 mg/ml), chloride (84 to 128 meq/l), and osmolality (263 to 326 mosmole/l). Plasma glucose and corticosteroid values increased within 15 min of handling or net confinement and typically returned to normal within 24 h after the stress was removed.

Depending on the degree of stress, chloride and osmolality changes may not be expressed until after the stress is removed, but may persist for 2 to 3 weeks. Consequently, careful monitoring of fish following any period of stress is important. Although a stress may have been removed, its effects may influence the health of the animal for several weeks.

Transporting fish for any length of time also causes changes in plasma characteristics. Chloride and osmolality values below 70 meq/l and 250 mosmole/l, respectively, are life-threatening and are likely to lead to substantial mortalities.

The use of saline solutions, mild anesthetics, bacteriostats, reduced temperatures, and prophylactic disease treatments reduce stress responses and subsequent mortality. Precapture anesthesia attenuated responses of salmonids to a second stressor (Strange and Schreck 1978) and has been recommended for use with striped bass hybrids (Tomasso et al. 1980) and LMB (Carmichael and Tomasso 1984a).

Sodium chloride solutions (0.3 to 1.0%), alone or with some anesthetic (e.g., MS-222 or etomidate), have been widely used in conjunction with hauling, generally with positive results (Collins and Hulsey 1963; Hattingh et al. 1975; Coetzee and Hattingh 1977; Murai et al. 1979; Tomasso et al. 1980). Hauling fish in salt solutions followed by posthaul tempering in salt during a recovery period reduces the stress response. Salt in recovery water lowers the osmotic gradient between plasma and the environment, thereby reducing the energy required for osmoregulation.

Cooled water has been used for fish transport with some success (Smith 1978, Leitritz and Lewis 1980). Transportation in chilled water reduces severity of the stress response and osmoregulatory dysfunction similar to that which occurs when salt is added to hauling water. However, the combination of chilling and adding salt provides even greater protection from stress and leads to increased survival.

The most effective method we have used for hauling 200 g LMB at a density of 180 g/l for up to 30 h is as follows:

1. Treat the fish prophylactically for disease with copper sulfate (10mg/l) 1 h daily for 10 d prior to hauling.
2. Withhold food from the fish for 72 h before hauling.
3. Anesthetize the fish before capture with MS-222 (50 mg/l) or an equivalent anesthetic.
4. Haul the fish in cool water (e.g., $16°C$) containing salts similar to those in fish plasma ($NaCl, KCl, KH_2PO_4, MgSO_4$), along with high calcium and bicarbonate (>100 mg/l $CaCO_3$), anesthetic (25 mg/l) MS-222, and add an antibacterial compound (10 mg/l active furacin).
5. Temper during recovery in salts similar to the plasma.
6. Use posthauling prophylactic disease treatment (e.g., copper sulfate treatment as described above).

Carmichael et al. (1984c) found that the described procedures virtually eliminated hauling stress mortality in LMB.

The most important water quality parameter for LMB during transportation and handling is DO concentration. Researchers at the SM NFH&TC measured the oxygen transfer efficiencies, under controlled conditions, of ten commercially available diffusers (Carmichael et al. 1991). The diffusers were tested within the methodologies used by those who haul fishes (Carmichael and Tomasso 1988). The diffusers selected for testing were micropore tubing (Micropor® 100), carbon stone (carbon rod), Bioweave® (Schramm hose), Leaky® pipe; 5-μm tubing; rubber diffuser (hose); fused silica; Plastipore (diffuser hose); Plastipore® diffuser; and Millionaire® (bubble hose). Selection depended on common use by fish culturists (Carmichael and Tomasso 1988). Surface areas of all diffusers were equalized to approximately 300 or 600 cm^2 and in some cases 900 or 1200 cm^2, as measured based on 0.7 to 2.8 cm^2 of diffuser area per liter of water. All diffusers were tested in both unaltered well water and water that had sodium chloride added. Salted water was tested because many fish culturists commonly add salt to hauling water (Carmichael and Tomasso 1988). The amount of salt used (0.6%) was suggested by Carmichael et al. (1984c). Oxygen flow rates were dependant on diffuser surface area and were regulated to supply adequate oxygen for hypothetical loads of fish.

All diffusers examined had oxygen transfer efficiencies of less than 15% (Table 6). Most diffusers were tested using five oxygen flow rates. As many as four sizes of diffusers were also examined. The three most efficient diffusers evaluated were Micropor® 100, carbon stone, and Schramm hose, although their relative efficiency values ranked differently depending on diffuser size and oxygen flow rates. Direct injection of oxygen into water circulating pumps was as efficient as most diffusers at adding oxygen to the water. Mino-saver surface agitators were also efficient at aerating hauling water.

TABLE 6
Efficiency Values for Various Aeration Systems[a]

	% Efficiency—300 cm^2			% Efficiency—600 cm^2		
Oxygen flow rates (l/min)	0.5	1.0	1.5	1.0	2.0	3.0
Diffuser name						
Micropore tubing (100)	8.7	7.2	7.6	13.8	11.1	10.1
Carbon stone	8.9	8.0	7.7	10.3	9.0	8.4
Schramm hose	7.0	8.9	8.5	5.2	4.7	4.6
Leaky pipe	6.2	4.6	4.8	5.3	4.8	4.7
5-µm tubing	6.1	5.4	5.5	5.5	5.3	4.9
Rubber diffuser hose	5.3	4.8	4.8	6.1	5.4	4.9
Fused silica stone	4.7	4.5	3.5	3.8	4.0	4.0
Plastipore diffuser hose	4.0	4.0	4.0	4.4	3.1	3.5
Plastipore diffuser	2.7	1.8	2.1	2.5	2.0	2.0
Millionaire bubble hose[b]	11.4	8.9	8.5	5.2	4.7	4.6
No diffuser[c]	—	—	—	14.7	11.4	8.0

[a]Efficiencies are reported as percentages of the oxygen added to the water divided by the amount
   of oxygen used (x 100).
[b]Variable quality; results of the best one tested.
[c]Oxygen injected directly into circulating pump.

Stress responses compared among the northern and Florida subspecies of LMB
and their reciprocal crosses revealed that the Florida strain is much more sensitive
to stress imposed by net confinement (Williamson 1986a, Williamson and Carmichael
1986). The northern strain was less sensitive and the hybrids were intermediate in
response. There was no heterosis expressed for this trait.

III. SMALLMOUTH BASS

Several culture regimes have been developed for SMB which may also be
applicable to Guadalupe and spotted bass. Two methods that have been successful
under different hatchery conditions include a raceway method developed at the
Tishomingo National Fish Hatchery, Tishomingo, OK and a pond method adapted
from Inslee (1975) at the A.E. Woods State Fish Hatchery, San Marcos, TX. Many
aspects of SMB culture are similar to those of the LMB.

The technique developed at Tishomingo involves raceway spawning and
mechanical hatching of eggs. Two advantages of raceway spawning include
control of spawning and the ability to obtain multiple spawns from the same
broodfish. This eliminates egg losses when males abandon their nests after a water
temperature drop. Smallmouth bass will renest and spawn after weather or
predators cause males to abandon their nests, though egg numbers in each
succeeding spawn are reduced, as in LMB.

The raceway technique involves holding adult SMB in concrete raceways for 2
to 3 weeks prior to normal spawning time, thus preventing uncontrolled spawning
in holding ponds. Wooden nests (40.6 × 10.6 × 10 cm) containing 2.5 to 7.6 cm rocks
are placed in the outdoor raceways. The nests are constructed from 2.5 × 10 cm pine

lumber with 0.32-cm hardware cloth covering the bottom. Weighted artificial turf can also be used as spawning substrate to reduce labor and facilitate egg handling and/or shipping.

Fish will usually spawn on the nests in 24 to 48 h. Nests are collected from the raceways daily or at least within 48 h of spawning. Daily observations of nests are made with a glass-bottomed tube made from 15.2-cm PVC pipe.

Mechanical hatching involves placing the entire nest in an indoor catfish-hatching facility, directly under the moving paddles. Compressed air forced through weighted Micropor® tubing is bubbled from under the nests. Fresh, heated (23°C) well water is circulated through the hatching troughs. Heated water improves survival to swim-up, and hatching occurs in 3 to 5 d. Heated water also reduces losses of eggs from fungus infection. Standard aluminum rearing troughs (3.7 × 55.9 × 25.4 cm), used without paddles, have led to good hatching success. Nests may be stacked in troughs, if necessary. Fry fall through the hardware cloth on the bottom of the nests into the troughs, are collected, and then stocked into rearing troughs at 100,000 per trough (about 850 g per trough). Nests and hatching troughs are disinfected with chlorine, rinsed, and air-dried prior to reuse.

Heated well water is used in the fry troughs when available, or well water can be mixed with pond water to raise the temperature. Air is supplied throughout the length of the troughs. Circulation provided by the air bubbles helps disperse the fry which have a tendency to aggregate. Vinyl-coated wire mesh (1.9 × 1.9 cm) may be placed on the bottom of the troughs. Fry will tend to occupy squares in the mesh and thus become more evenly distributed. After 3 to 5 d at 20 to 22°C, the fry swim-up. Subsequently, they are acclimated to pond water and stocked into rearing ponds for growout. Meat and bone meal has been broadcast into the ponds to supplement the natural zooplankton.

The box-nest pond-culture technique used by Pat Hutson, then manager of the A.E. Woods hatchery, offers an alternate method. Broodfish holding ponds are stocked with 225 kg/ha of adult SMB, and 7.2 kg of 5.0 to 7.6 cm goldfish are stocked for each kilogram of broodfish. Portable nests filled with rocks are used as spawning substrates. The nests consist of two boxes, one fitting within the other. The outer box (fry collection box) is 53.5 × 53.3 cm, constructed of 2.5 × 15.2 cm cypress lumber, and has a fiberglass window-screen bottom. The inner box (nest box) is also constructed of 2.5 × 10.2 cm cypress lumber and is 50.8 × 50.8 cm with a 0.6 cm mesh-hardware cloth bottom.

Spawning ponds should have extensive shorelines, silt bottoms with no visible rocks, and gradually sloping levees (3:1 slopes preferred). Pond bottoms are disk-harrowed, bladed, packed, and sprayed with Diuron® to prevent vegetation growth. Box nests are leveled around pond margins at a depth of 45 to 60 cm of water. They are placed 3 to 4 m apart. Rocks ranging from about 5 to 10 cm diameter are placed in the nests.

Ponds are filled 3 to 5 d prior to broodfish stocking. Broodfish are sexed, paired, and stocked when water temperatures are consistently above 16°C. Numbers of males stocked equals the number of nests in the pond. One third more females than males are stocked.

Visual observations of the nests are made daily using an underwater viewing scope constructed from 10.2-cm PVC pipe, 1.5 m long, with a lens from a diving mask attached at one end. After eggs are observed, they are viewed daily until they disappear, indicating hatch. Observations continue until black fry are seen on the rocks. Harvest of the fry is then achieved by slowly raising the entire nest structure to the water surface and gently raising and lowering the nest box. Water currents wash the fry from the rocks, through the hardware cloth, and into the collection box. The fry are then rinsed from the collection box into a wash tub for transfer to rearing ponds. Nests are cleaned and reset for subsequent spawns. Ponds are lowered about 1.0 m and refilled with fresh water to initiate spawning activity.

Rearing ponds are stocked with 125,000 to 187,500 fry per hectare. Since fry are nonswimming when stocked, they are held in $0.6 \times 0.9 \times 1.2$ m window-screen wire boxes within rearing ponds until they swim up. Boxes are placed where new water can be added for aeration. When fry are free-swimming, they are released. Rearing to 3.8 cm requires 20 to 30 d.

SMB are sensitive to handling and hauling stress. In tests designed to measure stress in the species (Carmichael et al. 1983), plasma glucose levels quickly increased after both hauling and handling, indicating that stress had occurred. After the fish were handled and loaded into a hauling unit, plasma sodium and chloride levels dropped. Hauling for 2.5 h caused further declines in both ions, indicating osmoregulatory dysfunction. Plasma glucose returned to baseline levels within 4 to 5 d after handling, but chloride remained low. The extended hypochloremia resulted from a relatively modest stress; recovery from more harsh hauling conditions may require much longer periods.

IV. OTHER SPECIES

Less research has been conducted on propagation of crappie and the various species of sunfishes and their hybrids than on the basses. Much of the information presented here on sunfishes was compiled by Flickinger and Williamson (1984).

A. SUNFISHES

Generally, persons involved with rearing sunfishes have used extensive culture methods. Most literature references pertain to bluegill and redear sunfish (*Lepomis microlophus*) culture, but the techniques suggested should apply to most other species and their hybrids.

Bluegills 2 years old or more (Higginbotham et al. 1983), totaling at least 18 cm TL and 100 to 150 g, are excellent broodfish (Huet 1970). Smaller fish will spawn, but production is reduced. Sunfish spawn several times after the water temperature reaches 21°C. Males defend small excavated nesting depressions which are often close together.

Broodfish should be stocked before water temperatures reach 21°C. The number stocked on an area basis determines the size of fingerlings obtained because young and adult sunfish are generally left together in the spawning-rearing ponds. Blosz (1948) found that 247 broodfish per hectare yielded 370,500 fingerlings per hectare

which averaged 2.5 cm, while 74 broodfish per hectare produced 123,000 fingerlings per hectare which averaged 660 fish per kilogram. Surber (1948) reported a
harvest of 676,780 small fingerling bluegill per hectare from a stock of 198
broodfish per hectare. Higgenbotham et al. (1983) noted that 375,000 fry can be
obtained from 100 broodfish. Huet (1970) generalized that stocking 200 to 300
pairs per hectare would produce 400,000 fingerlings per hectare. Stocking equal
numbers of each sex has been recommended (Huet 1970; Higgenbotham, et al.
1983), but Davis (1953) suggested using a ratio of 2:3 (males to females).

Sexes may be sorted according to methods described by McComish (1968). Egg
incubation requires less than 5 d at temperatures above 21°C (Huet 1970).

Since sunfish are fragile and not cannibalistic at small sizes, the young are reared
with adults. Spawning-rearing ponds should be fertilized to improve yields.
Bluegills accept pelleted feed (this may not be true for all sunfish species and
hybrids), and prepared feed can be used to supplement natural forage, leading to an
increase in fingerling production (Davis 1953; Higgenbotham et al. 1983) and
helping with maintenance of broodfish. Pellets should be broadcast throughout the
pond. Fines (feed dust) from large pellets can be used as food and fertilizer in
spawning-rearing ponds.

B. HYBRID SUNFISHES

Similar culture techniques can be used to produce hybrid sunfishes. Critical to
their production is broodfish care and selection. Crossing unpure broodfish does
not produce the proper progeny or expected sex ratios. Correct sorting of males and
females used in crosses is important as a single error may negate the entire effort.
Removing the opercular tabs on males may enhance hybridization success since
species recognition, which reduces the incidence of hybridization in nature, is
impaired (Lewis and Heidinger 1978). Young hybrid sunfish can be fed pelleted
feed or fines if bluegills are used as one of the parents in the cross (Lewis and
Heidinger 1971).

A hybrid xanthic color-mutant of green sunfish (*Lepomis cyanellus*) × orange
spotted sunfish (*L. humilis*) was located in central Texas in 1985 (Carmichael et al.
1987). The fish are yellow or golden in color and retain dark pigment (melanin) in
their eyes and on some in their skin. Pigments (including blue colored ones) are
present in the skin and fins. The common name "midas-morph hybrid" has been
given these fish.

White (1971) found and developed a hatchery stock of green sunfish with a
similar phenotypic appearance. His xanthic color mutant of green sunfish was found
to be highly prolific, precocious, and highly susceptible to predation. Dunham and
Childers (1980) examined offspring from six of White's (1971) golden color-
mutant green sunfish and found the mutant to be relatively slow-growing in Illinois
and highly susceptible to predation.

At the SM NFH&TC, 35 to 273 unsexed midas-morph hybrids (from 1.8 to
55.0 g) were stocked into 50 0.4-ha ponds. They produced 46,250 to 539,500
fingerlings per hectare in 68 to 201 d. In 21 months, the initial 59 fish and their
offspring produced at least 40,841 harvestable fingerlings using the traditional

open-pond spawning method of Snow (1975). Behavorial observations of the broodfish in spawning ponds indicate that the hybrid sunfish spawn similarly to other fishes within the genus. The hybrid fish are fertile and produce offspring when placed in ponds, even at very small sizes (Table 7). Fish spawned from April or May until at least September. Both the original captured generation, as well as their offspring, produced "color-true" fingerlings. Dunham and Childers (1980) reported that the golden color-morph in green sunfish was due to a recessive genetic trait.

Samples of the hybrid fingerlings (N = 12,508; weight was 0.08 to 0.52 g) were stocked in indoor culture tanks and were trained to consume formulated feed. They received 1.0 to 1.5 mm diameter semimoist pellets at 20% of the stocked biomass per day throughout 20 to 52 d of training and growth periods. Mean training success was greater than 96%. Growth rates during and after training were similar to those of various species of *Lepomis* and their hybrids. At about 1.3 g, sex-ratio differences favored males 63.3 to 36.7%. The results indicated that this highly visible hybrid-sunfish can be produced on fish hatcheries. They may be candidates for aquaculture, small pond management, the pet-fish industry, and for use as laboratory animals. Refugia populations exist, as of 1992, at the Private John Allen National Fish Hatchery, Tupelo, MS.

C. CRAPPIE

Black crappie (*Pomoxis nigromaculatus*) and white crappie (*P. annularis*) behave similarly under culture conditions. In management situations, however, they are dissimilar; 2-year-old crappie are used for spawning, though 1-year-old fish (12.7 cm) will spawn. Few crappie live longer than 4 years (Ming 1971). Large crappie feed on fish but will also consume invertebrates (Ming 1971). The addition

TABLE 7
Pond Production of Midas-Morph Hybrid Sunfish in 0.04 ha Ponds[a]

	Broodfish at stocking		Broodfish at harvest			Offspring at harvest	
Days	Number	Weight (g)	Number	Weight (g)	Survival (%)	Number	Weight (g)
68	59	2.0	35	24.6	59.3	1,850	0.45
133	35	24.6	24	104.6	68.6	19,730	0.63
117	273	1.8	173	25.5	63.4	4,316	0.96
97	100	20.0	78	55.0	78.0	4,349	1.08
90	99	55.0	92	87.3	92.9	10,596	0.30
201	59	2.0	197[b]	—	—	25,896	0.45–0.96
187	121[c]	20.0	92	—	—	14,945	0.30–1.08

[a] Weights of broodfish are presented as average values obtained from total weights divided by the total number of individuals. Offspring numbers and weights (g/fish) were obtained from total harvest weights divided by values derived from samples of 26 to 122 individuals.

[b] 24 original broodfish and 173 first generation broodfish.

[c] Includes 21 fish that were added after 60 d.

of fathead minnows to broodfish holding ponds may be desirable but is not necessary.

Smeltzer (1981) described a method to sex black crappie with nearly 100% accuracy. Portions of that method may also be applicable to white crappie. Basically, selection of fish with deep black coloration should yield adult males. Remaining fish are females and males not in spawning condition. The latter fish are gently stripped. Fish that do not yield eggs are unripe animals of both sexes.

Differences in the urogenital opening are not as distinct in crappie as they are in bluegill (McComish 1968). Fish of uncertain sex can be held without food for several days. Any distention of the abdomen above the urogenital opening indicates the presence of developing ovaries. Crappie spawn at water temperatures above 18°C, but males become dark in color at slightly cooler temperatures.

Generalization of stocking rates for broodfish is difficult because production has been variable. In many cases, both crappie species have been stocked together, and in some instances, they have been stocked with other species (Mraz and Cooper 1957; Smeltzer 1981) . Broodfish densities greater than 247 per hectare seem to retard production (Smeltzer 1981). Yields as high as 164,700 fingerlings per hectare have been reported from 125 adults (Smeltzer 1981; Smeltzer and Flickinger 1991).

Male crappie construct poorly defined nests. Crayfish may reduce production by eating eggs and should be controlled. Egg incubation requires 48 h at 19°C (Siefert 1968). Post-spawning adults are normally left in the ponds with fry.

Based on stomach analyses, cannibalism of adults on young is low (Smeltzer 1981). Addition of fathead minnows to crappie ponds seems to be detrimental (Smeltzer 1981).

One problem in crappie culture involves harvesting. Crappie are extremely susceptible to columnaris disease (Davis 1953) and shock (Smeltzer 1981). Crappie train well to pelleted feed by the method of Willis and Flickinger (1980). Better training success has been obtained with large (4.3 to 6.6 cm) than small (2.5 to 3.6 cm) fish (Smeltzer 1981). Smeltzer and Flickinger (1991) indicated that a 7-d training sequence beginning with carp eggs as a starter diet was more successful than when a pelleted ration was used as the initial feed. Those authors were able to train black crappie fry to a commercial pelleted ration with 85 to 95% success.

D. GUADALUPE AND SPOTTED BASS

In 1984, Carmichael and Williamson (1986) conducted studies at the SM NFH&TC to determine if Guadalupe bass could adapt to fish hatchery conditions. A total of 5 hormone-injected fish (1 female, 4 males), stocked in a 0.04-ha pond, produced about 1200 harvestable fingerlings. Fingerlings (22 to 45 mm TL) were trained to consume pelleted feed with 96.4% success. Growth rates during and following training were similar to other species of *Micropterus*. Cultured Guadalupe bass could be used in stocking programs as a means of supplementing depleted native populations.

Spotted bass have been reared in hatcheries using methods very similar to those used for LMB, SMB, and Guadalupe bass and need not be elucidated.

V. ENVIRONMENTAL REQUIREMENTS

Many environmental requirements of centrarchids are known from either controlled experimentation or general experience gained during culture operations. However, the environmental requirements, even for LMB, are not completely understood. The known environmental requirements for commonly cultured centrarchids are summarized here.

A. LARGEMOUTH BASS

LMB eggs spawned at 17 to 21°C exhibit good survival (>75%) on exposure to 10 to 24°C, but survival is reduced at temperatures above 24°C. Acclimation at the rate of 0.2°C/h increases the tolerance range to 10 to 29°C (Kelly 1968). Fry fed mixed zooplankton and small invertebrates exhibit maximum growth at 27°C (Coutant and DeAngelis 1983). Mortality during cold weather ("winterkill") has been prevented in laboratory studies by using a salmon diet which is high in choline (Roem and Stickney 1989). The authors, however, did not positively identify choline as the ameliorative agent.

Recent studies indicate that the northern and Florida subspecies of LMB differ in their abilities to resist extreme temperatures. Fingerlings of both northern and Florida subspecies can survive a temperature decrease (1°C/d) from 21 to 1°C, if followed by immediate warming (Carmichael et al. 1988). However, substantial mortality in the Florida subspecies will occur if the fish are held at 2 to 3°C for several days (Guest 1982; Carmichael et al.1988). Mortalities increase as the rate of temperature decline increases (Guest 1982). The northern subspecies appears to not be detrimentally affected by chronic exposure to temperatures of 2 to 4°C (Guest 1982; Carmichael et al. 1988). The Florida subspecies, on the other hand, survived higher temperatures (50% mortality at 39.2°C) as compared to the northern subspecies (50% mortality at 37.3°C) when temperatures were increased 1°C/d (Fields et al. 1987).

DO concentrations as low as 2.8 mg/l have no significant effect on survival of LMB eggs (Dudley and Eipper 1975). Food conversion efficiency decreases in LMB at DO levels below 4.0 mg/l. Growth is positively correlated with oxygen concentrations up to saturation but decreases under supersaturated conditions (Stewart et al. 1967). LMB acclimated to 25, 30, and 35°C require a minimum of 0.92, 1.19, and 1.40 mg/l DO for survival (Moss and Scott 1961). When faced with a DO concentration gradient, LMB avoid levels below 3.0 mg/l (Whitmore et al. 1960).

LMB eggs tolerate less than 3.5 ppt salinity. Fingerlings should be kept in < 3.5 ppt, although levels up to 12 ppt can be tolerated for short periods of time (Tebo and McCoy 1964). A population of LMB has recently been identified that grows well in brackish waters with salinities up to 8 ppt. LMB from freshwater populations show reduced growth in similar salinities (Meador and Kelso 1990a, 1990b).

The total ammonia-nitrogen (ammonia-N) 72-h median lethal concentration (LC_{50}) value for fingerling LMB was 15.3 mg/l (pH 8, temperature 25°C, alkalinity 232 mg/l, hardness 272 mg/l) (Tomasso and Carmichael 1982). This value corresponds to an un-ionized ammonia-N value of 0.82 mg/l. Similar values were reported by Roseboom and Richey (1977). Although no direct studies on the effects of ammonia on growth of LMB have been reported, estimates may be made by comparing LC_{50} and growth data available for other species. Based on information available, it is estimated that continuous exposure to 0.05 mg/l unionized ammonia may be expected to reduce growth in LMB. LMB are quite resistant to environmental nitrite when compared to other species (Tomasso 1986). The 96-h LC_{50} for LMB is 140.2 mg/l nitrite-N (Palachek and Tomasso 1984).

B. SMALLMOUTH AND GUADALUPE BASS

SMB fry grow best at 25 to 26°C (Coutant and DeAngelis 1983). Yearling SMB grow satisfactorily at temperatures no higher than 32 to 33°C, but tolerate 35°C for short periods (Wrenn 1980).

Hatching and survival of SMB are reduced after exposure to DO concentrations less than 4.2 mg/l at 20°C, and no fish survived for 2 weeks at 2.8 mg/l (Siefert et al. 1974). SMB larvae become more tolerant of low DO 11 d after hatch, with most animals surviving a 3-h exposure to 1 mg/l (Spoor 1984).

Young SMB are very sensitive to low pH levels (Holtz and Hutchinson 1989). During a 34-d study in which the pH was gradually adjusted to 7.4 (control), 5.7, or 5.0, survivals of fry in the pH 5.7 and 5.0 treatments were 43 and 4% of the controls, respectively (Hill et al. 1988).

SMB appear to be slightly more resistant to ammonia than LMB with a 96-h LC_{50} of 1.2 mg/l un-ionized ammonia-N (Broderius et al. 1985) determined at a pH of 7.7. The un-ionized ammonia-N LC_{50} decreases with decreasing pH, indicating some toxic effect of the ionized ammonia present in the environment. Guadalupe bass appear to be the most susceptible to ammonia of the black basses tested thus far with a 96-h LC_{50} of 0.6 mg/l unionized ammonia-N. Both SMB and Guadalupe bass are highly resistant to nitrite with 96-h LC_{50} values above 150 mg/l nitrite-N (Tomasso and Carmichael 1986).

C. BLUEGILL

Juvenile bluegill (initial weight 1.8 to 8.0 g) grow well in temperatures ranging from 22 to 34°C, with optimal growth in 30°C (Lemke 1977). Cold tolerance of common bluegill (*Lepomis macrochirus*) and coppernose bluegill (*L. purpurescens*) are similar (Sonski et al. 1988). Bluegill × green sunfish hybrids exhibit better winter growth than bluegill (Brunson and Robinette 1983).

The minimum DO concentrations tolerated by bluegill are 0.75 and 1.23 mg/l at 25 and 35°C, respectively (Moss and Scott 1961). When faced with a DO gradient, bluegill avoid concentrations below 3.0 mg/l (Whitmore et al. 1960).

Bluegill eggs tolerate salinities of 1.75 ppt. Fingerling survival is not affected below 10.50 ppt; however, it is suggested that bluegill be maintained at <3.5 ppt (Tebo and McCoy 1964).

Bluegill acclimated to pH 7.5 exhibited no mortalities due to exposures to pH as low as 4.0 (Ellgard and Gilmore 1984), and oxygen consumption is not affected (Ultsch 1978).

LC_{50} values (at 48 and 96 h) for un-ionized ammonia to bluegill range from 0.5 to 4.6 mg/l (Emery and Welch 1969; Roseboom and Richey 1977). Like all centrarchids tested, bluegill are relatively resistant to the toxic effects of nitrite with a 96-h LC_{50} >80 mg/l nitrite-N (Tomasso 1986).

VI. CONCLUSIONS

Historically, centrarchid culture techniques have evolved slowly, but recently developed techniques now allow successful large-scale rearing of perhaps the most popular warmwater game fish family in North America.

VII. RECOMMENDATIONS FOR FUTURE STUDY

Several lines of research would produce useful information for black bass and sunfish culture. These include, in no particular order: (1) "condition" factors and "fitness", (2) diet or nutrition and reproductive condition, (3) significance of "fatty liver" syndrome, (4) energy budgets for growth and reproduction, (5) "stress management" protocols, (6) genetic character heritability and domestication of broodfish, (7) dependable production and harvest of zooplankton, (8) reproduction physiology and control by photoperiod and temperature, and (9) economics and effects of lined ponds on culture.

REFERENCES

Anderson, R.J., Feeding artificial diets to smallmouth bass, *Prog. Fish-Cult.*, 36, 145, 1974.
Anderson, R.J., Kienholz, E.W., and Flickinger, S.A., Protein requirements of smallmouth bass and largemouth bass, *J. Nutr.*, 111, 1085, 1981.
Anon., Manual of Fish Culture, U.S. Fish Commission, Washington, D.C., 1900.
Anon., Spawn black bass in concrete raceway, *Aquacul. Mag.*, 10(6A), 46, 1984.
Anon., Annual Fish and Fish Egg Distribution Report of the National Fish Hatchery System, U. S. Fish and Wildlife Service, Washington, D.C., 1989, 1.
Badenhuizen, T.R., Effect of Incubation Temperature on Mortality of Embryos of the Largemouth Bass, *Micropterus salmoides* (Lacepede), Master's thesis, Cornell University, Ithaca, NY, 1969, 1.
Benz, G.W. and Jacobs, R.P, Practical field methods of sexing largemouth bass, *Prog. Fish-Cult.*, 48, 221, 1986.
Bishop, H., Largemouth bass culture in the southwest, in Proc. North Central Warmwater Fish Culture Workshop, Ames, IA, 1968, 24.
Blosz, H., Fish production program, 1947, in the southeast, *Prog. Fish-Cult.*, 10, 61, 1948.
Bowling, C.W., Rutledge, W.P., and Geiger, J.G., Evaluation of ryegrass cover crops in rearing ponds for Florida largemouth bass, *Prog. Fish-Cult.*, 46, 55, 1984.
Brandenburg, A.M., Ray, M.S., and Lewis.W.M., Use of carp eggs as a feed for fingerling largemouth bass, *Prog. Fish-Cult.*, 41, 97, 1979.
Brandt, T.M. and Flickinger, S.A., Feeding largemouth bass during cool and cold weather, *Prog. Fish-Cult.*, 49, 286, 1987.
Brandt, T.M., Jones, R.M., Jr., and Koke, J.R., Corneal cloudiness in transported largemouth bass, *Prog. Fish-Cult.*, 48, 199, 1986.
Brandt, T.M., Jones, R.M., Jr., and Anderson, R.J., Evaluation of prepared feeds and attractants for largemouth bass fry, *Prog. Fish-Cult.* 49, 198, 1987.

Brauhn, J.L., Holz, D., and Anderson, R.O., August spawning of largemouth bass, *Prog. Fish-Cult.*, 34, 207, 1972.

Broderius, S., Drummond, R., Fiandt, J., and Russom, C., Toxicity of ammonia to early life stages of the smallmouth bass at four pH values, *Environ. Toxicol. Chem.*, 4,87, 1985.

Brunson, M.W. and Robinette, H.E., Winter growth of bluegills and bluegill × green sunfish hybrids in Mississippi, *Proc. S.E. Assoc. Fish Wildl. Agen.*, 37, 343, 1983,.

Buck, D.H., Baur, R.J., and Rose, C.R., An experiment in the mixed culture of channel catfish and largemouth bass, Prog. Fish-Cult., 35, 19, 1973.

Buck, H., Baur, R.J., and Rose, C.R., Polyculture of Chinese carps in ponds with swine wastes, in *Culture of Exotic Fishes Symposium Proceedings*, Fish Culture Section of the American Fisheries Society, Auburn, AL, 1978, 144.

Campbell, E.A. and Johnson, D.L., Intraperitoneal injection and hyperosmotic infiltration for administering antibiotic to largemouth bass, *Prog. Fish-Cult.*, 51, 29, 1989.

Carlson, A.R., Induced spawning of largemouth bass, *Micropterus salmoides* (Lacepede), *Trans. Am. Fish. Soc.*, 102, 442, 1973.

Carlson, A.R. and Hale, J.G., Successful spawning of largemouth bass *Micropterus salmoides* (Lacepede) under laboratory conditions, *Trans. Am. Fish. Soc.*, 101, 539, 1972.

Carmichael, G.J., Seasonal variation in condition factors and plasma chemistry of hatchery-reared largemouth bass (Abstr.), *Proc. Tex. Chap. Am. Fish. Soc.*, 1987.

Carmichael, G.J. and Tomasso, J.R., Use of formalin to separate tadpoles from largemouth bass fingerlings after harvesting, *Prog. Fish-Cult.*, 45, 105, 1983.

Carmichael, G.J. and Tomasso, J.R., Characterization and alleviation of hauling-induced stress in largemouth bass, *Proc. Tex. Chap. Am. Fish. Soc.*, 6, 22, 1984a.

Carmichael, G.J. and Tomasso, J.R., Swim bladder stress syndrome in largemouth bass, *Tex. J. Sci.*, 35, 315, 1984b.

Carmichael, G.J. and Tomasso, J.R., Survey of fish transportation equipment and techniques, *Prog. Fish-Cult.*, 50, 155, 1988.

Carmichael, G.J. and Williamson, J.H., Differential response to handling stress by Florida, northern, and hybrid largemouth bass, *Trans. Am. Fish. Soc.*, 115, 756, 1986.

Carmichael, G.J., Williamson, J.H., and Graves, K.G., unpublished data, 1987.

Carmichael, G.J., Jones, R.M., and Morrow, J.C., Comparative efficiency of oxygen diffusers in a fish- hauling tank, *Prog. Fish-Cult.*, in press.

Carmichael, G.J., Wedemeyer, G.A., McCraren, J.P., and Millard, J.L., Physiological effects of handling and hauling stress on smallmouth bass, *Prog. Fish- Cult.*, 45, 110, 1983.

Carmichael, G.J., Graves, K.G., Bishop, H., and Jones, R.M., Problems concerning sampling techniques and analysis in pond fertilization and aeration studies (Abstr.), Annu. Meet. Texas Acad. of Sci., 1984a.

Carmichael, G.J., Tomasso, J.R., Simco, B.A., and Davis, K.B., Confinement and water quality-induced stress in largemouth bass, *Trans. Am. Fish. Soc.*, 113, 767, 1984b.

Carmichael, G.J., Tomasso, J.R., Simco, B.A., and Davis, K.B., Characterization and alleviation of stress associated with hauling largemouth bass, *Trans. Am. Fish. Soc.*, 113, 778, 1984c.

Carmichael, G.J., Williamson, J.H., Schmidt, M.E., and Morizot, D.C., Genetic marker identification in largemouth bass with electrophoresis at low-risk tissues, *Trans. Am. Fish. Soc.*, 115, 460, 1986.

Carmichael, G.J., Williamson, J.H., Caldwell-Woodward, C.A. and Tomasso, J.R., Responses of northern, Florida, and hybrid largemouth bass to low temperature and low DO, *Prog. Fish-Cult.*, 50, 225, 1988.

Chastain, G.A. and Snow, J.R., Nylon mats as spawning sites for largemouth bass, *Micropterus salmoides* (Lacepede), *Proc. S.E. Assoc. Game Fish Comm.*, 19, 405, 1966.

Chew, R.L., The failure of largemouth bass, *Micropterus salmoides floridanus* (LeSueur), to spawn in eutrophic, over-crowded environments, *Proc. S.E. Assoc. Game Fish Comm.*, 26, 306, 1974.

Coetzee, N. and Hattingh, J., Effects of sodium chloride on the freshwater fish, *Labeo capensis*, during and after transportation, *Zool. Afr.*, 12, 244, 1977.

Colgan, P.W., Brown, J.A., and Orsatt, S.D., Role of diet and experience in development of feeding behavior in largemouth bass *Micropterus salmoides, J. Fish Biol.,* 28, 161, 1986.

Collins, J.L. and Hulsey, A.M., Hauling mortality of threadfin shad reduced with MS-222 and salt, *Prog. Fish-Cult.,* 25, 105, 1963.

Coutant, C.C. and DeAngelis, D.L., Comparative temperature-dependent growth rates of largemouth and smallmouth bass fry, *Trans. Am. Fish. Soc.,* 112, 416, 1983.

Crumpton, J.E., Smith, S.L., and Moyer, E.J., Spawning of year class zero largemouth bass in hatchery ponds, *Fla. Sci.,* 40, 125, 1977.

Davies, W.D., Shelton, W.L., and Malvestuto, S.P., Prey-dependent recruitment of largemouth bass: a conceptual model, *Fisheries,* 7, 12, 1982.

Davis, H.S., *Culture and Diseases of Game Fishes,* University of California Press, Los Angeles, 1953, 1.

Driscoll, D.P., Sexing the largemouth bass with an otoscope, *Prog. Fish-Cult.,* 31, 183, 1969.

Dudley, R.G. and Eipper, A.W., Survival of largemouth bass embryos at low DO concentrations, *Trans. Am. Fish. Soc.,* 104, 122, 1975.

Dunham, R.A. and Childers, W.F., Genetics and implications of the golden color morph in green sunfish, *Prog. Fish-Cult.,* 42, 160, 1980.

Ellgard, E.G. and Gilmore, J.Y., III, Effects of different acids on the bluegill sunfish, *Lepomis macrochirus* Rafinesque, *J. Fish Biol.,* 25, 133, 1984.

Emery, R.M. and Welch, E.B., The Toxicity of Alkaline Solutions of Ammonia to Juvenile Bluegill Sunfish (*Lepomis macrochirus* Raf.), Water Quality Branch, Division of Health and Safety, Tennessee Valley Authority, Chattanooga, 1969, 1.

Erickson, D. L., Hightower, J. E., and Grossman, G. D., The relative gonadal index: an alternative index for quantification of reproductive conditions. *Comp. Biochem. Physiol.,* 81A, 117, 1985.

Farquhar, B.W., Evaluation of fertilization techniques used in striped bass, Florida largemouth bass, and smallmouth bass rearing ponds, *Proc. S.E. Assoc. Fish Wildl. Agen.,* 38, 346, 1984.

Farquhar, B.W., Comparison of granular and liquid inorganic fertilizers used in striped bass and smallmouth bass rearing ponds, *Prog. Fish-Cult.,* 49, 21, 1987.

Fields, R., Lowe, S.S., Kaminski, C., Whitt, G.S., and Phillip, D.P., Critical and chronic thermal maxima of northern and Florida largemouth bass and their reciprocal F1 and F2 hybrids, *Trans. Am. Fish. Soc.,* 116, 856, 1987.

Flickinger, S.A. and Williamson, J.H., Information presented at the U.S. Fish and Wildlife Service Warmwater Fish Culture Workshop, San Marcos National Fish Hatchery and Technology Center, San Marcos, TX, 1984.

Flickinger, S.A., Anderson, R.J., and Puttmann, S.J., Intensive culture of smallmouth bass, in *Black Bass Biology and Management,* Clepper, H., Ed., Sport Fishing Institute, Washington, D.C., 1975, 373.

Geiger, J.G., A review of pond zooplankton production and fertilization for the culture of larval and fingerling striped bass, *Aquaculture,* 35, 353, 1983.

Grant, N., Metabolic effects of adrenal glucocorticoid hormones, in *The Adrenal Cortex,* Eisenstein, A.B., Ed., Little, Brown, Boston, 1967, 269.

Graves, K.G. and Morrow, J.C., Method of harvesting large quantities of zooplankton from hatchery ponds, *Prog. Fish-Cult.,* 50, 184, 1988a.

Graves, K.G. and Morrow, J.C., Tube sampler for zooplankton, *Prog. Fish-Cult.,* 50, 182, 1988b.

Guest, W.C., Survival of adult, Florida and northern largemouth bass subjected to cold temperatures, *Proc. S.E. Assoc. Fish Wildl. Agen.,* 36, 332, 1982.

Harvey, W.D. and Campbell, D.L., Retention of passive integrated transponder tags in largemouth bass brood fish, *Prog. Fish-Cult.,* 51, 164, 1989.

Harvey, W.D., Greenbaum, I.F., and Noble, R.L., Electrophoretic evaluation of five hatchery stocks of largemouth bass in Texas, *Proc. Tex. Chap. Am. Fish. Soc.,* 3, 49, 1980.

Harvey, W.D., Noble, R.L., Neill, W.H., and Marks, J.E., A liver biopsy technique for electrophoretic evaluation of largemouth bass, *Prog. Fish- Cult.,* 46, 87, 1984.

Hattingh, J., LeRoux Fourie, F., and Van Vuren, J.H.J., The transport of freshwater fish, *J. Fish Biol.,* 7, 447, 1975.

Heidinger, R.C., Synopsis of Biological Data on the Largemouth Bass *Micropterus salmoides* (Lacepede) 1802, FAO Fisheries Synopsis No. 115, Food and Agriculture Organization, Rome, 1976, 1.

Higginbotham, B.J., Noble, R.L., and Rudd, A., Culture techniques of forage species commonly utilized in Texas waters, in Proc. Fish Farm. Conf. Annu. Conv. Catfish Farmers of Texas, Texas A&M University, College Station, 1983, 5.

Hill, J., Foley, R.E., Blazer, V.S., Werner, R.G., and Gannon, J.E., Effects of acidic water on young-of-the-year smallmouth bass (*Micropterus dolomieui*), *Environ. Biol. Fish.*, 21, 223, 1988.

Holtz, K.E. and Hutchinson, H.J., Lethality of low pH and Al to early life stages of six fish species inhabiting PreCambrian Shield waters in Ontario, *Can. J. Fish. Aquat. Sci.*, 46, 1188, 1989.

Huet, M., *Textbook on Fish Culture; Breeding and Cultivation of Fish*, Fishing News, London, 1970.

Hutson, P.L., Smallmouth bass culture in Texas, *Prog. Fish-Cult.*, 45, 169, 1983.

Inslee, T.D., Increased production of smallmouth bass fry, in *Black Bass Biology and Management*, Clepper, H., Ed., Sport Fishing Institute, Washington, D.C., 1975, 357.

Jackson, U.T., Controlled spawning of largemouth bass, *Prog. Fish-Cult.*, 41, 90, 1979.

Johnson, P.M., The embryonic development of the swim bladder of the largemouth black bass, *Micropterus salmoides* (Lacepede), *J. Morph.*, 93, 45, 1953.

Kelly, J.W., Effects of incubation temperature on survival of largemouth bass eggs, *Prog. Fish-Cult.*, 30, 159, 1968.

Kliensasser, L.J., Williamson, J.H., and Whiteside, B.G., Growth and catchability of northern, Florida, and F1 hybrid largemouth bass in Texas ponds, *North Am. J. Fish. Man.*, 10, 462, 1991.

Kramer, R.H. and Smith, L.L., Jr., First-year growth of the largemouth bass, *Micropterus salmoides*, and some related ecological factors, *Trans. Am. Fish. Soc.*, 89, 222, 1960.

Kramer, R.H. and Smith, L.L., Jr., Formation of year classes in largemouth bass, *Trans. Am. Fish. Soc.*, 91, 29, 1962.

Langlois, T.H., The problem of efficient management of hatcheries used in the production of pond fishes, *Trans. Am. Fish. Soc.*, 61, 106, 1931.

Lawrence, J.M., Estimated size of various forage fishes largemouth bass can swallow, *Proc. S.E. Assoc. Game Fish Comm.*, 11, 220, 1958.

Lee, D.S., *Micropterus salmoides* (Lacepede), largemouth bass, in *Atlas of North American Freshwater Fishes*, Lee, D. S., Gilbert, C. R., Hocutt, C. H., Jenkins, R. E., McAllister, D. E., and Stauffer, J. R., Jr., Eds., North Carolina State Museum of Natural History, Raleigh, 1980, 608.

Leitritz, E. and Lewis, R.C., Trout and Salmon Culture (Hatchery Methods), California Sea Grant Program Publ., University of California, Berkeley, 1980, 1.

Lemke, A., Optimum temperature for growth of juvenile bluegills, *Prog. Fish-Cult.*, 39, 55, 1977.

Lewis, S. D., The effect of salt solutions on osmotic changes associated with surface damage to the golden shiner, *Notemigonus chrysoleucas, Diss. Abstr.*, Part 13, 6346, 1971.

Lewis, W.M. and Heidinger, R.C., Supplemental feeding of hybrid sunfish populations, *Trans. Am. Fish. Soc.*, 100, 619, 1971.

Lewis, W.M. and Heidinger, R.C., Use of hybrid sunfishes in the management of small impoundments, in *New Approaches to the Management of Small Impoundments*, Novinger, G.D. and Dillard, J.G., Eds., North Central Division of the American Fisheries Society, Special Publ. No. 5, 104, 139, 1978.

Littauer, G., Avian Predators: Frightening Techniques for Reducing Bird Damage at Aquaculture Facilities, U.S. Department of Agriculture, Cooperative State Research Service and Extension Service, U.S. Fish and Wildlife Service and the U.S. Department of Agriculture, Animal and Plant Health Inspection Service—Animal Damage Control, Southern Regional Aquaculture Center Publ. No. 401, Mississippi State University, Starkville, 1990, 1.

Lovshin, L.L. and Rushing, J.H., Acceptance by largemouth bass fingerlings of pelleted feeds with a gustatory additive, *Prog. Fish-Cult.*, 51, 73, 1989.

McComish, T.S., Sexual differentiation of bluegills by the urogenital opening, *Prog. Fish-Cult.*, 30, 28, 1968.

MacCrimmon, H.R. and Robbins, W.H., Distribution of the black basses in North America, in *Black Bass Biology and Management*, Clepper, H., Ed., Sport Fishing Institute, Washington, D.C., 1975, 1.

McCraren, J.P., Hatchery production of advanced largemouth bass fingerlings, *Proc. S.E. Assoc. Game Fish Comm.*, 54, 260, 1974.

McCraren, J.P., Feeding young bass, *Farm Pond Harvest*, 9, 10, 1975.

Manns, R.E., Jr. and Whiteside, B.G., Inaccuracy of sex determination in Guadalupe and largemouth bass on the basis of external characteristics, *Prog. Fish-Cult.*, 42, 116, 1980.

Mazeud, M.M., Mazeud, F., and Donaldson, E.M., Primary and secondary effect of stress in fish: some new data with a general review, *Trans. Am. Fish. Soc.*, 106, 201, 1977.

Meador, M.R. and Kelso, W.E., Growth of largemouth bass in low-salinity environments, *Trans. Am. Fish. Soc.*, 119, 545, 1990a.

Meador, M.R. and Kelso, W.E., Physiological responses of largemouth bass, *Micropterus salmoides*, exposed to salinity, *Can. J. Fish. Aquat. Sci.*, 47, 2358, 1990b.

Meehan, O.L., A method for the production of largemouth bass on natural food in fertilized ponds, *Prog. Fish-Cult.*, 47, 1, 1939.

Merriner, J.V., Development of intergenetic centrarchid hybrid embryos, *Trans. Am. Fish. Soc.*, 100, 611, 1971.

Ming, A., A Review of the Literature on Crappies on Small Impoundments, Missouri Department of Conservation Division of Fisheries Research Section Project No. F-1-R- 20, Study No. I-15, No. 1, 1971, 1.

Moorman, R.B., Reproduction and growth of fishes in Marion Co. Iowa farm ponds, *Iowa State Coll. J. Sci.*, 32, 71, 1957.

Morizot, D.C., Schmidt, M.E., Carmichael, G.J., Stock, D.W., and Williamson, J.H., Minimally invasive tissue sampling, in *Electrophoretic and Isoelectric Focusing Techniques in Fisheries Management*, Whitmore, D.H., Ed., CRC Press, Boca Raton, FL, 1991a, 143.

Morizot, D.C., Calhoun, S.W., Clepper, L.L., Schmidt, M.E., Williamson, J.H., and Carmichael, G.J., Multispecies hybridization among native and introduced centrarchid basses of central Texas, *Trans. Am. Fish. Soc.*, 120, 283, 1991b.

Moss, D.D. and Scott, D.C., DO requirements of three species of fish, *Trans. Am. Fish. Soc.*, 90, 377, 1961.

Mraz, D. and Cooper, W.L., Reproduction of carp, largemouth bass, bluegill, and black crappies in small rearing ponds, *J. Wildl. Man.*, 21, 127, 1957.

Murai, T., Andrews, J.W., and Muller, J.W., Fingerling American shad: effects of valium, MS-222, and sodium chloride on handling mortality, *Prog. Fish- Cult.*, 41, 27, 1979.

Nagle, T., Rearing Largemouth Bass Yearlings on Artificial Diets, Division Wildlife In Service Note 335, Ohio Department Natural Research, 1976, 1.

Nelson, J.T., Bowler, R.G., and Robinson, J.D., Rearing pellet-fed largemouth bass in a raceway, *Prog. Fish-Cult.*, 36, 108, 1974.

Nielsen, L. A. and Johnson, D. L., Eds., *Fisheries Techniques*, American Fisheries Society, Bethesda, MD, 1983, 1.

Norton, V.M. and Davis, K.B., Effect of abrupt change in the salinity of the environment on the plasma electrolytes, urine volume, and electrolyte excretion in channel catfish, *Ictalurus punctatus*, *Comp. Biochem. Physiol.*, 56A, 425, 1977.

Padfield, J.H., Jr., Age and growth differentiation between the sexes of the largemouth black bass, *Micropterus salmoides* (Lacepede), *J. Tenn. Acad. Sci.*, 26, 42, 1951.

Palachek, R.M. and Tomasso, J.R., Toxicity of nitrite to channel catfish, tilapia, and largemouth bass: evidence for a nitrite exclusion mechanism in largemouth bass, *Can. J. Fish. Aquat. Sci.*, 41, 1739, 1984.

Parker, W.D., Preliminary studies on sexing adult largemouth by means of an external characteristic, *Prog. Fish-Cult.*, 36, 55, 1971.

Parmley, D., Alvarado, G., and Cortez, M., Food habits of small hatchery-reared Florida largemouth bass, *Prog. Fish-Cult.*, 48, 264, 1986.

Philipp, D.P., Childers, W.F., and Whitt, G.S., Management implications for different genetic stocks of largemouth bass (*Micropterus salmoides*) in the United States, *Can. J. Fish. Aquat. Sci.*, 38, 1715, 1981.

Philipp, D.P., Childers, W.F., and Whitt, G.S., A biochemical genetic evaluation of the northern and Florida subspecies of largemouth bass, *Trans. Am. Fish. Soc.*, 112, 1, 1983.

Pic, P., Mayer-Gostan, N., and Maetz, J., Branchial effects of epinephrine in the seawater-adapted mullet. I. Water permeability, *Am. J. Physiol*, 226, 698, 1974.

Rach, J.J. and Bills, T.D., Crayfish control with traps and largemouth bass, *Prog. Fish-Cult.*, 51, 157, 1989.

Robbins, W.H. and MacCrimmon, H.R., The Black Bass in America and Overseas, Biomanagement and Research Enterprises, Sault Sainte Marie, 1974, 1.

Roem, A.J. and Stickney, R.R, Winterkill syndrome in largemouth bass produced under laboratory conditions, *J. World Aquacult. Soc.*, 20, 277, 1989.

Rogers, W.A., Food habits of young largemouth bass (*Micropterus salmoides*) in hatchery ponds, *Proc. S.E. Assoc. Game Fish Comm.*, 21, 543, 1968.

Roseboom, D.P. and Richey, D.L., Acute Toxicity of Residual Chlorine and Ammonia to Native Illinois Fishes, Report of Investigation 85, Illinois State Water Survey, Urbana, 1977, 1.

Rosenblum, P.M., Horne, H., Chatterjee, N., and Brandt, T.M., Influence of diet on ovarian growth and steroidogenesis in largemouth bass, in 4th Int. Symp. Reproductive Physiology Fish, University of East Anglia, Norwich, 1991, 265.

Salmon, T.P. and Conte, F.S., Control of Bird Damage at Aquaculture Facilities, Wildlife Management Leaflet No. 475, U.S. Fish and Wildlife Service and the University of California Cooperative Extension Service, University of California, Davis, CA, 1981, 1.

Schreck, C.B., Stress and Rearing Salmonids, Oregon State University Agricultural Experiment Station Tech. Pap. No. 5911, Corvallis, 1981, 1.

Selye, H., Stress and the general adaptation syndrome, *Br. Med. J.*, 1, 1383, 1950.

Siefert, R.W., Reproductive behavior, incubation, and mortality of eggs, and postlarval food selection in the white crappie, *Trans. Am. Fish. Soc.*, 97, 252, 1968.

Siefert, R.E., Carlson, A.R., and Herman, L.J., Effects of reduced oxygen concentrations on the early life stages of mountain whitefish, smallmouth bass and white bass, *Prog. Fish-Cult.*, 36, 186, 1974.

Smeltzer, J.F., Culture, Handling, and Feeding Techniques for Black Crappie Fingerlings, Master's thesis, Colorado State University, Fort Collins, 1981, 1.

Smeltzer, J.F. and Flickinger, S.A., Culture, handling, and feeding techniques for black crappie fingerlings, *N. Am. J. Fish. Man.*, 11, 485, 1991.

Smeltzer, J.F. and Flickinger, S.A., 1991.

Smith, C.E., U.S. Fish and Wildlife Service, Manual of Fish Culture, Section G, Fish Transportation, Washington, D.C., 1978, 1.

Snow, J.R., An exploratory attempt to rear largemouth black bass fingerlings in a controlled environment, *Proc. S.E. Assoc. Game Fish Comm.* 14, 253, 1960.

Snow, J.R., Results of further experiments on rearing largemouth bass fingerlings under controlled conditions, *Proc. S.E. Assoc. Game Fish Comm.*, 17, 303, 1963.

Snow, J.R., The Oregon moist pellet as a diet for largemouth bass, *Prog. Fish-Cult.*, 30, 235, 1968a.

Snow, J. R., Some progress in the controlled culture of the largemouth bass, *Micropterus salmoides* (Lacepede), *Proc. S.E. Assoc. Fish Wildl. Agen.*, 22, 380, 1968b.

Snow, J.R., Culture of largemouth bass, in Report of the 1970 Workshop on Fish Feed Technology and Nutrition, Resource Publication U.S. Bureau of Sport Fisheries and Wildlife 102, Washington, D.C., 1970, 86.

Snow, J.R., Fecundity of largemouth bass, *Micropterus salmoides* (Lacepede), receiving artificial food, *Proc. S..E. Assoc. Game Fish Comm.*, 24, 550, 1971.

Snow, J.R., Controlled culture of largemouth bass fry, *Proc. S.E. Assoc. Game Fish Comm.*, 26, 392, 1973.

Snow, J.R., Hatchery propagation of the black bass, in *Black Bass Biology and Management*, Clepper, J. E., Ed., Sport Fishing Institute, Washington, D.C., 1975, 344.

Snow, J.R. and Maxwell, J.I., Oregon moist pellet as a production ration for largemouth bass, *Prog. Fish-Cult.*, 32, 101, 1970.

Snow, J.R., Jones, R.O., and Rogers, W.A., Training Manual of Warmwater Fish Culture, Bureau of Sport Fisheries and Wildlife, Washington, D.C., 1964, 1.

Sonski, A.J., Kulzer, K.E., and Prentice, J.A., Cold tolerance in two subspecies of bluegills, *Proc. S.E. Assoc. Fish Wildl. Agen.*, 42, 120, 1988.

Spoor, W.A., Oxygen requirements of larvae of the smallmouth bass, *Micropterus dolomieui* Lacepede, *J. Fish Biol.*, 25, 587, 1984.

Stevens, R.E., Hormonal relationships affecting maturation and ovulation in largemouth bass (*Micropterus salmoides*, Lacepede), Ph.D. dissertation, North Carolina State University, Raleigh, 1970.

Stewart, N.C., Shymway, D.L., and Doudoroff, P., Influence of oxygen concentration on the growth of juvenile largemouth bass, *J. Fish. Res. Bd. Can.*, 24, 475, 1967.

Stickley, A.R., Jr., Avian Predators on Southern Aquaculture, U.S. Department of Agriculture, Cooperative State Research Service and Extension Service, U.S. Department of Agriculture Animal and Plant Health Inspection Service-Animal Damage Control, U.S. Fish and Wildlife Service and the Southern Regional Aquaculture Center, Publication No. 400, Mississippi State University, Starkville, 1990, 1.

Strange, R.J. and Schreck, C.B., Anesthetic and handling stress on survival and the cortisol response in yearling chinook salmon (*Oncorhynchus tshawytscha*), *J. Fish. Res. Bd. Can.*, 35, 345, 1978.

Surber, E.W., Chemical control agents and their effects on fish, *Prog. Fish-Cult.*, 10, 125, 1948.

Swanson, R., Bass propagation at the Las Animas Fish Hatchery, in *Midwest Black Bass Culture*, Hutson, P.L. and Lillie, J., Eds., Texas Parks and Wildlife Department, Austin, and Kansas Fish and Game Commissions, Pratt, 1982, 25.

Swingle, H.S., Improvement of fishing in old ponds, *Trans. North Am. Wildl. Conf.*, 10, 299, 1945.

Swingle, H.S., Relationships and Dynamics of Balanced and Unbalanced Fish Populations, Bull. 274, Alabama Polytechnical Institute Agricultural Experiment Station, Auburn, 1950, 1.

Swingle, H.S., Appraisal of methods of fish population study. IV. Determination of balance in farm fish ponds, *Trans. North Am. Wildl. Conf.*, 21, 298, 1956.

Swingle, H.S. and Smith, E.V., Management of Farm Fish Ponds, Agricultural Experiment Station Alabama Polytechnical Institute, Auburn, 1938, 1.

Swingle, H.S. and Smith, E.V., The Management of Farm Fish Ponds, Bull. 254, Alabama Agricultural Experiment Station, Auburn, 1942, 1.

Tebo, L.B. and McCoy, E.G., Effect of sea-water concentration on the reproduction and survival of largemouth bass, *Prog. Fish-Cult.*, 26 ,99, 1964.

Timmons, T.J., Shelton, W.L., and Davies, W.D., Gonad Development, Fecundity, and Spawning Season of Largemouth Bass in Newly Impounded West Point Reservoir, Alabama-Georgia, U.S. Fish and Wildlife Service Tech. Pap. 100, 1980.

Tomasso, J.R., Comparative toxicity of nitrite to freshwater fishes, *Aquat. Toxicol.*, 8, 129, 1986.

Tomasso, J.R. and Carmichael, J.R., Ammonia and nitrite toxicity to the largemouth bass, (Abstr.), Annu. Meet. Texas Acad. of Sci., 1982.

Tomasso, J.R. and Carmichael, G.J., Acute toxicity of ammonia, nitrite and nitrate to the Guadalupe bass, *Micropterus treculi, Bull. Environ. Contam. Toxicol.*, 36, 866, 1986.

Tomasso, J.R., Davis, K.B., and Parker, N.C., Plasma corticosteroid and electrolyte dynamics of hybrid striped bass (white bass × striped bass) during netting and hauling, *Proc. World Maricult. Soc.*, 11, 303, 1980.

Ultsch, G.R., Oxygen consumption as a function of pH in three species of freshwater fishes, *Copeia*, 1978, 272, 1978.

Wedemeyer, G., The role of stress in disease resistance of fishes, in *A Symposium on Diseases of Fishes and Shellfishes*, Snieszko, S.F., Ed., Special Publication No. 5, American Fisheries Society, Bethesda, MD, 30, 1970.

Westers, H. and Pratt, K.M., The rational design of fish hatchery for intensive culture based on metabolic characteristics, *Prog. Fish-Cult.*, 39, 157, 1977.

Wetzel, R.G., *Limnology*, Saunders, Philadelphia, 1975, 1.

White, B.L., Culture of Florida largemouth bass *Micropterus salmoides floridanus*, in *Midwest Black Bass Culture*, Hutson, P.L. and Lillie, J., Eds., Texas Parks and Wildlife Department, Austin, and Kansas Fish and Game Commission, Pratt, 1982, 146.

White, B L., Culture of Florida largemouth bass, in Proc. Fish Farm. Conf. Annu. Conv. Catfish Farmers Texas, Davis, J., Ed., Texas A&M University, College Station, 1985, 20.

White, G.E., The Texas golden green: a color mutation of the green sunfish, *Prog. Fish-Cult.*, 33, 155, 1971.

Whitmore, C.M., Warren, C.E., and Doudoroff, P., Avoidance reactions of salmonid and centrarchid fishes to low oxygen concentrations, *Trans. Am. Fish. Soc.*, 89, 17, 1960.

Wickstrom, G.A. and Applegate, R.L., Growth and food selection of intensively cultured largemouth bass fry, *Prog. Fish-Cult.*, 51, 79, 1989.

Wilbur, R.L. and Langford, F., Use of human chorionic gonadotropin (HCG) to promote gametic production in male and female largemouth bass, *Proc. S.E. Assoc. Fish Wildl. Agen.*, 28, 242, 1984.

Williamson, J.H., Comparing training success of two strains of largemouth bass, *Prog. Fish-Cult.*, 45, 3, 1983.

Williamson, J.H., An Evaluation of Florida, Northern and Hybrid Largemouth Bass, *Micropterus salmoides* under Intensive Culture Conditions, Ph.D. dissertation, Texas A&M University, College Station, 1986a, 1.

Williamson, J.H, unpublished data, 1986b.

Williamson, J.H. and Carmichael, G.J., Differential response to handling stress by Florida, northern, and hybrid largemouth bass, *Trans. Am. Fish. Soc.*, 115, 756, 1986.

Williamson, J.H. and Carmichael, G.J., The concept of optimality versus genetic variability in fish culture and management, (Abstr.), Annu. Meet. Am. Fish. Soc., 1989.

Williamson, J.H. and Carmichael, G.J., An aquacultural evaluation of Florida, northern, and hybrid largemouth bass, *Micropterus salmoides*, *Aquaculture*, 85, 247, 1990.

Williamson, J.H., Carmichael, G.J., Schmidt, M.E., and Morizot, D.C., New biochemical genetic markers for largemouth bass, *Trans. Am. Fish. Soc.*, 115, 460, 1986.

Willis, D.W. and Flickinger, S.A., Survey of private and government hatchery success in raising largemouth bass, *Prog. Fish-Cult.*, 42, 232, 1980.

Willis, D.W. and Flickinger, S.A., Intensive culture of largemouth bass fry, *Trans. Am. Fish. Soc.*, 110, 650, 1981.

Wrenn, W.B., Effects of elevated temperature on growth and survival of smallmouth bass, *Trans. Am. Fish. Soc.*, 109, 617, 1980.

Chapter 6

NORTHERN PIKE AND MUSKELLUNGE

Harry Westers and Robert R. Stickney

TABLE OF CONTENTS

8633-9/93/$0.00 + $.50
© 1993 by CRC Press, Inc.

I. INTRODUCTION

Members of the pike family, Esocidae, are widely revered as sport fishes. There are only five species, all in the genus *Esox*, all of which are strongly piscivorous. The culture of them is almost exclusively for purposes of augmenting natural populations and the sportfishing that is targeted on such populations. Because the fish are piscivorous, their production offers a special challenge to fish culturists.

The natural distribution of the popular northern pike (*Esox lucius*) is circumpolar throughout the freshwater temperate and arctic regions of the northern hemisphere. The muskellunge (*E. masquinongy*), which is sought by anglers as a trophy fish, has a rather limited distribution in the eastern U.S. and Canada. Interestingly, muskellunge have been propagated by fish culturists in the U.S. since the late 1800s (Sorenson et al. 1966). In recent years, a hybrid between the northern pike and muskellunge, the tiger muskellunge, has received a considerable amount of attention from fish culturists.

Northern pike often reach weights of 10 to 15 kg, though 20-kg fish have been caught. Most of the northern pike landed are smaller than those values. Muskellunge are typically caught in the 10 to 15 kg range (Eddy and Underhill 1974), with fish weighing over 30 kg having been taken occasionally.

In addition to the U.S., nations involved in esocid production include Canada, Germany, Austria, Switzerland, The Netherlands, and, to a lesser extent, France, Belgium, and Sweden (Huisman et al. 1980). Low hatchability, cannibalism, and severe production variability have been the main barriers to large-scale propagation (Huet 1970; Sorenson et al. 1966), though progress has been made in improving culture techniques and developing acceptable prepared diets (Sorenson et al. 1966; Graff and Sorenson 1970; Colesante 1977; Bender et al. 1979; Timmermans 1979; Pecor and Humphrey 1984). Development work is continuing, especially in the U.S., to improve upon prepared diets and develop intensive culture techniques.

II. REPRODUCTION

A. BROODSTOCK

In nearly all cases, esocid broodstock come from the wild rather than being produced from captive stocks. Broodfish are readily captured on their spawning grounds with either trapnets or pound nets. In some instances, electrofishing is also used. The sexes of both northern pike and muskellunge can be determined with a high degree of reliability from external examination. Muskellunge adults and fingerlings can be reliably sexed by external urogenital morphology (Lebeau and Pageau 1989).

Natural spawning of northern pike normally occurs at temperatures ranging from 5 to 10°C. Males typically arrive on the spawning grounds before the females, usually at the time of ice-out in the spring. Muskellunge tend to spawn later when the temperature of the water ranges from 10 to 14°C. Natural hybridization occasionally occurs; however, pike and muskellunge are not generally compatible because pike are detrimental to muskellunge. The earlier spawning of pike give them a size advantage over the muskellunge. Northern pike fingerlings are sufficiently large to feed on muskellunge fry. Large females of both northern pike and muskellunge may produce 100,000 eggs (Eddy and Underhill 1974).

The use of natural populations for broodstock precludes the development of a selective breeding program. In the case of esocids, a selection program favoring domestication for ease of artificial propagation may be desirable but can only be implemented by developing a captive broodstock. In North Dakota, muskellunge broodstock were maintained in a 0.3 ha pond on a diet of fathead minnows, perch, and white suckers. Although growth compared favorably with that of wild fish, initial spawning success was poor. Egg viability in 4-year-old females was less than 10% (Phillips and Graveen 1973).

B. SPAWNING AND EGG INCUBATION

Low hatchability has long been a barrier to large-scale production of northern pike and muskellunge. Procedures involved in obtaining eggs and milt, as well as incubation techniques, can contribute to the problem (Sorenson et al. 1966, Colesante 1977).

The incidence of broken eggs has been significantly reduced because of the use of anesthetics to immobilize large broodfish. The technique has led to greatly improved fertilization success and leads to reduced stress on broodfish. Employment of anesthetics also reduces the chance of injury to hatchery workers charged with handling the fish. One method used to anesthetize broodfish involves spraying a Chlorotone mixture over the gills (1 part trichloro-trimethyl-propanol to 4 parts ethanol). A dose of 1.5 to 2.0 full sprays with a spray bottle of the type used for household window cleaners can bring about deep anesthesia (Colesante 1977). MS-222 is another effective anesthetic. It is often employed at a concentration of 100 to 150 mg/l.

There have been those who question the use of anesthetics, because use of the chemicals may lull the people responsible for handling broodfish into a state of inattention during which a reflex muscular contraction by the fish may lead to the partial destruction of a pan of eggs. Instead of anesthetics, some experts advise the use of a sock-like bag to restrain broodfish during the collection of gametes (Klingbiel 1984).

Other refinements have been developed, such as the use of a catheter to collect uncontaminated milt, and employment of a sphygmomanometer (blood pressure cuff) to gently extrude eggs under a controlled pressure of 280 g/cm^2 (Sorenson 1966).

Hypophysation has been used to improve spawning success. Carp pituitary has been successfully employed to accelerate ovulation when it is injected at levels of from 4.5 to 9.0 mg/kg of broodfish body weight (Sorenson et al. 1966; Phillips and Graveen 1973; Jennings, 1982).

Since male esocids produce milt in sparse amounts, they have been injected experimentally with progesterone at levels of 10 and 100 mg/kg of body weight. Milt yield increased from two to threefold, with increasing hormone level within 48 h of injection. There was no adverse effect on fertilization with even the highest level of progesterone, and the yield of eyed eggs was 70% (de Montalembert et al. 1978).

The testes can be surgically removed and sperm obtained by squeezing the organs through cheesecloth (Sorenson et al. 1966). This requires sacrificing the male but enhances sperm yield. Milt collection from live fish typically results in the production of 0.2 to 0.3 ml per male. With care, 1 liter of eggs can be successfully fertilized with 75% success with as little as 0.02 ml of milt (Huisman et al. 1980).

Esocid sperm will remain viable for several days at 3 to 5°C, but once diluted with water, its activity ceases after 1 or 2 min, depending on water temperature. At 5°C, sperm motility after exposure to water will continue for up to 2 min (Huisman et al. 1980). Sperm motility is also affected by pH. No motility occurs below a pH of 5.4, while the time that sperm are motile increases up to at least a pH of 7.9 (Duplinsky 1982).

The duration of sperm motility can be more than doubled if the milt is diluted with a saline solution rather than with just freshwater. An additional benefit of use of saline diluent solutions is that fertilization rate is improved (Rieniets and Millard 1987).

Changes in incubation techniques have contributed more to improvements in hatcheries than have changes in the manner in which eggs and milt are collected (Colesante 1977). Esocid eggs should be handled very carefully. Fertilized eggs can be transported for 1 h following spawning, but must not be jarred.

Eggs are most commonly incubated in some type of jar incubator. It is recommended that eggs not be "rolled" during the first several days of incubation; thereafter, water flow should be increased by about 2.0 l/min to 5.5 l/min for a jar containing 1.0 to 2.5 l of eggs. In Europe, Zoug jars of 6 to 8 l are used to incubate 1 to 5 l of eggs in a flow of 2 to 6 l/min. Eggs are gently rolled during the latter portion of the incubation period.

Fertilization success ranges from 70 to 90% (Huet 1970). A good balance of flow must be maintained. If the water exchange-rate is too high, the fragile eggs will be damaged. If the flow rate is insufficient, the eggs may clump, depriving them of oxygen.

An alternative incubation method is to introduce the eggs into jars which have had small size gravel placed on their bottoms. The gravel will prevent egg rolling even when relatively high flow rates are employed. Up to 2 liters of eggs can be incubated in a 6-liter jar. Improvements in the percentage of eggs that reach the eyed stage of from 15 to 75% have been reported for northern pike and from 6 to 87% for muskellunge when the eggs are not rolled during incubation (Sorenson et al. 1966). Such dramatic improvements seem convincing, but other investigators have reported good hatching success when they have allowed eggs to roll during incubation (Huet 1970; Klingbiel 1984).

Artificial turf has been employed for incubation of muskellunge eggs in lakes. That technique was proposed for lakes in which inadequate natural spawning habitat is available (Dombeck 1987). It might also be useful in a hatchery situation.

The optimum incubation temperature for esocid eggs ranges from 9 to 13°C, but a range of 5 to 21°C can produce viable fry. Lethal temperatures for northern pike embryos and sac-fry have been observed at 3 and 24°C.

The best results can be expected when incubation occurs under conditions of relatively constant temperature with the optimum range. Dissolved oxygen (DO) should be maintained at near saturation. Northern pike eggs incubated at the lower end of the temperature range are reportedly more advanced morphologically than those incubated at higher temperatures. The result is larger than normal fry (Huisman et al. 1976).

A common practice during incubation is daily treatment with a 1:600 formalin solution for 17 min, or a 1:4000 to 1:6000 formalin solution for 1 h. Since formalin may harden egg shells, perhaps leading to reduced hatchability, some workers recommend the use of Diquat at 4 mg/l for 15 min as an alternative to formalin in combating fungus problems (Johnson 1958).

C. HATCHING AND CARE OF SAC-FRY

Approximately 120 degree-d (12 d at 10°C) are required to hatch northern pike eggs, and 180 degree-days (18 d at 10°C) are required to hatch muskellunge eggs. Another 160 to 180 degree-days are required for yolk-sac absorption. The time from initiation of incubation to feeding is from 300 to 360 degree-days, depending somewhat on water temperature and the species being produced (Johnson 1958; Huet 1970).

Since hatching usually encompasses a period of 6 h or more, it is often advantageous to induce hatching of esocid eggs. As soon as hatching commences, eggs can be placed in a container and the temperature rapidly elevated 5 to 7°C without negative effect. Within a short period of time, all the fry will hatch and can be separated from egg shells, dead embryos, and other debris.

A technique used in Michigan involves transfer of fry to Heath tray incubators immediately upon hatching. Heath incubators are designed for salmonids which have much larger eggs, so the egg baskets must be modified with finer mesh screens to prevent fry escapement. The smallest size window screen, approximately 6 × 7 meshes per square centimeter, is acceptable. Ideally, the mesh size should be 7 × 7 meshes per square centimeter. Retention of fish by the 6 × 7 mesh size can be assured if several coats of paint are sprayed on the screen.

The cover on the incubation basket should be tightly sealed, and several layers of artificial turf should be placed in the baskets to keep the fry from piling up and suffocating. From 30,000 to 35,000 fry can be placed in a single tray. A normal stack consisting of 16 Heath incubator trays can, therefore, accommodate over one half million fry during yolk-sac absorption. Flow rate should be maintained at 15 to 20 l/min.

The fry are maintained at a temperature of 16 to 17°C and remain in the incubator for 6 to 10 d depending on species. Once or twice during the first 5 d in the incubator, a cleaning operation is undertaken. During that operation the fry are transferred from their incubation tray into a pan and debris is removed from the tray.

While in the incubator, treatment of the fry for fungus is not required. A savings of about 80% labor can be obtained by employing Heath incubators as compared with rearing sac-fry in troughs or tanks. Sac-fry in Heath trays need a minimum amount of attention. The upwelling water and rapid exchange rates prevent the development of bacteria, a problem that is often associated with fry rearing in troughs. Ultraviolet (UV) light treatment of the entering water to control bacteria has been an effective means of reducing bacterial losses of muskellunge fry reared in tanks (Colesante et al. 1981).

In addition to reducing the incidence of bacterial growth, the accompanying problem of gill fungus is also avoided when Heath incubators are used. Fry handled in incubators suffer reduced mortality and appear to be healthier than those reared in troughs.

The fry of northern pike possess adhesive papillae on the anterior part of their heads just in front of the eyes. The papillae are used by the fish to attach themselves to substrates which are often provided in culture troughs in the form of cloth, aquatic plants, fir branches, wood, artificial turf, and similar objects. The hanging stage of fry development lasts about 9 d at 12°C, but clinging is not required to ensure survival. When fry that are kept in Heath incubator trays during the hanging stage are denied the opportunity to cling, good survival may still be obtained. After the hanging stage, the fry swim to the surface to fill their swim bladders and from that point on adopt a horizontal position in the water column (Timmermans 1979).

III. FINGERLING PRODUCTION

A. EXTENSIVE OR POND-REARING TECHNIQUES

The production of esocids has historically occurred in extensive culture systems, but the practice is changing to one that utilizes more intensive methods. However, most production continues to occur in ponds, and it is unlikely that pond culture will ever be entirely abandoned. It is believed that pond culture on natural foods is needed in certain instances for the production of large, high quality fingerlings which are required to satisfy certain management strategies (Klingbiel 1984).

While, in most cases, hatchery-produced fish are introduced to ponds for rearing, it is possible to stock adult esocids into relatively small ponds and obtain fingerlings through natural spawning. Bry and Souchon (1982) obtained yields of northern pike fingerlings of 0.9 to 2.6 fish per square meter after 50 days when they stocked 5 to 10 fry per square meter. Comparable production was obtained when the ponds were stocked with 1 northern pike female and 2 males in February (equivalent of about 60 eggs per square meter) and harvested in May.

Stocking rates for fry vary greatly. The reported range is from 25,000 to 250,000 fry per hectare (Braum 1964; Huet 1970). Stocking rates from 80,000 to 125,000 fry per hectare are recommended.

Another alternative to stocking ponds with fry is to rear them in cages. Lejolivet and Dauba (1988) maintained northern pike fry for 16 d in cages by feeding them zooplankton. Optimum food intake was 17 g per cage of plankton when the cages were initially stocked with 600 fry per cubic meter.

A number of factors affect the production of esocids in ponds. Those which are most difficult to control, or which are uncontrollable, relate to climate. The vagaries of spring weather can make production of zooplankton blooms difficult.

Ideally, muskellunge fry should be stocked when pondwater temperature is 16 to 17°C. Should the temperature drop to 8 or 9°C a few days after stocking, as sometimes occurs, the fry may die even in the presence of an abundant food supply (Klingbiel 1984). For tiger muskellunge 3 to 4 cm long, growth, production, and food conversion appear to be best at from 20 to 22°C. For fish in the 12- to 13-cm size range, best growth and food conversion seems to occur at 23°C (Meade et al. 1983). Maximum growth for northern pike in Ohio was found to occur at 25°C, while in Ontario the same fish species grew best at 19°C.

Climatically induced problems generally are most pronounced during the early culture phase; thus, the critical period during the culture of northern pike and muskellunge has been identified as that from swim-up to 8.0 cm (Klingbiel 1984). On the other hand, there are those who feel that the production of fingerlings larger than 10 cm is utopian (Timmermans 1979). The difference is influenced by whether production is viewed from a biological or an economic perspective.

Ponds should be properly prepared prior to stocking with esocid fry. Proper fertilization is required for the production of zooplankton blooms. It is recommended that culture ponds be kept dry during the winter. In spring, they should be filled and fertilized. Both organic and inorganic fertilizers are commonly used in combination. Rates and frequencies vary depending upon how a given pond responds to treatment. One method which has been used successfully in the midwestern U.S. involves the application of 300 to 400 kg/ha of alfalfa meal or pellets. Lanoiselee et al. (1986) found that northern pike production and survival could be increased when the rearing ponds received repeated applications of swine manure. Their best results were obtained at a manuring rate of 15 kg/ha of dry matter daily.

Feeding activity is temperature dependent in esocids as in other fishes, though optimum temperatures for growth may differ by as much as 6°C, depending on the geographical distribution of the species under culture. Not only does feeding activity increase as optimum temperatures are approached, but success of capture is also greater. For example, northern pike were found to strike at prey 6 times more often at 20°C than at 13°C, and their success in hitting the target was 7 times as good at the higher temperature (Hiner 1961).

Tiger muskellunge fry can be fed brine shrimp (*Artemia* spp.) and have been shown to consume fairy shrimp (*Streptocephalus seali*) in the hatchery (Meade and Bulkowski-Cummings 1987). They will also accept pelleted feed, though fish reared entirely on live food survive better than those which are converted from dry pellets to forage fish. Gillen et al. (1981) found that tiger muskellunge converted from pelleted feeds to bluegills (*Lepomis macrochirus*) more slowly than to minnows (*Notropis* spp. and *Pimephales promelas*). That could affect growth and survival of the esocids following stocking.

Little information on nutritional requirements of esocids is available. In one study in which tiger muskellunge were reared on diets containing 35, 45, or 55% crude protein (Lemm and Rottiers 1986), the best growth was obtained at 45 and 55% protein at 20 and 23°C. At 17°C, growth improved with increasing dietary protein level.

Cannibalism among esocids can be a major barrier to successful culture. Northern pike fry a few days of age can convert from zooplankton to fish, but they can exist on zooplankton until they reach 10 cm, provided the planktonic food is qualitatively and quantitatively correct (Huet 1970). In one study, it was determined that 2.2-cm northern pike consumed 600 copepods daily, and that by the time they reached 5.0 cm each would have ingested from 50,000 to 60,000 planktonic crustaceans. Careful observation of the food supply for young esocids is critical. Evaluation of the zooplankton bloom as many as four or five times daily is not unrealistic. Making such observations early in the morning is particularly important (Klingbiel 1984).

Once esocid fingerlings reach 8 to 10 cm, enormous numbers of forage fish are required as food. Forage must be continuously present in high concentrations. Some culturists recommend a predator to prey ratio of 1:10 to prevent cannibalism among the esocids. Once the predator to prey ratio drops below some critical level, heavy losses due to cannibalism will quickly begin to occur, and an irreversible trend may become established. It is the need for large numbers of forage fish of the proper size that makes the production of large esocid fingerlings quite expensive. For example, the production of some 11,000 muskellunge fingerlings averaging 20 cm required 60 million sucker fry and 13,860 kg of minnows (Menz 1978).

Fry of the white sucker (*Catastomus commersonii commersonii*) are commonly stocked on a daily basis at 10 fish for each esocid for as long as the sucker fry supply lasts. Thereafter, common carp (*Cyprinus carpio*) fry or, more reliably, minnows may be fed. A problem with carp fry is that those which are not quickly consumed become unavailable to the esocids since the carp grow more rapidly.

Heavy mortality of esocids have been reported in association with stocking. Various sources of stress are coincident with stocking, with thermal stress being perhaps one of the most important (Mather et al. 1986). Handling and moderate-density confinement appeared to be less important than significant rapid temperature change.

Extensive production of relatively large fingerlings in ponds wherein the forage base is replenished through natural recruitment ranges from only 4 to 100 kg/ha. Semi-intensive pond rearing, wherein forage fish are added at frequent intervals, can routinely produce yields of 2000 to 4000 kg/ha. In one case, 9000 kg/ha were produced under those conditions (Klingbiel 1984). The cost of fingerling production under semi-intensive culture conditions can easily reach $24/kg. Production under those conditions is virtually restricted to rearing of muskellunge for use in the management of highly valuable trophy sport fisheries. Large northern pike fingerlings could be produced equally as well under semi-intensive conditions if agencies were willing to underwrite the costs involved.

Pond rearing results are often highly variable. Survival can range from 0 to over 50%. A consistent survival rate of from 10 to 20% is considered satisfactory for the production of small fingerlings. Once the fish are over 8 cm, survival to 20 cm typically exceeds 60%, provided an abundance of forage continues to be available.

The muskellunge management program for Wisconsin has an annual cost of $190,000, but supports a specialized fishery which has an overall economic value of $3.5 million (Klingbiel 1983). The net value of the New York St.

Lawrence-Chautaugua-Niagara muskellunge fishery in 1976 was $1.7 million (Menz 1978). In 1983, that fishery contributed $3 to $4 million to the local economy.

B. INTENSIVE REARING TECHNIQUES

Extensive pond culture of esocids with dependence on natural foods leads to a high degree of unpredictable variability in production. In an attempt to avoid the problem, fish culturists have attempted to develop prepared rations for northern pike and muskellunge. The first breakthrough came with tiger muskellunge in the late 1960s (Graff and Sorenson 1970). Significant improvements in both rearing technology and feed formulation have been made since the initial successes were achieved. Today, tiger muskellunge and northern pike can be successfully reared in tanks on prepared feeds. Muskellunge have been more difficult to rear on prepared feed though some progress is being made (Jorgensen 1984).

Advantages of intensive fingerling culture include cost effectiveness. In the study of Jorgensen (1984), pond rearing of muskellunge was six times more costly than intensive culture. In theory, unlimited numbers of fingerlings can be produced under intensive culture with prepared diets since food should not be a limiting factor. Production stability and predictability are more completely assured with intensive culture than with extensive culture.

Fry are most susceptible to mortality during the period when they are being trained to accept prepared feed. It is important to begin training with only healthy, vigorous fry. Their health and vigor will relate, in large part, to how the fish were handled during incubation and yolk-sac absorption. Employment of the Heath tray method of incubation appears to be one excellent means of producing good quality swim-up fry.

At swim-up, the fry should be transferred to their rearing troughs or tanks. A rigorous sanitation program should be maintained, even though the intensity of labor required to maintain cleanliness is high, particularly during the early stages of culture.

The fish must be fed to excess to ensure that food is continually available to them. Automatic feed dispensers programmed to drop feed every 3 to 5 min over at least 15 h daily should be employed.

Systemic bacterial infections such as furunculosis and columnaris can be controlled with Terramycin provided the bacteria are not resistant to the antibiotic. *Pseudomonas* sp. and *Aeromonas hydrophila*, which can lead to heavy losses in muskellunge fry, have been successfully controlled with UV treatment of lake and well-water supplies during egg incubation and yolk-sac absorption (Colesante et al. 1981). In Europe, northern pike eggs are disinfected with Wescodyne (50 mg active I_2 per liter) for 10 min to control rhabdovirus (Timmermans 1979). Disinfection of muskellunge eggs can be accomplished safely with povidone-iodine (1% active I_2) in a 1:100 solution with 10-min exposure (Schachte 1979).

In general, diseases have not posed a serious problem to intensive esocid culturists. In most instances, epizootics can be successfully prevented or controlled, though it is important that the culturist remain constantly alert for signs of disease.

Cannibalism may occur, but can be controlled. Proper feeding levels and regimes are the best deterrent to cannibalism, but even when such programs are in force it may be necessary to remove obvious cannibals from the culture system daily, particularly in the case of highly aggressive northern pike. If cannibalism is not detected and circumvented early in its occurrence, high levels of mortality can occur over a short period of time.

Cannibalism can be reduced or prevented during the fry feeding stage by employing the technique of feeding at 3- to 5-min intervals. Once the fish are established on feed, the frequency can be reduced to 10-min intervals. When the fish reach 10 cm or larger, further reduction in feeding intervals to 15 min can be imposed. Feed should be offered for at least 16 h daily (dawn to dusk), though feeding over 24 h is not recommended (Bender et al. 1979). Two types of diets developed by U.S. Fish and Wildlife Service personnel and used successfully are the W-Series (14-16) coolwater fish diet and the Abernathy Salmon diet (Tables 1 and 2).

Esocids grow much more rapidly than salmonids; thus, it is necessary to adjust feeding levels more frequently than is common for salmonid fishes. Preferred feeding temperatures range from 18 to 22°C, with the lower part of the range most suitable for northern pike. It is best to work with a relatively constant temperature. Most modern hatcheries which are involved with intensive esocid culture are equipped with water-temperature control apparatus. Though temperatures higher than those recommended may lead to somewhat more rapid growth rates, it is advisable to operate below optimum growth temperatures (23 to 24°C) to avoid

TABLE 1
Coolwater Diet Composition
(Percentage of Diet Basis)

	Diet		
	W14	W15	W16
Ingredient			
Herring meal	50.0	50.0	50.0
Blood meal	10.0	10.0	20.0
Shrimp meal	5.0	5.0	5.0
Soy flour	23.0	20.0	10.0
Fish oil (not anchovy)	9.0	12.0	12.0
Vitamin premix	0.6	0.6	0.6
Mineral premix	0.025	0.025	0.025
Ascorbic acid	0.13	0.13	0.13
Choline chloride (50%)	0.225	0.225	0.225
Pellet binder	2.0	2.0	2.0
Proximate composition			
Crude protein	58.2	56.8	61.1
Digestible protein	48.7	47.5	49.8
Crude fat	13.6	16.1	16.0
Metabolizable energy (kcal/g)	3944	4122	4176

TABLE 2
Abernathy Diet S8-2(84)

Ingredient	%
Herring meal	58.0
Dried whey	10.0
Dried blood or blood meal	10.0
Condensed fish solubles	3.0
Poultry byproduct meal	1.5
Vitamin premix	1.5
Choline chloride	0.58
Ascorbic acid	0.1
Trace mineral mix	0.0005
Pellet binder	2.0
Proximate composition	
Crude protein	48.0
Fish meal protein	40.5
Crude fat	17.0
Moisture	10.0

problems with pathogens which also grow most efficiently in the optimum temperature range of the fish.

The intensive culture method offers the culturist an opportunity to produce large numbers of fingerlings in a limited amount of space, often indoors. Water quality in the culture system can be controlled to a large extent. Production levels per unit area are high and can be expressed in kilograms per cubic meter rather than kilograms per hectare, as is the case in ponds. To put the difference in perspective, assume a 1-ha pond with an average depth of 1 m. Such a pond would have a volume of 10,000 m^3. Production might be 400 kg of 20-cm fingerlings, which represents a density of 0.04 kg/m^3. In intensive culture, a rearing density of 32 kg/m^3 is not uncommon. Thus, the ratio of extensive to intensive culture production can be 1:800 or more with respect to space requirements. Production from one 60 m^3 raceway could equal that from 4 to 6 ha of ponds.

One Michigan fish hatchery utilizes the following rearing program with a target production figure of 200,000, 18-cm fingerling esocids. Initially, swim-up fry numbers well in excess of the target figure are taken directly from Heath incubators and placed into two rearing raceways, each with a volume of 6 m^3. Water temperature is maintained at 18°C. Each raceway is stocked with approximately 150,000 fry.

The fingerling production raceways are fitted with automatic feeders which, when activated, cover about 90% of the water surface with food. A feeding regime of one activation of the feeders every 5 min is initiated with a starter diet of 0.4- to 0.6-mm diameter particles. Daily, or twice daily, raceway cleaning is undertaken to limit bacterial growth. Frequent observations are made to determine fry condition, feeding activity, and the incidence of cannibalism. Once the training period has been completed (10 to 15 d at 18°C), the feeding frequency is reduced to every 10 min.

When the fish appear to be vigorous and healthy and are actively feeding, a complete inventory is obtained. Numbers are then reduced, if necessary, to 240,000, a level 20% above the target figure. The fish are redistributed among sufficient raceways to complete the raceway-rearing phase. The raceways should be equipped with baffles to aid in self-cleaning (Boersen and Westers 1986).

At 18°C, esocid fingerlings are expected to grow at least 0.2 cm daily. A hatchery constant of 90 is recommended (Westers 1981). The percent body weight (BW) to feed is calculated by the formula HC/L, where HC is the hatchery constant (90) and L represents fish length in centimeters. Thus, a 10-cm fish should be fed at 9% of body weight daily.

As the fish grow, feed particle size is increased, as is flow rate through the rearing tanks. Ultimately, a maximum flow rate of 400 l/min is reached. That flow rate provides one water exchange every 15 min. If that much water is not available, maximum loading of the tank is reduced. Maximum loading (kilogram of fish per liter per minute) can be determined on the basis of the metabolic characteristics of the species, in particular its oxygen requirement (Westers and Copeland 1983). Data indicate that tiger muskellunge consume 110 g of DO per kilogram of dry feed offered over each 16-h period. Thus, the maximum loading can be expressed as:

$$l / min / kg \ food \ = \ 110 / available \ DO \tag{1}$$

The available DO is defined as the level in the incoming water less that selected for the effluent. If the incoming water contains 8.0 mg/l and the culturist wishes to maintain 5.0 mg/l in the outflow, 3.0 mg/l are available to the fish. Thus, the liters per minute flow required for each kilogram of feed offered is $110/3 = 36.7$ l/min. Translating that into kilograms of fish per liter per minute, the loading equation becomes:

$$kg \ fish / l / min \ = \ available \ DO / (1.1 \times \%BW) \tag{2}$$

and since $\%BW = 90/8 = 11.25$ in this case:

$$kg \ fish / l / min = available \ DO \ (1.1 \times 11.25) = 0.24 \tag{3}$$

A rearing tank receiving 400 l/min can support a maximum of 400×0.24 kg = 97 kg of 8.0-cm fingerlings. If the condition factor (K) of the fingerlings is 0.005 (it may range from 0.0042 to 0.0055), the weight of an 8.0 cm fish is $0.005 \times 8^3 = 2.56$ g. At maximum loading, a tank receiving 400 l/min can accommodate 97,000 g/2.56 g = 38,000 fish of 8.0 cm.

Redistribution of the fish after initial stocking is required well before they reach 8.0 cm if overloading of the tanks is to be avoided. The best procedure is to redistribute fish after the first complete inventory. In Michigan, final rearing from 8.0 to 18 cm takes place in outdoor raceways, so the initial redistribution is 40,000 fish per tank. Subsequent redistribution is then unnecessary prior to stocking the fish into outdoor raceways.

The raceways have rearing volumes of 60 m^3 and maximum inflow rates of 4000 l/min. Final rearing to 18 cm leads to fish of 29.16-g average (assuming a K of 0.005, the calculation is $0.005 \times 18^3 = 29.16$). The percent BW to feed for those fish is 90/18 = 5%. Maximum loading is $3(1.1 \times 5.0) = 0.55$ kg fish per liter per minute. With a maximum flow of 4000 l/min, the maximum weight of 18-cm fish that can be supported by each raceway is $4000 \times 0.55 = 2200$ kg. The maximum number of 18-cm fish which can be supported by each raceway is 2,200,000 g/29.16 = 75,446 fish.

To produce 200,000 fingerlings, only 3 such raceways are required. The maximum weight attained, 2200 kg, will result in a rearing density of 37 kg/m^3. In Michigan, rearing densities of up to 60 kg/m^3 are attainable (Klingbiel 1983).

From the above description, it is obvious that intensive culture of esocids parallels that for salmonids. There is a need to determine more accurately the nutritional requirements of esocids so diets can be improved. Also, esocids have not been domesticated, so genetic selection holds potential as a means of improving upon the results obtained to date. Rearing techniques, especially the difficult and labor-intensive early-rearing phase, must be improved. The quality of intensively reared fingerlings remains to be fully evaluated, particularly with respect to the utility of such fish in management situations.

Sufficient progress has been made during the past several years that intensive culture of esocids on prepared feeds is practical, both biologically and economically. The technique can be expected to be more widely applied and expanded to additional species in the future.

REFERENCES

Bender, T.R., Jr., Mudrak, V.A., and Hood, S.E., Final Report on Intensive Culture of Tiger Muskellunge Using Prepared Diets, Pennsylvania Fish Commission, Harrisburg, 1979, 1.

Boersen, G.L. and Westers, H., Solids control in a baffled raceway, *Prog. Fish-Cult.*, 48, 151, 1986.

Braum, E., Experimentelle Untersuchungen zur ersten Nahrungsaufnahme und Biologie am Jungfischen von Blaufelchen (*Coregonus wartmanni*, Bloch), Weissfelchen (*Coregonus fera*, Jurine) und Hechten (*Esox lucius*, L.), *Arch. Hydrobiol.*, Suppl. 28, 183, 1964.

Bry, C. and Souchon, Y., Production of young northern pike families in small ponds: natural spawning versus fry stocking, *Trans. Am. Fish. Soc.*, 111, 476, 1982.

Colesante, R.J., Improvement in Esocid Culture Techniques, Report of 1976 Studies, New York Department of Environmental Conservation, 1977, 1.

Colesante, R.J., Engstrom-Heg, R., Ehlinger, N., and Youmans, N., Cause and control of muskellunge fry mortality at Chautauqua Hatchery, New York, *Prog. Fish-Cult.*, 43, 17, 1981.

de Montalembert, G., Bry, C., and Billard, R., Control of reproduction in northern pike, in *Selected Coolwater Fishes of North America*, Special Publ. No. 11, Kendall, R.L., Ed., American Fisheries Society, Bethesda, MD, 1978, 217.

Dombeck, M.P., Artificial turf incubators for raising muskellunge to swim-up fry, *North Am. J. Fish. Manage.*, 7, 425, 1987.

Duplinsky, P.D., Sperm motility of northern pike and chain pickerel at various pH values, *Trans. Am. Fish. Soc.,* 111, 768, 1982.

Eddy, S. and Underhill, J.C., *Northern Fishes,* University of Minnesota Press, Minneapolis, 1974, 1.

Gillen, A.L., Stein, R.A., and Carline, R.F., Predation by pellet-reared tiger muskellunge on minnows and bluegills in experimental systems, *Trans. Am. Fish. Soc.,* 110, 197, 209.

Graff, D.R. and Sorenson, L., The successful feeding of a dry diet to esocids, *Prog. Fish-Cult.,* 32, 31, 1970.

Hiner, L.E., Propagation of northern pike, *Trans. Am. Fish. Soc.,* 90, 298, 1961.

Huet, M., *Textbook on Fish Culture: Breeding and Cultivation of Fish,* Fishing News (Books), London, 1970, 1.

Huisman, E.A., Skjervold, H., and Richter, C.J.J., Aspects of Fish Culture and Fish Breeding, Misc. Pap. 13, Agricultural University of Wageningen, The Netherlands, 1976, 1.

Huisman, E.A., Richter, C.J.J., and Hogendorn, H., *Visteelt,* Agricultural University of Wageningen, The Netherlands, 1980, 1.

Jennings, T., Hatchery Branch Production Report, Iowa Conservation Commission, 1982, 1.

Johnson, L.J., Pond Culture of Muskellunge in Wisconsin, Wisconsin Conservation Department Technical Bulletin, Madison, 1958, 1.

Jorgensen, W., Iowa culture of muskellunge on artificial diets, in *Managing Muskies. A Treatise on the Biology and Propagation of Muskellunge in North America,* Hall, G.E., Ed., Proc. Int. Symp., LaCrosse, WI, April 4-6, Spec. Publ. No. 15, American Fisheries Society, Bethesda, MD, 1984, 285.

Klingbiel, J.H., Ed., Warmwater Stocking and Propagation Audit, Wisconsin Department of Natural Resources, Madison, 1983, 1.

Klingbiel, J.H., Culture of purebred muskellunge, in *A Treatise on the Biology and Propagation of Muskellunge in North America,* Hall, G.E., Ed., Proc. Int. Symp., LaCrosse WI, April 4-6, Spec. Publ. No. 15, American Fisheries Society, Bethesda, MD, 1984, 273.

Lanoiselee, B., Billard, R., and de Montalembert, G., Organic fertilization of nursery ponds for pike (*Esox lucius*), *Aquaculture,* 54, 141, 1986.

Lebeau, B. and Pageau, G., Comparative urogenital morphology and external sex determination in muskellunge, *Esox masquinongy* Mitchill, *Can. J. Zool.,* 67, 1053, 1989.

Lejolivet, C. and Dauba, F., Croissance et comportement alimentaire d'alevins de brochet, (*Esox lucius,* L.) eleves en cages dans le reservoir de Pareloup, *Ann. Limnol.,* 24, 183, 1988.

Lemm, C.A. and Rottiers, D.V., Growth of tiger muskellunge fed different amounts of protein at three water temperatures, *Prog. Fish-Cult.,* 48, 101, 1986.

Mather, M.E., Stein, R.A., and Carline, R.F., Experimental assessment of mortality and hyperglycemia in tiger muskellunge due to stocking stressors, *Trans. Am. Fish. Soc.,* 115, 762, 1986.

Meade, J.W. and Bulkowski-Cummings, L., Acceptability of fairy shrimp (*Streptocephalus seali*) as a diet for larval fish, *Prog. Fish-Cult.,* 49, 217, 1987.

Meade, J.W., Krise, W.F., and Ort, T., Effect of temperature on production of tiger muskellunge in intensive culture, *Aquaculture,* 32, 157, 1983.

Menz, F.C., An Economic Study of the New York State Muskellunge Fishery, New York Sea Grant Institute, Albany, 1, 37, 1978.

Pecor, C.H. and Humphrey, R., Tiger Muskies at Michigan's Platte River Hatchery in 1980, 82, and 83, Michigan Department of Natural Resources, Lansing, 1984, 1.

Phillips, R.A. and Graveen, W.J., A domestic muskellunge broodstock, *Prog. Fish-Cult.,* 35, 176, 1973.

Rieniets, J.P. and Millard, J.L., Use of saline solutions to improve fertilization of northern pike eggs, *Prog. Fish-Cult.,* 49, 117, 1987.

Schachte, J.H., Iodophor disinfection of muskellunge eggs under intensive culture in hatcheries, *Prog. Fish-Cult.,* 41, 189, 1979.

Sorenson, L., Buss, K., and Bradford, D., The artificial propagation of esocid fishes in Pennsylvania, *Prog. Fish-Cult.*, 28, 133, 1966.

Timmermans, G.A., Culture of Fry and Fingerlings of Pike, *Esox lucius*, EIFAC/T35, Suppl. 1, 177, 1979.

Westers, H., Principles of Intensive Fish Culture, Michigan Department of Natural Resources, Lansing, 1981, 1.

Westers, H. and Copeland, J., The Art of Intensive Esocid Culture in Michigan, Michigan Department of Natural Resources, Lansing, 1983, 1.

Chapter 7

YELLOW PERCH

Roy C. Heidinger and Terrence B. Kayes

TABLE OF CONTENTS

8633-9/93/$0.00 + $.50
© 1993 by CRC Press, Inc.

I. INTRODUCTION

Historically, the North American yellow perch (*Perca flavescens*) and the Eurasian perch (*P. fluviatilis*) were considered distinct species. A study by Svetovidov and Dorofeeva (1963) concluded that there was a single, circumpolar species with three subspecies. That taxonomic status was accepted by some North American authors (Scott and Crossman 1967; McPhail and Lindsey 1970), but not by others (Robins 1991). Since the recommendation of Svetovidov and Dorofeeva (1963), Collette and Banarescu (1977) found that the predorsal bone in *P. flavescens* extends between the first and second neural spine while it is anterior of the first neural spine in *P. fluviatilis*, that morphological difference clearly separates the two species. Even though Thorpe (1977) concluded that the two species are biologically equivalent, the present review is limited to yellow perch.

Yellow perch can be distinguished from all other North American members of the family Percidae by the following characteristics: the mouth is large with the maxilla extending at least to the midpoint of the eye; canine teeth are absent, but the preopercle is serrate; and the anal fin has two spines and six to eight soft rays. Yellow perch are intermediate-sized percids that seldom exceed 0.5 kg, though individuals exceeding 1.9 kg have been reported (Scott and Crossman 1973).

Originally, yellow perch occurred in eastern North America from Labrador to Georgia and west to the Mississippi River (Scott and Crossman 1973, Collette and Banarescu 1977). The presently established range includes much of the U.S. and most of Canada. On the east coast of North America, it extends from Nova Scotia south to Florida. The western edge of the range extends from northern Missouri and eastern Kansas, northwest to Montana and north to Great Slave Lake (63°N latitude), then southeast to James Bay, Quebec, and New Brunswick. In the U.S., yellow perch have been introduced into nearly all states west and south of its original range. Its extensive range and diversity of habitat is a reflection of the great ecological adaptability of the species.

In North America, the yellow perch is valuable as a sport, forage, and commercial food fish. By the early 1900s, various agencies cultured yellow perch to the fingerling stage for stocking. Eggs were obtained from the wild and incubated in jars, screened-bottom floating boxes, or wire baskets suspended in streams (Leach 1928; Muncy 1962). Leach (1928) reported that in 1927, 15 states stocked 12 million eggs, 194 million fry, and 1.25 million fingerlings produced in U.S. governmental hatcheries. Due to those early stockings, the yellow perch became naturalized over a large area, and as a result, the demand for fish to be stocked decreased so that by 1984 only 12.7 million eggs, fry, and fingerling yellow perch were distributed in the U.S. (Conover 1986).

Demand for cultured yellow perch depends upon season of the year, price in relation to competing marine species such as cod and ocean perch, and the commercial catch. Traditionally, people in Wisconsin have consumed 75% (9 to 11 million kg) of yellow perch annually. Those fish are captured by commercial fishermen in the Great Lakes (Follett 1975). Nearly all commercially caught yellow perch come from Lake Erie and Green Bay, WI. Commercial harvest of wild stocks in the Great Lakes region does not meet the demand of the local markets (Starr

1991). In general, commercially caught fish are not available from December through March.

Following the record 17-million kg commercial harvest of 1969, the mean ex-vessel price of yellow perch in the U.S. and Canada was only $0.25/kg; in 1974, it increased to $0.71/kg, and in 1976, to $1.64/kg (Lesser and Vilstrup 1979). During 1983 and 1984, the price ranged from approximately $2.00 to $3.50/kg and unbreaded fillets sold for $6.00 to $13.00/kg wholesale and $8.00 to $17.50 retail. In 1991, fillets sold for $19.00 to $30.00/kg. As the market demand for yellow perch began to exceed the supply, interest in developing an economical means for culturing the species increased (Downs 1975).

II. SPAWNING

Yellow perch are annual spawners with synchronous oocyte growth during fall through winter, culminating in a spring spawning season of approximately 2 weeks. Just prior to spawning, the gonadosomatic index (GSI) of mature females ranges from 20 to 31% (Lagler et al. 1962; Hutchinson 1974; Brazo et al. 1975; Clugston et al. 1978). GSI is calculated by dividing the weight of the gonad by the weight of the fish and expressing the number obtained as a percentage. The GSI of male yellow perch ranges from 8 to 15% (Lagler et al. 1962).

West and Leonard (1978) found that males mature at between 98 and 165 mm (mean is 108 mm) and females mature from 140 to 191 mm (mean is 158 mm). Clugston et al. (1978) reported that a 92 mm male and a 129 mm female were sexually mature. Relative fecundity ranges from 79 to 233 eggs per gram of female body weight (Sheri and Posner 1969; West and Leonard 1978). Based on the regression equation of Clugston et al. (1978) [log fecundity = $-4.21565 + 3.58816$ log total length (TL)], a 130 mm female would have approximately 3000 eggs and a 250 mm female 109,000 eggs. In the wild, a few males mature after 1 year of age and some females at 2 years (Herman et al. 1959; Muncy 1962), while most 3- and 4-year-old fish are mature (Clady 1976).

In the spring, males move to the shoreline first, followed by females. Harrington (1947) described the actual spawning behavior for yellow perch in the wild; Hergenrader (1969) and Kayes (1977) described it for fish held in aquaria and tanks. A female is accompanied by 2 to 25 males as she drags the unique transparent, gelatinous, accordion-folded hollow egg tube through the milt. The egg tube (Figure 1) unfolds from the female like a long concertina. It may be several meters long and 10 cm wide (Mansueti 1964). Access to the eggs by sperm and aeration is partially accomplished by water circulation through holes in the gelatinous matrix of the egg tube to the central canal (Worth 1892). It takes several minutes at 14 to 15°C and up to several days of temperatures below 5°C to extrude the entire egg mass (Kayes 1977). No protection is given the egg mass or young by either parent, but the egg mass appears to contain some compound that deters predation by other fish (Newsome and Tompkins 1985).

Spawning has been reported at temperatures ranging from 2.8 to 19.9°C (Hokanson and Kleiner 1974; Hokanson 1977b). Optimum incubation temperatures from fertilization to hatching are considered to be from 10 to 20°C (Table 1).

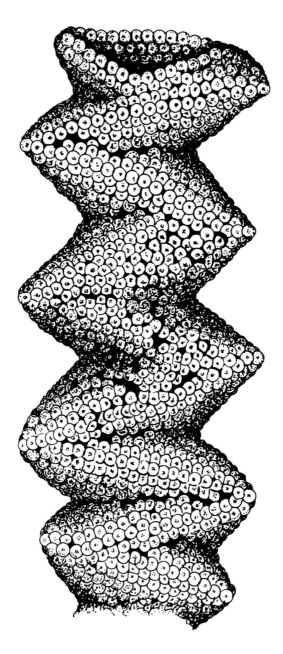

FIGURE 1.   Section of fertilized yellow perch egg mass.

A temperature rise of approximately 0.5°C per day maximizes survival and hatch. Within the range of 13 to 17°C, 85 to 90% of the eggs can be expected to hatch and 70 to 75% of the fry will reach the swim-up stage (Figure 2A). Embryos can be incubated at up to 22°C after the neural keel is formed (Figure 2B).

TABLE 1
Yellow Perch Temperature Requirements for
Various Life Stages

Life stage	Temperature (°C)	
	Optimum	Tolerance range
Maturation	3.9–6.1	<11.1
Spawning	7.8–11.1	2.8–18.9
Cleavage embryo	7.8–12.2	3.9–21.1
Embryo	12.2–16.1	7.7–22.8
Fertilization to hatch	10.0–20.0	6.7–20.0
Hatch to swim-up	20.0–23.9	2.8–27.8
Feeding larvae	20.0–23.9[a,b]	10.0–30.0[a]
Juvenile (survival)	23.9–27.8	0.0–33.3
Juvenile (growth)	23.9–27.8	6.1–31.1[b]

[a] Denotes best estimate based on culture experience.
[b] Results obtained on excess ration. Restricted rations or mass culture situations will result in lower growth optimum temperature and upper growth limit which will be dependent on feeding regime.

From Hokanson, K. E. F., *Perch Fingerling Production for Aquaculture*, Soderberg, R. W., Ed., University of Wisconsin Sea Grant Advisory Rep. 421, Madison, 1977a, 24.

Yellow perch appear to require a cooling period (chill period) for late stage yolk deposition and final egg maturation. Hokanson (1977b) stated that the minimum chill period is 160 d at approximately 10°C or less. He did not obtain viable eggs when the perch were held at 12°C. In yellow perch obtained from Minnesota waters, the optimum chill period was 185 d at 6°C or lower (Figure 3).

Attempts to change the natural spawning cycle of yellow perch by manipulating temperature and light have met with limited success. Of perch exposed to thermal regimes ranging from 4 to 10°C for 120 to 240 d, those exposed to the higher temperatures spawned only slightly earlier than fish exposed to colder temperatures (Figure 4). Kayes and Calbert (1979) were not able to significantly change the time of spawning by taking perch from a lake in the winter, warming them from 9 to 13°C, and increasing the photoperiod to either 13.5 or 18 h of light. Those authors, as well as Hokanson (1977b), postulated that within a certain temperature range the onset of springtime spawning depends more on the intrinsic maturation state of the gonads than on photoperiod or temperature cues.

Hatchability of eggs collected from tank-spawned fish frequently ranges from 40 to 85%. In tanks, females may drop their eggs in the absence of males, or the male, if present, may not fertilize them (Kayes 1977). Kayes (1977) was able to obtain 80% hatch using the wet and 83% hatch using the dry-stripping method. Fertility was increased by 97% by stripping eggs and sperm into a 0.5% sodium chloride solution. He found that the fertility of eggs exposed to water was reduced from 80% when fertilization was done within 1 min, to 15% when fertilization was delayed by 5 min. Nearly all the larvae produced from eggs fertilized after 5 min were

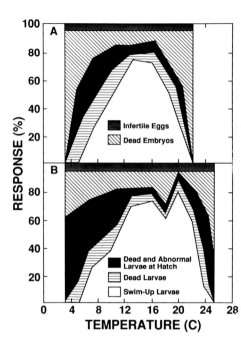

FIGURE 2.   Effect of constant incubation temperatures on percentage hatch, normal hatch, and swim-up larvae of known aged yellow perch embryos. (A) Embryos incubated at test temperatures from fertilization to larval swim-up stage; (B) embryos incubated at 12°C until neural keel formation, then at test temperatures to larval swim-up stage. (From Hokanson, K. E. F. and Kleiner, C. F., *The Early Life History of Fishes,* Blaxter, J. S., Ed., Springer-Verlag, Heidelberg, 1974. With permission.)

deformed. To avoid deformities and increase fertility, Kayes (1977) recommended mixing the sperm and eggs within 20 s after the eggs are stripped. This is usually not a problem because male perch generally have copious amounts of milt.

No attempts to store yellow perch eggs or sperm have been reported in the literature; however, the basic composition of the milt has been determined (Koenig et al. 1978). Perch milt contained from 1.14 to 3.02×10^{10} sperm per milliliter. The pH of the milt was 8.5 and osmotic pressure was 316.7 mOsm. The seminal plasma contained 2.64, 0.46, 0.13, and 0.14 mg of sodium, potassium, calcium, and

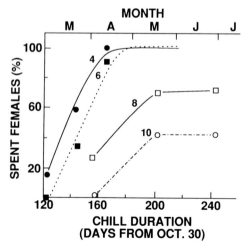

FIGURE 3.   Percentages of female yellow perch that spawned during exposure to 4 chill temperatures (4, 6, 8, and 10°C) of different durations (123 to 242 d from October 30). Temperature was increased at the rate of 2°C per week to a maximum of 20°C after termination of the exposure to various chill temperatures. (From Hokanson, K. E. F., *J. Fish. Res. B. Can.,* 34, 1524, 1977b. With permission.)

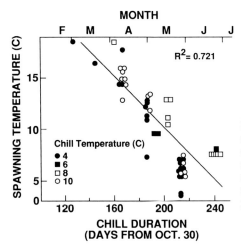

FIGURE 4. Spawning temperature observed at the time of laboratory spawning of yellow perch from Minnesota that had been exposed to four chill temperatures for various periods. (From Hokanson, K. E. F., *J. Fish. Res. Bd. Can.*, 34, 1524, 1977b. With permission.)

magnesium per gram, respectively. Corresponding cationic concentrations for spermatozoa were 1.81, 1.65, 0.07, and 0.23 mg/g.

The use of hormones to spawn yellow perch is not well-documented. *In vitro* germinal vesicle breakdown did not occur using carp pituitary extracts or mammalian gonadotropins (Goetz and Bergman 1978b). However, *in vitro* treatment with various steroids caused germinal vesicle breakdown (Goetz and Bergman 1978a). Compounds that raise the intraoocyte level of cAMP can block steroid-induced germinal vessicle breakdown (DeManno and Goetz 1987).

Before water hardening, the diameter of fertilized eggs ranges from 1.6 to 2.1 mm; after hardening, the eggs increase in diameter to 1.7 to 4.5 mm (Mansueti 1964). Hatching-time varies greatly, depending on temperature. Eggs require 51, 27, 13, and 6 d to hatch at 5.4, 8.0, 16.0, and 19.7°C, respectively (Worth 1892; Mansueti 1964; Wiggins et al. 1983).

Prior to hatching, the egg mass, which initially has a specific gravity slightly greater than water, loses rigidity and becomes flaccid; a gold iris pigmentation surrounds the melanin in the eyes of the larvae; larval movement decreases; and bubbles accumulate in the egg mass, giving it a tendency to rise. This floating tendency can be a problem if the eggs are being incubated in jars (West and Leonard 1978). For that reason, Heath incubators have been used to hatch yellow perch eggs. They contain the egg mass, but do not completely eliminate the floating problem. Hatching occurs within 24 h after mouth and opercular movements are synchronized in a regular breathing fashion (Worth 1892).

III. LARVAL DEVELOPMENT

Newly hatched postlarvae are 4.7 to 6.6 mm TL (Mansueti 1964; Scott and Crossman 1973; Hokanson and Kleiner 1974). The literature contains conflicting information on when gas bladder inflation occurs. Mansueti (1964) indicated that it occurs when the prolarvae are approximately 7 mm long, while Hokanson (1977b) believed that gas bladder inflation takes place during the swim-up stage, which at water temperatures above 13°C occurs on the day of hatching. At temperatures below 13°C, swim-up occurs within 2 d of hatching. Hokanson (1977b) further postulated that the prolarvae must fill their gas bladders with air at the water surface. However, Ross et al. (1977) found that at temperatures above 20°C the gas bladder inflates in 7 to 10 d after hatching. These differences cannot be attributed to salinity. Mansueti (1964) worked with broodfish obtained from a brackish water population, but both Hokanson (1977b) and Ross et al. (1977) used broodfish obtained from fresh water.

Prolarvae reach the larval stage at 13 to 14 mm. Although all of the fins are present within that size range, they are not complete until the larvae are 25 to 30 mm TL (Mansueti 1964). Soon after hatching, the larvae move into the limnetic zone. When they reach 25 to 30 mm TL they return to the littoral zone (Whiteside et al. 1985; Post and McQueen 1989). The median period between swim-up and mortality of unfed larvae increases with a decrease in temperature (Figure 5). In one study, mortality occurred in 9 d at 19.8°C and after 21 d at 10.5°C (Hokanson and Kleiner 1974).

Optimum water velocities for rearing yellow perch larvae in intensive culture systems have not been defined. However, Houde (1969) determined that perch larvae are better swimmers than walleye larvae for length classes less than 9 mm; swimming ability of the 2 species is equal between 9 and 15 mm. Velocities that yellow perch larvae under 9.5 mm can sustain are less than 3 cm/s. Thus, they are poor swimmers. Larger larvae can sustain current velocities of 3 to 4 body lengths per second for at least an hour.

Optimum temperatures for rearing and feeding larval perch are between 20.0 and 23.9°C (Table 1). This range corresponds closely with their thermal preference (McCauley and Read 1973; Ross et al. 1977). Researchers have been relatively unsuccessful in trying to train larval perch less than 18 mm to accept prepared diets, and brine shrimp nauplii do not appear to be an adequate food source for larvae less

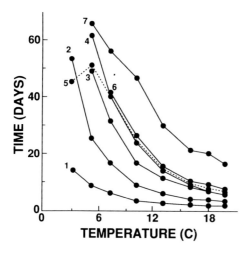

FIGURE 5. Median times for development to various stages of yellow perch embryos and larvae at various temperatures. (1) Neural keel, (2) heart beat, (3) retinal pigmentation, (4) branchial respiration, (5) mass hatch, (6) swim-up larvae, (7) mortality of unfed larvae. (From Hokanson, K. E. F. and Kleiner, C. F., *The Early Life History of Fishes*, Blaxter, J. S., Ed., Springer-Verlag, Heidelberg, 1974. With permission.)

than 13 mm long (Mansueti 1964). However, yellow perch larvae have been reared experimentally on mixed zooplankton obtained from lakes. According to Hale and Carlson (1972), 250 żooplanktonic organisms per larval yellow perch daily are required to obtain 50% survival during the first 3 weeks of feeding. Those authors recommended feeding at least four times daily using fry tanks with dark bottoms. Egestion time for larval perch held between 20 and 23°C is less than 1 h (Hokanson 1977a).

The thermal tolerance of successive embryonic and larval stages of yellow perch increases with morphological differentiation (Hokanson and Kleiner 1974). The optimum temperature for feeding and rearing juvenile perch is 23.9 to 27.8°C (McCormick 1974; McCormick 1976). This is slightly higher than the 20 to 24°C range found by Huh (1975). The upper incipient lethal temperature for juveniles is 29.1°C and for adults is 33°C (Hokanson 1977b). Deformities occurred as 32°C, and all fish died within 7 d at 34°C. Little growth occurred below 8°C (McCormick 1976).

West and Leonard (1978) were able to train 0.38-g yellow perch to accept prepared feed with 38% survival. Stairs (1977) successfully trained an unspecified percentage of 18- to 25-mm yellow perch to accept a prepared ration. Survival is directly related to fingerling size at the beginning of the training period. Fewer than 50% of larvae less than 16 mm can be expected to survive, while 80% of larvae 18 mm long and 98% of those longer than 31 mm have been shown to survive when started on the Spearfish W-7 starter mash and #1 crumble (600 to 800 μm) according to Best (1981).

IV. GROWTH

Mean growth rate of yellow perch from natural populations in North America, calculated from data presented by Carlander (1950), indicates that fish reach 7.4, 13.3, 18.0, 20.4, and 22.9 cm, respectively, during their first 5 years of life. The current market size for yellow perch is about 20 cm (150 g). Based on experimental rearing trials, Calbert and Huh (1976) postulated that 1.0 to 1.5 g yellow perch will reach market size in 9 to 11 months when cultured at 21°C, maintained at 16-h light, and fed 3 to 4% of body weight daily. Those authors estimated that the feed conversion ratio would be about 1.5 on a high quality feed. In terms of growth rate and feed conversion in fish larger than 12 g, they found little difference among feeds containing 27, 40, and 50% protein. However, protein-energy ratios were not held constant in their study.

Photoperiod, in addition to temperature, appears to have a significant influence on growth rate in yellow perch (Huh 1975; Huh et al. 1976). Growth of perch reared at the same temperature, but with a short photoperiod (8 h of light) was only about one third that of fish exposed to photoperiods of 16 or 20 h light (Huh et al. 1976). Perch of 9 to 17 g, reared under 16 h of light at 22°C and fed at 3% of body weight daily on the Spearfish W-3 diet with 3% added gelatin, assimilated 80% of the daily ration and converted 67% of the total food energy into body substance (Huh 1975).

In terms of food intake at 25°C, yellow perch appear to be rather tolerant of oxygen levels down to 3.8 mg/l, as well as to diurnally fluctuating oxygen levels (Carlson et al. 1980). The mean routine oxygen consumption for 1.8 g perch was 0.153, 0.250, and 0.426 mg O_2/h/g at 15, 20, and 25°C (log Y = 1.486 + 0.47x). Active oxygen consumption was approximately 0.500, 0.540, 0.610, and 1.500 mg O_2/h/g at 15, 20, 25, and 30°C (Huh 1975). In tanks, perch over 50 g are not very active.

Female yellow perch grow considerably faster and reach larger ultimate sizes than males (Leach 1928; Carlander 1950). In a laboratory study reported by Schott et al. (1978), females began growing faster than males at 110 mm (15 g). Since the difference in growth rate expresses itself long before the fish reach marketable size, there has been some interest in reversing the phenotypic sex of genetically male yellow perch with hormones. Treatment of 20 to 35 mm perch with estradiol-17β induced complete germ-cell sex inversion in most males. Similarly, 17σ-methyltestosteroine induced spermatogenesis and formation of ovotestes in females (Malison et al. 1986). Malison et al. (1988) presented evidence that estrogens promote growth in yellow perch by stimulating food consumption. Female perch outgrew males because of both greater food consumption and higher food conversion efficiency, and differences in growth between the sexes were not a consequence of intersexual competition for food.

V. CULTURE

Yellow perch fingerlings can be raised in predator-free fertilized rearing ponds for stocking or subsequent growout as food fish. Various methods have been used to fertilize ponds, depending on the location of the culture unit and the availability of inorganic and organic fertilizers. The goal has been to produce a dense bloom of zooplankton without the production of excess vegetation. Few actual fingerling production values appear in the literature; in particular, information on trials wherein attempts have been made to maximize production is almost unavailable. West and Leonard (1978) stocked a 0.23 ha pond with 111,000 fertilized yellow perch eggs in anchored, floating screens (30 × 30 × 3.8 cm) made of 6 mm mesh. They harvested 35,789 fingerlings averaging 0.38 g.

Eggs, obtained from the wild by tanking, spawning, or stripping, can also be incubated in trays such as those found in Heath incubators or incubators of the upwelling type. After hatching, the fry may be stocked into fertilized rearing ponds.

Fingerlings can be harvested from rearing ponds by seining or drawing the pond down to the harvest basin; also, since young perch are positively phototaxic (Schumann 1963), harvest can be accomplished at night with the aid of a light and a lift net. Manci et al. (1983) found that perch from 8 to 50 mm are attracted to light. They successfully removed 61% of the fingerings (23,000) from a 0.08 ha pond with a 132 × 366 cm rectangular lift net constructed of 3-mm white knotless nylon mesh.

As is the case with other fish species, commercial production of yellow perch to marketable size on natural foods is not economically feasible. Thus, it is necessary to train fingerlings to accept prepared feeds. After fingerling yellow perch reach 18 mm in length, 80% of them can be trained to such rations. The training procedure for perch is essentially the same as that used for largemouth bass (Heidinger 1976) or striped bass (Lewis et al. 1981). Significant components of the training procedure include concentrating the fingerlings, removing the natural food source, elevating the temperature to ensure an aggressive feeding response, feeding frequently, and grading fingerlings to reduce cannibalism. As in the case with largemouth bass and striped bass, an initial feed with a soft texture (such as the Oregon Moist Pellet) appears to improve training success. After the fish are trained, a hard pellet may be substituted for the soft diet. Fingerling yellow perch can consume 0.84 to 1.19 mm diameter pellets (Reinitz and Austin 1980).

Trained fingerlings can be placed in ponds or tanks and raised to marketable size. Preliminary research demonstrated that at 20 to 24°C, perch grew at 0.8 to 1.1 cm/month and had monthly food conversion ratios from 1.3 to 3.5 (Starr 1991). In the north central U.S., 2 to 3 years are required for the fish to reach marketable size in ponds on natural foods. A bioenergetics model developed by Kitchell et al. (1977) predicted that yellow perch under ambient (Michigan-Wisconsin) temperature conditions, including winter, should reach marketable size in 15 to 18 months if food is not limiting. In recirculating water systems where temperature and light can be controlled, the rearing period might be reduced by as much as 50%. According to Garling (1991), loading rates in flow-through systems for yellow perch are approximately 1.5 times those for rainbow trout.

Large production studies which include economics are lacking for yellow perch reared in ponds. West and Leonard (1978) estimated that in 1977 dollars it would cost $2.31/kg to produce 23,700 kg of yellow perch in a tank system. Kocurek (1979) estimated that cost at $8.65/kg. Based on the increase in the wholesale price index, those costs in 1985 dollars would have approximately doubled. The economics are not favorable, since the 1984 market value of perch ranged from $1.98 to $3.52/kg. Either the selling price would have to increase significantly or production costs must be reduced.

One problem associated with reducing production costs is that genetically the yellow perch is a relatively small fish, and even though it can be marketed at a small size, it may go through its rapid growth phase before it reaches the desired 150 g. Further, even if one assumes 100% trainable fingerlings and 100% survival, each kilogram of perch produced represents approximately 6.5 fingerlings. Thus, the cost of fingerlings accounts for a considerable percentage of the selling price per kilogram of fish produced.

This argument carries through to the restaurant menu. By hand filleting wild fish and leaving the belly flap intact (butterfly fillet), dress-out yields that average 45% are obtained. With very diligent grading and constant equipment adjustment, machine processing can yield dress-outs of 42% (Lesser 1978), though 37 to 40% is more realistic. The dress-out weight of cultured fish is approximately 5% higher than that of wild fish (Calbert and Huh 1976). Thus, if the mean weight of a serving is 123 g and the mean weight of a butterfly fillet is 64 g, 2 fish are required per serving. Assuming a training success of 80% and 80% survival of trained fish, the cost of producing 3 fingerlings must be reflected in each serving.

VI. CONCLUSION

State and federal fish hatcheries can produce yellow perch fry and fingerlings for use in fish management. The economic feasibility of commercially culturing yellow perch for the food fish market has not been demonstrated in scientific literature. It is doubtful that the demand for yellow perch will elevate the price sufficiently to make commercial culture economically feasible in recirculating water systems; however, neither the market, nor the price that the fish would command outside the states bordering the Great Lakes, has been tested. Also, the feasibility and costs associated with rearing yellow perch to market size in ponds or cages under various ambient temperature conditions remains to be examined.

REFERENCES

Best, C. D., Initiation of Artificial Feeding and the Control of Sex Differentiation in Yellow Perch, *Perca flavescens*, Master's thesis, University of Wisconsin, Madison, 1981, 1.

Brazo, D. C., Tack, D. I., and Liston, C. R., Age, growth, and fecundity of yellow perch, *Perca flavescens*, in Lake Michigan, *Trans. Am. Fish. Soc.*, 104, 726, 1975.

Calbert, H. E. and Huh, H. T., Culturing yellow perch *Perca flavescens* under controlled environmental conditions for the upper midwest market, *Proc. World Maricult. Soc.*, 7, 137, 1976.

Carlander, K. D., *Handbook of Freshwater Fisheries Biology*, Wm. C. Brown, Dubuque, IA, 1950, 1.

Carlson, A. R., Blocker, J., and Herman, L. J., Growth and survival of channel catfish and yellow perch exposed to lowered constant and diurnally fluctuating dissolved oxygen concentrations, *Prog. Fish-Cult.*, 42, 73, 1980.

Clady, M. D., Influence of temperature and wind on the survival of early stages of perch, *Perca flavescens*, *J. Fish. Res. Bd. Can.*, 33, 1187, 1976.

Clugston, J. P., Oliver, J. L., and Ruell, R., Reproduction, growth, and standing crops of yellow perch in southern reservoirs, in *Selected Coolwater Fishes of North America*, Special Publ. No. 11, Kendall, R. L., Ed., American Fisheries Society, Washington, D.C., 1978, 89.

Collette, B. B. and Banarescu, P., Systematics and zoogeography of the family Percidae, *J. Fish. Res. Bd. Can.*, 34, 1450, 1977.

Conover, M. C., Stocking cool-water species to meet management needs, in *Fish Culture in Fisheries Management*, Stroud, R.H., Ed., American Fisheries Society, Bethesda, Maryland, 1986, 31.

DeManno, D. A. and Goetz, F. W., The effects of of forskolin, camp, and cyanoketone on steroid-induced meiotic maturation of yellow perch (*Perca flavescens*) oocytes *in vitro*, *Gen. Comp. Endocrinol.*, 66, 233, 1987.

Downs, W., Wisconsin: the dairy state takes a look at fish farming. Raising perch for the midwest market, *Comm. Fish. Farm.*, 1(5), 27, 1975.

Follett, R., Raising Perch for the Midwest Market, Advisory Report 13, University of Wisconsin Sea Grant Program, Madison, 1975, 1.

Garling, D. L., NCRAC research programs to enhance the potential of yellow perch culture in the North Central Region, in Proc. North Central Regional Aquaculture Conf., Kalamazoo, MI, March 18 to 21, 1991, 253.

Goetz, F. W. and Bergman, H. L., The effects of steroids on final maturation and ovulation of oocytes from brook trout (*Salvelinus fontinalis*) and yellow perch (*Perca flavescens*), *Biol. Reprod.*, 18, 293, 1978a.

Goetz, F. W. and Bergman, H. L., The *in vitro* effects of mammalian and piscine gonadotropin and pituitary preparations on the final maturation in yellow perch (*Perca flavescens*) and walleye (*Stizostedion vitreum*), *Can. J. Zool.*, 56, 348, 1978b.

Hale, J. G. and Carlson, A. R., Culture of the yellow perch in the laboratory, *Prog. Fish-Cult.,* 34, 195, 1972.

Harrington, R. W., Observations on the breeding habits of the yellow perch, *Perca flavescens* (Mitchill), *Copeia,* 1947, 199, 1947.

Heidinger, R. C., Synopsis of Biological Data on the Largemouth Bass *Micropterus salmoides* (Lacepede) 1802, FAO Fisheries Synopsis No. 115, Food and Agriculture Organization, Rome, 1976, 1.

Hergenrader, G.L., Spawning behavior of *Perca flavescens* in aquaria, *Copeia,* 1969, 839, 1969.

Herman, E., Wisly, W., Wiegert, L., and Burdick, M., The Yellow Perch: Its Life History, Ecology, and Management, Wisconsin Conservation Department Publ. 228, 1959, 1.

Hokanson, K. E. F., Optimum culture requirements of early life phases of yellow perch, in Perch Fingerling Production for Aquaculture, Soderberg, R. W., Ed., University of Wisconsin Sea Grant Advisory Rep. 421, Madison, 1977a, 24.

Hokanson, K. E. F., Temperature requirements of some percids and adaptations to the seasonal temperature cycle, *J. Fish. Res. Bd. Can.,* 34, 1524, 1977b.

Hokanson, K. E. F. and Kleiner, C. F., Effects of constant and rising temperatures on survival and developmental rates of embryonic and larval yellow perch, *Perca flavescens* (Mitchill), in *The Early Life History of Fish,* Blaxter, J. S., Ed., Springer-Verlag, New York, 1974, 437.

Houde, E. D., Sustained swimming ability of larvae of walleye (*Stizostedion vitreum*) and yellow perch (*Perca flavescens*), *J. Fish. Res. Bd. Can.,* 26, 1647, 1969.

Huh, H. T., Bioenergetics of Food Conversion and Growth of Yellow Perch (*Perca flavescens*) and Walleye (*Stizostedion vitreum vitreum*) Using Formulated Diets, Ph.D. dissertation, University of Wisconsin, Madison, 1975, 1.

Huh, H. T., Calbert, H. E., and Stuiber, D. A., Effects of temperature and light on growth of yellow perch and walleye using formulated feed, *Trans. Am. Fish. Soc.,* 105, 254, 1976.

Hutchinson, B., Yellow Perch Egg and Prolarvae Mortality, Fish and Wildlife Restoration Project Report, F-17-R-17, Cornell University, Ithaca, NY, 1974, 1.

Kayes, T., Reproductive biology and artificial propagation methods for adult perch, in Perch Fingerling Production for Aquaculture, Advisory Report 421, Soderberg, R. W., Ed., University of Wisconsin Sea Grant Program, Madison, 1977, 6.

Kayes, T. B. and Calbert, H. E., Effects of photoperiod and temperature on the spawning of yellow perch (*Perca flavescens*), *Proc. World Maricult. Soc.,* 10, 306, 1979.

Kitchell, J. F., Steward, D. J., and Weininger, D., Applications of a bioenergetics model to yellow perch (*Perca flavescens*) and walleye (*Stizostedion vitreum vitreum*), *J. Fish. Res. Bd. Can.,* 34, 1922, 1977.

Kocurek, D., An Economic Study of a Recirculating Perch Aquaculture System, Master's thesis, University of Wisconsin, Madison, 1979, 1.

Koenig, S. T., Kayes, T. B., and Calbert, H. E., Preliminary observations on the sperm of yellow perch, in *Selected Coolwater Fishes of North America,* Special Publ. No. 11, Kendall, R. L., Ed., American Fisheries Society, Washington, D.C., 1978, 177.

Lagler, K. F., Bardach, J. E., and Miller, R. R., *Ichthyology,* John Wiley & Sons, New York, 1962, 1.

Leach, G. C., Propagation and Distribution of Food Fishes, Fiscal Year 1927, Report to the U.S. Commissioner of Fisheries, U. S. Government Printing Office, Washington, D.C., 1928, 683.

Lesser, W. H., Marketing Systems for Warm Water Aquaculture Species in the Upper Midwest, Ph.D. dissertation, University of Wisconsin, Madison, 1978, 1.

Lesser, W. H. and Vilstrup, R., The Supply and Demand for Yellow Perch 1915—1990, University of Wisconsin College of Agriculture Life Science Res. Bull. R3006, 1979, 1.

Lewis, W. M., Heidinger, R. C., and Tetzlaff, B. L., Tank Culture of Striped Bass, Fisheries Research Laboratory, Southern Illinois University, Carbondale, 1981, 1.

Malison, J. A., Kayes, T. B., Best, C. D., Amundson, C. H., and Wentworth, B. C., Sexual differentiation and use of hormones to control sex in yellow perch (*Perca flavescens*), *Can. J. Fish. Aquat. Sci.,* 43, 26, 1986.

Malison, J. A., Kayes, T. B., Wentworth, B. C., and Amundson, C.H., Growth and feeding response of male versus female yellow perch (*Perca flavescens*) treated with extradiol-17β, *Can. J. Fish. Aquat. Sci.,* 45, 1943, 1988.

Manci, W. E., Malison, J. A., Kayes, T. B., and Kuczynaki, T. E., Harvesting photopositive juvenile fish from a pond using a lift net and light, *Aquaculture,* 34, 157, 1983.

Mansueti, A. J., Early development of the yellow perch, *Perca flavescens, Chesapeake Sci.,* 5, 46, 1964.

McCauley, R. W. and Read, L. A. A., Temperature selection by juvenile and adult yellow perch (*Perca flavescens*) acclimated to 24°C, *J. Fish. Res. Bd. Can.,* 30, 1253, 1973.

McCormick, J. H., Temperatures Suitable for the Well-Being of Juvenile Yellow Perch During their First Summer-Growing Season, Annual Report, ROAP-16AB1, National Water Quality Laboratory, Duluth, MN, 1974, 1.

McCormick, J. H., Temperature Effects on Young Yellow Perch, *Perca flavescens* (Mitchill), EPA-600/3-76-057, U.S. Environmental Protection Agency Ecology Research Service, Washington, D.C., 1976, 1.

McPhail, J. D. and Lindsey, C. C., Freshwater fishes of northwestern Canada and Alaska, *J. Fish. Res. Bd. Can.,* 173, 1, 1970.

Muncy, R. J., Life history of the yellow perch, *Perca flavescens,* in estuarine waters of Seven River, a tributary of Chesapeake Bay, Maryland, *Chesapeake Sci.,* 6, 545, 1962.

Newsome, G. E. and Tompkins, J., Yellow perch egg masses deter predators, *Can. J. Zool.,* 63, 2882, 1985.

Post, J. R. and McQueen, D. J., Ontogenetic changes in the distribution of larval and juvenile yellow perch (*Perca flavescens*): a response to prey or predators?, *Can. J. Fish. Aquat. Sci.,* 45, 1820, 1989.

Reinitz, G. and Austin, R., Experimental diets for intensive culture of yellow perch, *Prog. Fish-Cult.,* 42, 29, 1980.

Robins, R., *A List of Common and Scientific Names of Fishes from the United States and Canada,* American Fisheries Society, Bethesda, MD, 1991, 1.

Ross, J., Powles, P. M., and Berrill, M., Thermal selection and related behavior in larval yellow perch (*Perca flavescens*), *Can. Field Nat.,* 91, 406, 1983.

Schott, E. F., Kayes, T. B., and Calbert, H. E., Comparative growth of male versus female yellow perch fingerlings under controlled environmental conditions, in *Selected Coolwater Fishes of North America,* Special Publ. No. 11, Kendall, R. L., Ed., American Fisheries Society, Washington, D.C., 1978, 181.

Schumann, G. O., Artificial light to attract young perch: a new method of augmenting the food supply of predacious fish fry in hatcheries, *Prog. Fish-Cult.,* 25, 171, 1963.

Scott, W. B. and Crossman, E. J., Provisional Checklist of Canadian Freshwater Fishes, Information Leaflet, Department Ichthyology and Herpetology, Royal Ontario Museum, July, 1967, 1.

Scott, W. B. and Crossman, E. J., Freshwater fishes of Canada, *Fish. Res. Bd. Can. Bull.,* 184, 1973, 1.

Sheri, A. N. and Posner, G., Fecundity of the yellow perch, *Perca flavescens* (Mitchill), in the Bay of Quinte, Lake Ontario, *Can. J. Zool.,* 47, 55, 1969.

Stairs, S., Experience in perch fingerling production from the Lake Mills National Fish Hatchery, in Perch Fingerling Production for Aquaculture, Soderberg, R. W., Ed., University of Wisconsin Sea Grant Advisory Report 421, Madison, 1977, 58.

Starr, C. J., Commercial production of yellow perch (*Perca flavescens*), in proc. North Central Regional Aquaculture Conferences, Kalamazoo, MI, March 18 to 21, 1991, 256.

Svetovidov, A. N. and Dorofeeva, E. A., Systematics, origin, and history of the distribution of the Eurasian and North American perches and pike-perches (genera *Perca, Lucioperca,* and *Stizostedion*), *Vopr. Ikhtiol.,* 3, 625, 1963.

Thorpe, J. E., Morphology, physiology, behavior, and ecology of *Perca fluviatilis* L. and *P. flavescens* Mitchill, *J. Fish. Res. Bd. Can.,* 34, 1504, 1977.

West, G. and Leonard, J., Culture of yellow perch with emphasis on development of eggs and fry, in *Selected Coolwater Fishes of North America*, Kendall, R. L., Ed., Special Publ. No. 11, American Fisheries Society, Washington, D.C., 1978, 172.

Whiteside, M. C., Swindoll, C. M., and Doolittle, W. L., Factors affecting the early life history of yellow perch, *Perca flavescens, Environ. Biol. Fish.*, 12, 47, 1985.

Wiggins, T. A., Bender, T. R., Mudrak, V. A., and Takacs, M. A., Hybridization of yellow perch and walleye, *Prog. Fish-Cult.*, 45, 131, 1983.

Worth, S. G., Observations on hatching of yellow perch, *Bull. U.S. Fish Comm. for 1890*, 10, 331, 1892.

Chapter 8

WALLEYE

John G. Nickum and Robert R. Stickney

TABLE OF CONTENTS

8633-9/93/$0.00 + $.50

I. INTRODUCTION

In North America, two fishes within the family Percidae have been the focus of a considerable amount of research by fish culturists. One of those, the yellow perch (*Perca flavescens*), was considered in Chapter 7. The other, the walleye (*Stizostedion vitreum*), is considered here. Two subspecies of *S. vitreum*, *S. v. vitreum* (walleye) and *S. v. glaucum* (blue pike) have been recognized (Robins 1991), but aquaculture attention has been directed only on the walleye. Culture interest has also developed on the sauger (*S. canadense*) and saugeye (hybrid between *S. canadense* and *S. v. vitreum*), though relatively little information is available on those fishes. Other common names for the walleye are yellow walleye, pickerel, yellow pickerel, pikeperch, yellow pikeperch, walleye pike, and yellow pike (Colby et al. 1979).

Historically, the distribution of the walleye was in the freshwaters of Canada from Quebec eastward. In the U.S., walleye were native from the St. Lawrence River south to the Gulf coast of Alabama, but not east of the Appalachians. The western extent of the range of the species in the U.S. was east of a line from the western border of Alabama through the northeastern corner of Montana, essentially following the northern boreal and central and southern hardwood forests (Scott and Crossman 1973).

The walleye is one of the most valuable and sought-after fishes in North America, both as a commercial species and as a game fish. Commercial production from Lake Erie alone totalled nearly 14,000 tons in the late 1950s (Baldwin and Sealfeld 1962). States such as Minnesota, Wisconsin, and Michigan that feature walleye fishing consistently rank near the top in both resident and nonresident angling licenses sold. Walleye have been introduced into lakes throughout much of the U.S. Most culture has been by state and federal hatcheries for the purpose of stocking recreational fishing waters. Much of the emphasis on walleye production will continue in that vein; however, the popularity of walleye with consumers implies that a foodfish market could also be established if economical culture of edible fish becomes possible.

Techniques for taking and incubating walleye eggs stripped from wild-cultured fish are well established. Although the origins of that aspect of walleye culture have not been recorded with certainty, it is known that fry have been used in stocking programs for over 100 years (Webster et al. 1978). Cobb (1923) described walleye propagation in Minnesota nearly 70 years ago, and Nevin (1887) discussed techniques for hatching the species at the 16th Annual Meeting of the American Fisheries Society over 100 years ago. The general procedures employed by early culturists have not changed substantially in the intervening decades.

Management needs in some areas call for the production of fingerling walleye of large sizes. Until recently, the only system available for rearing walleyes to fingerling size was pond culture. However, neither large numbers, nor even predictable production can be obtained from that method. In spite of somewhat erratic pond production results, most states continue to produce walleyes in that type of culture environment.

The demand for more and larger walleyes for sport fishery management, as well as the potential for commercial fish farming, have produced a strong interest in intensive walleye culture. That methodology holds potential for rearing walleyes to any desired size, though certain breakthroughs in technology will be required before economical intensive culture becomes a reality. Basic techniques for intensive culture were described by Nickum (1978), and experimental work has continued in the interim.

Since successful rearing of first-feeding fry under intensive culture conditions has not been economically accomplished, all intensive walleye culture has been based on pond rearing until the fish reach at least 2.5 cm. Thereafter, the fish may be offered formulated feeds.

Although it was predicted that intensive walleye culture would be well established by 1988 (Nickum 1978), that objective was not reached, even though interest in walleye culture has remained high, and some advances have been achieved. We remain optimistic that further advances, which should be possible over the next several years, can raise large-scale walleye culture from the level of fry production only to that wherein fish of various larger sizes can be routinely produced.

II. LIFE HISTORY

A. ENVIRONMENTAL CONDITIONS

Walleye juveniles and adults can be found in lakes above the thermocline. They often occur over relatively deep waters during summer and move inshore in early autumn as water temperatures begin to fall (Johnson 1969; Kelso 1976). The species can be found over a temperature range from 0 to 30°C, though the preferred range is from about 20 to 23°C (Ferguson 1958). Like other percid fishes, walleye appear to have a life cycle which is adapted to temperate climates (Hokanson 1977).

In one study (Johnson et al. 1988), it was determined that walleyes occurred primarily over coarse sediments in the fall and fine ones during early summer. The same pattern was followed by saugeye. In addition, both walleye and saugeye showed movement offshore when temperatures were above 22°C.

Adult walleye are negatively phototaxic (Scherer 1976) and may spend most of the daylight period in contact with the substrate or hiding under various objects in the water (Ryder 1977). Walleyes feed primarily at night in shallow water (Ali and Anctil 1968; Ryder 1977) and migrate to shoal areas diurnally.

Walleye are able to tolerate a wide range of turbidity (Ryder 1977) and often occur in highly colored lakes that are rich in humic acids (Baldwin and Sealfeld 1962). They tolerate dissolved oxygen (DO) concentrations as low as 2.0 mg/l in the laboratory (Scherer 1971; Petit 1973), though in nature walleyes are generally found at DO levels above 3.0 mg/l (Dendy 1948).

The effects of ammonia on walleye have been examined in a number of studies (Alexander et al. 1986; Mayes et al. 1986; Arthur et al. 1987; Hermanutz et al. 1987). The 96-h LC_{50} value for unionized ammonia was placed at 1.06 mg/l by Mayes et al. (1986). Based on studies by Alexander et al. (1986), Michigan placed the acute water quality criterion for unionized ammonia in the waters of that state at 1.0 mg/l. Walleye typically occur over a pH range of 6 to 9 (Scherer 1971).

B. REPRODUCTION

Maturity in walleyes depends on water temperature and may also be a function of food availability (Baldwin and Sealfeld 1962). In general, males mature at 2 to 4 years and females between 3 and 6 years of age. Sizes at maturity are >279 mm total length (TL) in males and 356 to 432 mm TL in females (Scott and Crossman 1973).

Spawning behavior in nature may vary considerably and has been described by Eschmeyer (1950) for lakes, Ellis and Giles (1965) for streams, and Priegel (1970) for marshes. Group spawning appears to be a common phenomenon for walleye.

Spawning occurs in shallow water — often in <1 m of water in lakes, but in up to 6.1 m in at least one Mississippi River pool — and over various bottom types (Eschmeyer 1950; Pitlo 1989). Milt and eggs are spawned into the water column, where fertilization subsequently occurs (Colby et al. 1979).

The number of eggs per unit body weight in females is relatively constant within a given population of fish, but has been found to vary from 28,000/kg (Smith 1941; Arnold 1960) to over 120,000/kg (Muench 1966; Wolfert 1969). Values of approximately 60,000 eggs per kilogram body weight are typical throughout the walleye's range.

The spawning season may begin as early as January (Cook 1959) and in some regions may not be completed until June (Scott and Crossman 1973), depending upon latitude. In some northern parts of their range, walleye may not spawn during years when temperatures are unfavorably cold (Scott and Crossman 1973). In general, spawning begins shortly after the ice breaks up in spring (Colby et al. 1979). A so-called chill period may be required to induce spawning as walleyes not exposed to certain minimum temperatures — below about 10°C in at least some instances (Colby et al. 1979) — will not reproduce.

Walleye eggs are adhesive when spawned. The adhesiveness lasts for an hour or more (Nelson et al. 1965; Niemuth et al. 1966) during which the eggs become water-hardened (Niemuth et al. 1966). Development rate is dependent upon temperature. As reviewed by Colby et al. (1979), 10 d are required for hatching at 12.8°C, while the eggs will hatch in 4 d at 23.9°C. Best hatching-success rates have been achieved at intermediate temperatures, though incubation at the low end of the above range leads to a higher percentage hatch than at the high end. Koenst and Smith (1976) obtained greatest hatching success at 6.0°C, though no fry survived the period of yolk-sac absorption. Those authors obtained best survival from fertilization through yolk-sac absorption at temperatures between 8.9 and 12°C.

A great deal of variability in walleye egg viability has been reported. For example, instances of hatch rates as low as 3.4% (Baker and Scholl 1969) and as high as 100% (Johnson 1961) have been observed. Some of the observed variation undoubtedly relates to water quality. Prentice and Dean (1977) found that water temperatures above 12°C led to reduced hatching rates. However, hatching rates were significantly increased if eggs that were fertilized at temperatures above 12°C

were chilled to 7.2°C for incubation. On the other hand, Smith and Koenst (1975) indicated that while optimum fertilization temperatures ranged from 6 to 12°C, optimum incubation temperatures ranged from 12 to 15°C.

Other water quality variables that can influence hatching success include DO. Siefert and Spoor (1973) found that embryos and larvae of walleye could tolerate levels of 50% DO saturation, but that mortality increased sharply when the DO saturation fell to 35%. Oseid and Smith (1971) found that the time from incubation to hatching was reduced as DO dropped from 7 to 2 ppm. They also indicated that the influence of DO on time to hatching is greater at low incubation temperatures.

Smith and Oseid (1972) suggested that the safe level of hydrogen sulfide to which walleye eggs and fry can be safely exposed should be less than 0.006 mg/l, though the eggs could tolerate a higher level. High mortality of walleye eggs (90.5%) has been reported when incubation occurred at pH 5.4. Lower mortalities (25.5 to 33.5%) occurred in water of pH 6.0 (Hulsman et al. 1983).

C. NATURAL FOOD

Walleye fingerlings and adults are highly piscivorous except during late spring and early summer when invertebrates may be important food for fingerlings (Colby et al 1979). As reviewed by Colby et al. (1979), it has been repeatedly demonstrated that young-of-the-year yellow perch are a primary prey species in instances where perch are available. In the absence of yellow perch, various other fish species are consumed.

Limited information is available on first-feeding walleye fry, but there have been reports that fry up to 7 or 8 mm consume phytoplankton extensively in Lake Erie (Hohn 1966; Paulus 1969). Mathias and Li (1982), on the other hand, reported that the primary food of postlarval walleye was crustacean zooplankton and that the gill rakers on the fish were not sufficiently developed to filter food until the fish were from 20 to 30 mm in length. As reviewed by Colby et al. (1979), studies in rearing ponds have shown that various types of zooplankton are the primary food of walleyes from 5 to 9 mm in total length. Selectivity of zooplankton by walleye fry was studied in the laboratory by Raisaen and Applegate (1983), as well as by Mathias and Li (1982).

III. CULTURE

A. SPAWNING AND HATCHING

Captive walleye broodstock have yet to be produced for routine use by culturists, so most hatcheries obtain their broodfish from the wild. Spawners are obtained most often from trap-nets set in selected locations in streams or lakes (Colby et al. 1979). Spawning generally occurs at water temperatures from 6.7 to 8.9°C (Scott and Crossman 1973), so temperature recommendations for setting nets to trap walleye broodstock (7.2 to 10.0°C) which date back to the first quarter of the present century (Cobb 1923) are generally appropriate.

Collections of spawners are made at sites known to be natural spawning grounds. Walleyes typically spawn in relatively shallow water over substrates of gravel and/or rubble. Water as deep as 4.5 m and substrates of sand (Eschmeyer 1950) have been reported as spawning sites, however. The nature of the bottom and water movement and exchange must be sufficient to provide adequate oxygen for the developing embryos; thus, spawning seldom occurs over silt or mud bottoms.

Males generally arrive over the spawning grounds before females and may remain for several days after the females have departed. Even collections of adults made at the peak of spawning activity are often dominated by males.

Broodfish are commonly stripped at the site of capture, though if unripe, they may be held in pens or tanks until sexual maturity is reached (Colby et al. 1979). Many hatcheries have developed facilities in which to hold recently captured fish so that spawning at the peak of development can be achieved. Holding broodfish also ensures that sufficient numbers of both sexes are available.

Experienced walleye culturists can identify mature females on the basis of general body shape, particularly the distended, somewhat softer belly as compared with males. However, no absolutely reliable external characteristics for separation of the sexes have been identified.

No special procedures beyond normal care to minimize stress are used in the transport and holding of walleye spawners. While most culturists do not employ hormone injections and no widely accepted methodology has been developed, ovulation in wild-captured prespawning adult females can be induced by injections of acid-dried carp pituitary (Nelson et al. 1965; Lessman 1978) and human chorionic gonadotropin (HCG) as reported by Lessman (1978). Injections of 7.7 to 22 mg/kg of carp pituitary at 72-h intervals or HCG at the rate of 152 IU/kg are effective in inducing ovulation (Hearn 1980). Pond-reared walleyes can also be induced with HCG (Hearn 1980). If attempts are made to force ovulation with hormones too far in advance of the time of natural spawning, the eggs will flow freely but will not be viable (Heidinger 1985).

Eggs and sperm may be mixed in a dry pan (dry method) or the pan may be dipped in water and shaken relatively free of water film before eggs are added (wet method). The milt of two or more males is added to the pan of eggs in either method and the mixture is stirred. Olson (1971) reported that maximum fertilization could be achieved by mixing the eggs with water before the milt is added.

To prevent the eggs from clumping due to their adhesiveness, compounds such as starch, bentonite, tannic acid, and protease solutions can be added to the eggs during the water-hardening period (Davis 1953; Dumas and Brand 1972; Waltemyer 1975, 1977; Colesante and Youmans 1983; Krise et al. 1986; Krise 1988). Fuller's earth may be added at 35 to 40 g/l of water in a slurry that is poured over the eggs and stirred for approximately 5 min, after which it is rinsed free. Waltemyer (1975) recommended adding tannic acid after fertilization and an initial rinse to remove excess milt. Stirring during exposure for 2 min to tannic acid was found to enhance the effect of the chemical (Waltemyer 1977). Colesante and Youmans (1983) recommended exposing walleye eggs for 2 to 3 min to a 400 mg/l tannic acid solution. Water-hardening in a 0.01% protease solution has been shown effective (Krise et al. 1986), and if the concentration of protease is increased to 0.1%, treatment time can be reduced from 30 to 5 min (Krise 1988). Exposure for 1 min to a protease solution of 1.0% is lethal to walleye eggs (Krise 1988).

Once the eggs become water-hardened (1 to 2 h), they are rinsed and placed in hatching jars such as Downing jars (Colesante and Youmans 1983). Water flows (3 to 6 l/min) are adjusted to produce a gentle "rolling" movement of eggs throughout the 3.8-l jars. Hatchery workers in Pennsylvania add a layer of gravel to the bottom of each jar, which produces a diffuse upwelling of water without causing the eggs to roll. Hatching rates using that technique are similar to those obtained in other hatcheries. The volume of eggs added to each jar is largely a matter of personal preference; however, jars are seldom filled to over 75% of total volume. The number of eggs present in each jar can be estimated by volumetric procedures, though walleye eggs seem to vary slightly in size from stock to stock and as a function of female size and condition; therefore, counts of known volumes from each lot of eggs should be made if accurate estimates of egg numbers are desired. Values of 30,000 to 35,000 eggs per liter are typical.

The length of incubation time required for hatching is temperature dependent, as previously discussed. The range is from about 3 weeks at 10°C to 1 week at 20°C (Koenst and Smith 1976). Many walleye hatcheries use surface water supplies with variable temperatures. Other local conditions may also affect the length of time required for hatching; therefore, no fully reliable guide can be offered. Colby et al. (1979) included information derived from the literature on incubation times, but did not develop standards based on temperature or other universal units. Hatching success rates of 55 to 70% are typical. Methods used to estimate egg and fry numbers have been less accurate than may be desirable, so while hatching rates approaching 100% have been reported, they have not be verified.

Walleye eggs do not require substantial care during incubation. It is not standard practice to remove dead eggs from hatching jars, though clumps of dead eggs held together by the hyphae of fungi should be manually removed. Some culturists apply routine prophylactic treatments with various fungicides to reduce egg loss. For example, workers in Pennsylvania hatcheries employ a 1:600 formalin bath for 17 min daily as a means of controlling fungus on eggs. However, no specific procedure or treatment can be recommended for general use. Jars should be observed several times daily for flow rate, egg condition, and the stage of larval development.

Larvae may be transferred from hatching jars to holding tanks. In some hatcheries, they are allowed to swim out of the jars and into holding tanks. Larvae may be transported immediately after hatching and are often stocked into rearing ponds or lakes at swim-up. Some managers prefer to wait 1 to 3 d following hatching before stocking walleye. Larval walleye must obtain adequate food within 3 to 5 d posthatch or they will reach a level of starvation that is irreversible even if abundant supplies of food are subsequently provided (Jahncke 1981). The time required for yolk-sac absorption and initiation of feeding is related to temperature; for example the critical time for onset of feeding is 3 d at 20°C.

B. POND CULTURE TECHNIQUES

Culture of walleyes in ponds is dependent upon the production of adequate supplies of food for the young fish. Walleye fry generally feed on zooplankton (Smith and Moyle 1945; Dobie and Moyle 1956; Houde 1967); however, diatoms have also been found as major stomach content items in fry recovered from Lake

Erie (Hohn 1966; Paulus 1969). Walleye often become piscivorous when they reach lengths of 6 to 8 cm, though Walker and Applegate (1976) found that the fish would remain planktivorous up to 9 cm or larger if abundant supplies of zooplankton were available.

Management of ponds to produce and maintain large zooplankton populations is still a mixture of art and science. Many biotic and abiotic variables interact to affect plankton community dynamics, so fish culturists must be able to analyze, interpret, and then manipulate conditions in their ponds on a continuing basis. Dobie and Moyle (1956) and Dobie (1971) discussed fertilization techniques that substantially increased walleye production in drainable Minnesota ponds. The techniques were based primarily on maintaining more than 4% organic matter in pond bottom soils. The use of organic fertilizers such as sheep and other barnyard manures was recommended. Fox et al. (1989) examined walleye performance in ponds treated with fermented soybean meal at both a constant and a progressively reduced rate of application. They found that chironomid larvae and pupae were the dominant prey in terms of biomass in walleye larger than 22 ml TL. Fish in the ponds that received constant addition of the fertilizer (32 g/m^3 weekly) performed best because of the greater abundance of chironomids. Mean zooplankton density did not vary significantly between the treatments.

Walleye culturists responsible for pond-rearing often use various combinations of inorganic and organic fertilizers in a rather subjective manner to manipulate plankton densities. Each set of ponds, and even individual ponds within a series, seem to require independent management. The techniques employed for individual ponds are often dependent upon the experience of the culturist. Even after many years of experience, the culturist is often unable to predict fingerling production from year to year or from pond to pond. The latest techniques for pond fertilization, including the inoculation of desirable zooplankton, were reviewed by Buttner (1989).

Cannibalism is a major source of fry loss in walleye ponds. Cannibalism begins in larval walleye (Loadman et al. 1986) and increases as zooplankton populations diminish and as differential growth creates substantial variation in fish size within ponds. Harvest before fingerlings reach 6 to 8 cm can increase survival and size uniformity.

Walleye fry typically reach 5 to 6 cm in 6 weeks or less, depending upon stocking density and food supply in the pond. Lengths of 20 to 25 cm can be obtained in 12 weeks; however, the numbers of fish harvested of that size are generally quite low.

Ponds are usually stocked with at least 125,000 fry per hectare. Stocking rates as high as 375,000 fry per hectare have been used, but growth rates and survival tend to be poor when the higher stocking rate is employed. Survival to harvest can exceed 50% when pond fertility is high and the fish are harvested before they reach 5 cm. Survival to harvest of fish 10 cm or larger can be less than 1%, particularly when pond fertility is poor. Li and Ayles (1981) reported survival in 11 ponds ranged from 1.2 to 18% in fish ranging from 10.2 to 16.4 cm. Cheshire and Steel (1972) documented a negative logarithmic relationship between survival and fish length at harvest for pond-reared walleye.

A technique whereby newly hatched walleye fry were reared in circular net enclosures placed in a lake was described by Brugge and McQueen (1991). Survival to 50 d of age ranged from 25.4 to 45.3%. The authors indicated that the technique might be particularly applicable to private groups interested in walleye enhancement and restoration.

Harvest of walleye fingerlings from ponds usually takes place during the summer. Considerable care is required to minimize stress and prevent both severe immediate and delayed losses resulting from handling. The following recommendations are based upon unpublished experiences and observations at hatcheries in Iowa, New York, and Pennsylvania.

1. Ponds, whether drainable or not, should be free of filamentous algae and macrophytes. Many fish tend to become trapped in vegetation during draining and seining operations.
2. Harvest operations should be conducted when water temperatures are below 20°C, if possible. If fish are concentrated in a catch basin, a flow of water through the basin will greatly reduce stress.
3. Harvested fish should be immediately placed in a solution of 0.5% NaCl, and an appropriate antibacterial agent should be added to the water.
4. DO concentrations in all tanks, tubs, pails, or other transport containers should be maintained at 4 mg/l or above.

Given the above conditions, walleyes can be harvested and transported for up to 6 h with no appreciable losses according to studies conducted by the Iowa Cooperative Fishery Research Unit. The importance of extreme care in the harvest and transportation of walleyes has not been fully appreciated and cannot be overemphasized. Too often it has been assumed that as long as the fish were alive when stocked, they would survive. There is no established basis for such an assumption.

C. INTENSIVE CULTURE METHODS

A comprehensive review with respect to intensive culture of walleyes was prepared by Nickum (1978). The discussion that follows draws from that material and is supplemented, as cited, with more recent information. The discussion also incorporates the results of unpublished studies that have been conducted since the original review.

The National Task Force for Public Fish Hatchery Policy (1974) identified "the inability to rear the tiny delicate larvae of species like striped bass and walleye on artificial diets [as] the most critical bottleneck in the national fish-culture program." Over a decade later, Krise and Meade (1986) indicated that intensive walleye culture could soon be a reality as a result of research on larval development. Yet, while improvements are being made each year many of the techniques still utilized today were developed in the 1970s (McCauley 1970; Graves 1974; Nagel 1974; Beyerle 1975; Nagel 1976; Huh et al. 1976).

1. Fry Culture

Although progress has been made toward understanding the factors that affect feeding and the general culture requirements of walleye larvae, survival rates have commonly reached only a few percent, and frequently 1% or less, through the first month posthatch when prepared feeds have been used. A variety of factors seem to affect the feeding of intensively reared fry, but one can only speculate on the mechanisms that control unsatisfactory growth and poor survival. Rearing units, stocking densities, feeds, feeding practices, and the physical conditions that exist have all been implicated as important considerations.

a. Physical Conditions and Facilities

A variety of rearing units have been evaluated in intensive walleye culture situations. They range from so-called "standard" start-troughs (about $3 \times 0.4 \times 0.15$ m) to hatching jars. No particular unit can be recommended at this time, since no practical method for intensive culture has been developed. There are, however, a number of physical factors that have been demonstrated to affect larval walleyes.

Corazza and Nickum (1983) found that walleye larvae are so strongly attracted to the sides of light-colored rearing units that they ignore all forms of live or formulated feeds. Fry in uniformly lighted units with dark or neutral-colored sides become more uniformly distributed and feed actively. Light manipulation and the use of tanks with darkened sides have been found to significantly improve survival of walleye fry reared on brine shrimp nauplii and zooplankton (Colesante 1989).

Walleye fry feed most actively when they are within the water column of their rearing unit rather than on the surface or near the bottom. They also seem to have a relatively short search-radius in which they can recognize and ingest food particles. It seems important, therefore, to use rearing units designed to maintain feed particles in suspension. Various upwelling systems, usually with perforated baffle plates in the bottom, have been used to accomplish feed suspension without producing strong currents that might cause battering of the fish.

The outlets of rearing units should be screened with a fine-meshed material so larval walleye (7 to 8 mm long) do not escape. Kindschi and Barrows (1991) recommended using screens with no more than 710 µm between threads and no more than 53% open area for the containment of young walleye being fed formulated feed. Since feed and fecal material rapidly accumulate on such screens, frequent cleaning may be required to prevent clogging and subsequent overflow with resultant loss of fish. Water exchange rates of at least twice hourly seem to reduce fouling and disease problems.

No controlled, replicated studies have been conducted to test the effects of temperature on the feeding behavior of walleye fry. In natural environments fry may be assumed to begin feeding at 10 to 15°C if a normal warming trend follows spawning. It seems reasonable to suggest similar temperatures for walleye fry culture. Temperatures in the range of 20°C have generally led to poor results. Temperatures above 10 to 15°C lead to increased metabolic rates and apparently cause fry to exhaust their nutrient reserves before they can adapt to prepared feeds, thus causing massive starvation and cannibalism. Temperatures above 20°C have been found desirable for walleye fingerling rearing, however (Siegwarth and Summerfelt 1990).

Stocking density and feeding intervals have also been thought to influence the feeding of walleye under intensive culture conditions. No density guidelines can be offered as yet, but a typical density is 200,000 fry per standard trough. Some workers believe that crowding aids initial feeding, though no systematically gathered evidence supports that conclusion. Frequent feeding (e.g., intervals of 2 to 5 min) seem to improve feed intake and survival, perhaps due to the fact that the feed particles are fresh and are within a short distance of the fish.

b. Feeds and Feeding

Numerous diets and environmental conditions have been used when feeding walleye fry under intensive culture conditions. Brine shrimp nauplii have been a standard live food source. Various experimental diets, including the Oregon Moist Pellet, chicken egg yolk, salmon starter diets, liver slurry, farina slurry, mixed zooplankton, and some of the more recently produced commercially available proprietary larval fish diets have been tested under intensive culture conditions.

Failures with both dry and live feeds have tended to take one of two forms. Immediate failures to accept feed have been followed by cannibalism, "tail-biting," and early starvation. When the feed has been accepted by a proportion of the fry, a phenomenon commonly called the "dwindles" by hatchery workers has been repeatedly observed. Fish of 2 to 3 weeks of age die in substantial numbers for no apparent reason and the population may fall to near zero. Malnutrition is the most probable explanation for the phenomenon. Failure of intensively cultured fry to inflate their swim bladders may also contribute to the phenomenon (Davis 1953).

Cuff (1977) concluded that greater food availability significantly reduced cannibalism in walleye less than 20 d of age, primarily because feeding fish more successfully avoided attacks by cannibals in the population. Jahncke (1981) found that feeding a nutritionally balanced diet must begin within 2 to 5 d of hatching, depending upon temperature, or the fry will develop an energy/nutrition deficit from which they will not recover.

The feeding behavior of walleye fry on natural foods may provide insights into the development of prepared feeds and the manner in which they should be presented (Li and Mathias 1982; Mathias and Li 1982; Raisanen and Applegate 1983). The live foods most commonly utilized typically have one axis less than 0.4 mm. Tests with various pelleted diets also indicate that walleye fry will ingest particles of 0.2 to 0.4 mm. Some evidence points to greater acceptance of orange and red particles over other colors, but that speculation has not been confirmed. It appears important to maintain a high density of feed particles whether living or prepared feeds are used.

Colesante et al. (1986) found that walleye fry reared intensively would ingest prepared feed, but all the fish died after just over a month. Walleye fry reared on brine shrimp before conversion to formulated feed, and those fed brine shrimp and zooplankton prior to conversion to formulated feed, had survival rates of 3.9 and 6.0% through 65 d of culture.

Successful rearing of walleye fry for 20 d on proprietary formula prepared feeds with good survival was accomplished by Kindschi and MacConnel (1989). Survival was up to 54.1%. Source of the fry appeared to affect the results. Loadman et al. (1989) extended the period of reasonable survival to 30 d of intensive culture. They fed an experimental diet manufactured commercially and obtained 28% survival at 19°C. Survival at 24°C was only 4%, but those fish experienced high mortality as a result of an oxygen depletion.

c. Pathology

Pathological problems in walleye fry culture have not been widely investigated. Problems related to the presence of myxobacteria and fungi have been reported (but not confirmed) at several hatcheries. A fungal infection involving the oral cavity and gills was prevalent in one set of rearing trials conducted in New York during 1977, and a similar problem has been reported in Pennsylvania. It is not known whether the infection was primary or secondary. The flush rate in all troughs was doubled in later trials and no further problems were observed.

Preliminary tests of the toxicity of standard therapeutic agents have been inconclusive. Walleye fry seem unusually sensitive to many of the chemicals commonly employed by fish culturists. If that observation is sustained, the necessity for maintaining high water quality standards will become even more important, since even chemicals approved by the U.S. Food and Drug Administration may be toxic to walleye at levels necessary for efficacy with other species.

d. Suggested Methods for Walleye Fry Culture

The following suggestions, while still preliminary, may be helpful to culturists involved in fry rearing:

1. Cylindrical rearing units with large, finely screened outlets are preferred over standard start troughs. In either type of unit, water flow should be directed in a manner which aids in suspending food particles. Rearing units should be of a neutral color and should be uniformly lighted.
2. At least 2 exchanges per hour of 10 to 15°C water should be provided. Once feeding is established, the temperature may be raised to 20°C.
3. Live feed such as brine shrimp nauplii should be provided in large quantities (1000 nauplii per fish per day seems appropriate) until swim bladder inflation occurs. Conversion to prepared feeds should be accomplished abruptly; that is, no gradual weaning should be used. Fry reared on brine shrimp nauplii will ignore prepared feeds so long as live food is available.
4. If dry feeds are used as the initial diet, they should be fed at 2 to 5 minute intervals, 24 h/d. Recently available commercial fry feeds appear to meet the nutritional requirements of walleye. If the culturist elects to formulate a walleye ration, the nutrient content should be similar to that of unfertilized walleye eggs. Since feed particles must be small (0.2 to 0.4 mm), the solubility of nutrients in water must be considered and appropriate adjustments in diet formulation made. A ring to confine feed particles and oils from the feed to a portion of the rearing unit surface may improve survival.

2. Fingerling Culture

Intensive culture of pond-reared walleye fingerlings is relatively new, but methods developed in laboratory systems have been adapted for use in several production hatcheries. Methods are similar to those described by Nickum (1978), but a number of modifications have been introduced. Most of those modifications are based on research and development studies conducted within the last few years. Therefore, the material that follows incorporates recent experience and conversations with other walleye culturists.

a. Physical Conditions and Facilities

Supplies of fingerling walleyes for intensive culture can be reliably produced in ponds using the methods described above. Highest production is obtained when the fish are harvested before reaching a length of 5 cm. Fish as small as 2.5 cm convert readily to formulated feeds if harvest and handling stress are minimized and the methods presented below are followed.

Rearing units of various sizes and shapes have been successfully used for intensive culture of walleye fingerlings. Experience indicates that the most important consideration involving rearing units is that the flow of water should be directed in a manner which will keep feed particles in suspension. Upwelling systems or water flows introduced perpendicular to the long axis of the culture unit have been successfully employed. Water exchange rates should be adjusted to reflect the oxygen demand of the fish and the feeding rate. Heavy stocking densities and high feeding rates may require three to five exchanges each hour; however, a single exchange is satisfactory for lower stocking rates. At present, 25 kg/m^3 is considered maximum stocking density.

Walleye fingerlings should be harvested and transported at temperatures below 20°C, though they will survive and grow in intensive systems at temperatures up to 30°C. Temperatures of 20 to 25°C seem optimal for good growth and low disease susceptibility. When the temperature falls below 18°C, growth rate is reduced. Growth ceases when the temperature is less than 10°C.

Dim light seems to enhance feeding. The fish are less excitable if they are kept in covered units where disturbance by passersby is reduced. Photoperiods of 16 to 24 h of light daily seem appropriate.

Waters of widely varying quality have been used for walleye culture. Water clarity may be of some importance, but other factors do not seem to influence walleye growth and survival so long as certain limits are not exceeded. Water of low turbidity is generally associated with fewer losses from cannibalism; however, it is not known whether that result is related to changes in fish behavior or to the ability of the culturist to more closely observe the behavior of cannibalistic individuals. Clear, clean water is also recommended because of reduced incidence.

b. Feeds and Feeding

Pond-reared walleye can be returned to raceways and easily trained to accepted formulated feeds when they reach about 5 cm in length (Cheshire and Steele 1972). It is now considered desirable to start pond-reared fingerlings that have been moved to intensive culture systems directly on dry pellets with no weaning period.

Starter-size granules of such diets as W-14 and W-16 (see Chapter 6, Table 1) supplied through automatic feeders at 2 to 10 min intervals, 16 to 24 h/d, have been readily accepted by walleye fingerlings. Within a week, 60 to 80% (survival as high as 99% through 1 month has been obtained) of 3 to 5 cm fish can be expected to accept such prepared feeds. If nonfeeding fish are removed and isolated in separate tanks, many of them will learn to accept feed. Particle size should be increased as the fish grow.

Published information on feeding rates and feed conversion rates is not available. Experimental programs in which walleye fingerlings have been fed to satiation have typically resulted in feeding rates of 5 to 10% of body weight daily. Feed conversions as low as 1.5 (weight of feed offered per body weight increase) have been obtained in such studies, but many variables are known to influence the values obtained.

The W-series of diets developed at U.S. Fish and Wildlife Service laboratories contain protein levels in excess of 55% and are, therefore, quite expensive. Preliminary results from studies conducted by the Iowa Cooperative Fishery Research Unit indicate that protein percentage can be reduced to 45% with no sacrifice of growth. Substantial modifications in diet specifications can be expected now that it is possible to reliably produce pellet-fed fingerling walleyes for such studies.

Variations in feed color and texture, as well as flavor enhancers, have been used to increase the acceptance of dry feeds developed for walleye fingerlings; however, most studies have lacked adequate replication and controls. It is the opinion of most workers that soft feeds are more acceptable than hard pellets. No consistent effects attributable to flavor enhancers have been reported.

c. Pathology

Myxobacterial infections, particularly those attributable to *Flexibacter columnaris*, have been the primary pathological problems associated with walleye fingerling culture. *Ichthyophthirius multifiliis*, *Trichodina* sp., *Scyphidia* sp., bacterial gill disease, furunculosis, fin rot, and fungal infections have also been reported (Hnath 1975), as have the parasites *Diplostomum* sp., *Spiroxys* sp., and *Camallanus* sp. (Muzzall et al. 1990). The incidence of dermal sarcoma in fingerling walleye has also recently been reported (Bowser et al. 1990; Martineau et al. 1990).

Various treatments have been employed to combat pathological problems in walleye culture. Nagel (1976) controlled outbreaks of *F. columnaris* with 10-s dip treatments in 500 mg/l copper sulfate and 1 min baths in copper sulfate at 30 mg/l. Hyamine 3500 at 2 mg/l for 45 min and Diquat at 16 mg/l have also been reported as successful treatments against *F. columnaris* (Hnath 1975). Malachite green, formalin, Acriflavin, potassium permanganate, and Furanace have been used with variable success. Hnath (1975) reported that Roccal at 2 mg/l for 1 h effectively controlled bacterial gill disease. He also indicated that Acriflavin at 5 mg/l for 1 h or Hyamine 3500 at 2 mg/l for 1 h controlled fin rot and that Terramycin in feed controlled the symptoms of furunculosis. Formalin treatments for *Ichthyophthirius multifiliis* were not successful. Nagel (1974) controlled bacterial gill disease with Roccal at 2 mg/l and fungus disease with formalin at 1:6000.

Disease problems with fingerling walleyes have apparently been reduced by minimizing handling when water temperatures exceed 20°C, by strict sanitation, and by maintaining high water quality and rapid rates of flushing. Dietary insufficiencies may contribute to disease problems; however, systematic studies of pathology in association with diet have yet to be conducted.

d. Suggested Method for Fingerling Production

The methods outlined below should lead to acceptable survival and growth of walleye fingerlings under intensive culture conditions. However, as with any new technology, modifications of the methods will be required to reflect advances made by researchers and practicing culturists.

1. Walleye fingerlings should be harvested from ponds when they reach 2.5 to 3.5 cm, a range in which starvation and cannibalism losses are low. Harvest should be undertaken when water temperature is below 20°C, and stress should be minimized through careful handling and the use of 0.5% NaCl and a bacteriostatic agent in all transportation units.
2. Rearing tanks designed to maintain feed particles in suspension should be used. Water flows of one or two exchanges hourly of 20°C water are recommended. Covered tanks or troughs and dim lights for at least 16 h daily are also recommended.
3. Feed with formulations similar to W-14 or W-16 should be provided at 2 to 10 min intervals at least 16 h/d. A feeding rate of 10% of body weight daily may be needed during initial feeding, but 3 to 6% daily should be adequate once the fish are actively feeding.
4. The fish should be carefully observed at frequent intervals. Cannibals should be removed and the fish graded at regular intervals if differential growth is observed.

3. Post-Fingerling Culture

Walleyes will continue to grow year round if water temperature is maintained above 20°C. Growth will continue at temperatures down to 10°C, but becomes very slow below 15°C. Length increases of 5 cm monthly and doublings in weight at 2-week intervals have been obtained, but should not be expected on a sustained basis. Two-year-old fish with well-developed gonads have been produced under laboratory conditions.

The methods used to rear walleyes past the fingerling stage are essentially the same as those used for fingerling production. Pellets of 8 mm diameter will be accepted by walleyes of all sizes beyond 20 cm (and by some smaller individuals). All types of rearing units, including floating net-pens (cages) have been used on an experimental basis. Specific optimal rearing conditions and diets have not been developed, but the diet, facilities, and conditions used for fingerling production are generally adequate.

REFERENCES

Alexander, H. C., Latvaitis, P. B., and Hopkins, D. L., Site-specific toxicity of un-ionized ammonia in the Tittabawassee River at Midland, Michigan: overview, *Environ. Toxicol. Chem.,* 5, 427, 1986.

Ali, M. A. and Anctil, M., Corrélation entre la structure rétinienne et l'habitat chez *Stizostedion vitreum vitreum* et *S. canadense, J. Fish. Res. Bd. Can.,* 25, 2001, 1968.

Arnold, B. B., Life History Notes on the Walleye, *Stizostedion vitreum vitreum* in a Turbid Water Utah Lake, Master's thesis, Utah State University, Logan, 1960, 1.

Arthur, J. W., West, C. W., Allen, K. N., and Hedtke, S. F., Seasonal toxicity of ammonia to five fish and nine invertebrate species, *Bull. Environ. Contam. Toxicol.,* 38, 324, 1987.

Baker, C. T. and Scholl, R. L., Walleye Spawning Area Study in Western Lake Erie, Fish and Wildlife Restoration Project Report, unpublished, 1969, 1.

Baldwin, N. S. and Sealfeld, R. W., Commercial Fish Production in the Great Lakes, 1867-1960, Great Lakes Fishery Commission, Ann Arbor, MI, 1962, 1.

Beyerle, G. B., Summary of attempts to raise walleye fry and fingerlings on artificial diets, with suggestions on needed research, and procedures to be used in future tests, *Prog. Fish-Cult.,* 37, 103, 1975.

Bowser, P. R., Martineau, D., and Wooster, G. A., Effects of temperature on experimental transmission of dermal sarcoma in fingerling walleyes, *J. Aquat. Anim. Health,* 2, 157, 1990.

Brugge, G. T. and McQueen, D. J., *In situ* enclosure culture of walleye, *Prog. Fish-Cult.,* 53, 91, 1991.

Buttner, J. K., Culture of fingerling walleye in earthen ponds. State of the art 1989, *Aquacult. Mag.,* 19(2), 37, 1989.

Cheshire, W. F. and Steele, K. L., Hatchery rearing of walleyes using artificial food, *Prog. Fish-Cult.,* 34, 96, 1972.

Cobb, E. W., Pike-perch propagation in northern Minnesota, *Trans. Am. Fish. Soc.,* 53, 95, 1923.

Colby, P. J., McNicol, R. E., and Ryder, R. A., Synopsis on Biological Data on the Walleye *Stizostedion v. vitreum* (Mitchill 1818), FAO Fish. Synop. 119, Food and Agriculture Organization, Rome, 1979, 1.

Colesante, H., Improved survival of walleye fry during the first 30 days of intensive rearing on brine shrimp and zooplankton, *Prog. Fish-Cult.,* 51, 109, 1989.

Colesante, H. I. and Youmans, N. B., Water-hardening walleye eggs with tannic acid in a production hatchery, *Prog. Fish-Cult.,* 45, 126, 1983.

Colesante, H. I., Youmans, N. B., and Ziolkoski, B., Intensive culture of walleye fry with live food and formulated diets, *Prog. Fish-Cult.,* 48, 33, 1986.

Cook, F. A., Freshwater Fishes of Mississippi, Mississippi Game and Fish Commission, Jackson, 1959, 1.

Corazza, L. and Nickum, J. G., Rate of food passage through the gastrointestinal tract of fingerling walleye, *Prog. Fish-Cult.,* 45, 183, 1983.

Cuff, W. R., Initiation and control of cannibalism in larval walleyes, *Prog. Fish-Cult.,* 39, 29, 1977.

Davis, H.S., *Culture and Diseases of Game Fishes,* University of California Press, Los Angeles, 1953, 1.

Dendy, J. S., Predicting depth distribution of fish in three TVA storage type reservoirs, *Trans. Am. Fish. Soc.,* 75, 65, 1948.

Dobie, J., Minnesota walleye nursery ponds and transportation of fingerling walleye, in Proc. North Central Warmwater Fish Culture-Management Workshop, Iowa Coop. Fish. Res. Unit, Ames, 1971, 133.

Dobie, J. and Moyle, J. B., Methods Used for Investigating Productivity of Fish Rearing Ponds in Minnesota, Spec. Publ. 5, Minnesota Department of Conservation, St. Paul, 1956, 1.

Dumas, R. F. and Brand, J. S., Use of tannin solution in walleye and carp culture, *Prog. Fish-Cult.,* 34, 7, 1972.

Ellis, D. V. and Giles, M. A., The spawning behavior of the walleye, *Stizostedion vitreum* (Mitchill), *Trans. Am. Fish. Soc.,* 94, 358, 1965.

Eschmeyer, P. H., The life history of the walleye, *Stizostedion vitreum vitreum* (Mitchill), in Michigan, *Bull. Mich. Dep. Conserv. Inst. Fish. Res.*, 3, 1, 1950.

Ferguson, R. G., The preferred temperature of fish and their midsummer distribution in temperate lakes and streams, *J. Fish. Res. Bd. Can.*, 15, 607, 1958.

Fox, M. G., Keast, J. A., and Swainson, R.J., The effect of fertilization regime on juvenile walleye growth and prey utilization in rearing ponds, *Environ. Biol. Fish.*, 26, 129, 1989.

Graves, G., They said it couldn't be done, *Farm Pond Harvest*, 8(1), 6ff, 1974.

Hearn, M. C., Ovulation of pond-reared walleyes in response to various injection levels of human chorionic gonadotropin, *Prog. Fish-Cult.*, 42, 228, 1980.

Heidinger, R.C., personal communication, 1985.

Hermanutz, R. O., Hedtke, S. F., Arthur, J. W., Andrew, R. W., and Allen, K. N., Ammonia effects on microinvertebrates and fish in outdoor experimental streams, *Environ. Pollut.*, 47, 249, 1987.

Hnath, J. G., A summary of fish diseases and treatments administered in a coolwater diet testing program, *Prog. Fish-Cult.*, 37, 106, 1975.

Hohn, M. H., Analysis of plankton ingested by *Stizostedium* (sic) *vitreum vitreum* (Mitchill fry and concurrent vertical plankton tows from southwestern Lake Erie, May 1961 and May 1962, *Ohio J. Sci.*, 66, 193, 1966.

Hokanson, K. E. F., Temperature requirements of percids and adaptations to seasonal temperature cycle, *J. Fish. Res. Bd. Can.*, 34, 1524, 1977.

Houde, E. D., Food of pelagic young of the walleye, *Stizostedion vitreum vitreum* in Oneida Lake, New York, *Trans. Am. Fish. Soc.*, 96, 17, 1967.

Huh, H. T., Calbert, H. E., and Stuiber, D. A., Effects of temperature and light on growth of yellow perch and walleye using formulated feed, *Trans. Am. Fish. Soc.*, 105, 254, 1976.

Hulsman, P. F., Powles, P. M., and Gunn, J. M., Mortality of walleye eggs and rainbow trout yolk-sac larvae in low-pH waters of the LaCloche Mountain area, Ontario, *Trans. Am. Fish. Soc.*, 112, 680, 1983.

Jahncke, M. L., Selected Factors Influencing Mortality of Walleye Fry in Intensive Culture, Master's thesis, Cornell University, Ithaca, NY, 1981, 1.

Johnson, B. L., Smith, D. L., and Carline, R. R., Habitat preferences, survival, growth, foods, and harvests of walleyes and walleye × sauger hybrids, *North Am. J. Fish. Manage.*, 8, 292, 1988.

Johnson, F. H., Walleye egg survival during incubation on several types of bottom in Lake Winnibigoshish, Minnesota, and connecting waters, *Trans. Am. Fish. Soc.*, 90, 312, 1961.

Johnson, F. H., Environmental and species associations of the walleye in Lake Winnibigoshish and connected waters, including observations on food habits and predator-prey relationships, *Minn. Fish. Invest.*, 5, 5, 1969.

Kelso, J. R. M., Diel movement of walleye, *Stizostedion vitreum vitreum*, in West Blue Lake, Manitoba, as determined by ultrasonic tracking, *J. Fish. Res. Bd. Can.*, 33, 2070, 1976.

Kindschi, G. A. and F. T. Barrows, Optimal screen mesh size for restraining walleye fry, *Prog. Fish-Cult.*, 53, 53, 1991.

Kindschi, G. A. and MacConnell, E., Factors influencing early mortality of walleye fry reared intensively, *Prog. Fish-Cult.*, 51, 220, 1989.

Koenst, W. M. and Smith, L. L., Jr., Thermal requirements of the early life history stages of walleye, *Stizostedion vitreum vitreum* and sauger *Stizostedion canadense*, *J. Fish. Res. Bd. Can.*, 33, 1130, 1976.

Krise, W. H., Optimum protease exposure time for removing adhesiveness of walleye eggs, *Prog. Fish-Cult.*, 50, 126, 1988.

Krise, W. H. and Meade, J. W., Review of the intensive culture of walleye fry, *Prog. Fish-Cult.*, 48, 81, 1986.

Krise, W. H., Bulkowski-Cummings, L., Shellman, A. U., Kraus, K. A., and Gould, H. W., Increased walleye egg hatch and larval survival after protease treatment of eggs, *Prog. Fish-Cult.*, 48, 95, 1986.

Lessman, C. A., Effects of gonadotropin mixtures and two steroids on inducing ovulation in the walleye, *Prog. Fish-Cult.*, 40, 3, 1978.

Li, S. and Ayles, G. B., Preliminary experiments on growth, survival, production and interspecific interactions of walleye (*Stizostedion vitreum vitreum*) fingerlings in constructed earthen ponds in the Canadian prairies, Can. Tech. Rep. Fish. Aquat. Sci. No. 1041, 1981, 1.

Li, S. and Mathias, J. A., Causes of high mortality among cultured larval walleyes, *Trans. Am. Fish. Soc.*, 111, 710, 1982.

Loadman, N. L., Mathias, J. A., and Moodie, G. E. E., Methods for the intensive culture of walleye, *Prog. Fish-Cult.*, 51, 1, 1989.

Loadman, N. L., Moodie, G. E. E., and Mathias, J. A., Significance of cannibalism in larval walleye (*Stizostedion vitreum*), Can. J. Fish. Aquat. Sci., 43, 613, 1986.

Martineau, D., Bowser, P. R., Wooster, G. A., and Armstrong, L. U., Experimental transmission of a dermal sarcoma in fingerling walleyes (*Stizostedion vitreum vitreum*), Vet. Pathol., 27, 230, 1990.

Mathias, J. A. and Li, S., Feeding habits of walleye larvae and juveniles: comparative laboratory and field studies, *Trans. Am. Fish. Soc.*, 111, 722, 1982.

Mayes, M. A., Alexander, H. C., Hopkins, D. L., and Latvaitis, P. B., Acute and chronic toxicity of ammonia to freshwater fish: a site-specific study, *Environ. Toxicol. Chem.*, 5, 437, 1986.

McCauley, R. W., Automatic food pellet dispenser for walleyes, *Prog. Fish-Cult.*, 32, 42, 1970.

Muench, K. A., Certain Aspects of the Life History of the Walleye, *Stizostedion vitreum vitreum* in Center Hill Reservoir, Tennessee, Master's thesis, Tennessee Technical University, Cookeville, 1966, 1.

Muzzall, P. M., Sweet, R. D., and Milewski, C. L., Occurrence of *Diplostomum* sp. (Trematoda: Diplostomatidae) in pond-reared walleyes from Michigan, *Prog. Fish-Cult.*, 52, 53, 1990.

Nagel, T. O., Rearing of walleye fingerlings in an intensive culture using Oregon Moist Pellets an an artificial diet, *Prog. Fish-Cult.*, 36, 59, 1974.

Nagel, T. O., Intensive culture of fingerling walleyes on formulated feeds, *Prog. Fish-Cult.*, 38, 90, 1976.

National Task Force for Public Fish Hatchery Policy, Report of the National Task Force for Public Fish Hatchery Policy, U.S. Fish and Wildlife Service, Washington, D.C., 1974, 1.

Nelson, W. R., Hines, N. R., and Beckman, L. G., Artificial propagation of saugers and hybridization with walleyes, *Prog. Fish-Cult.*, 27, 216, 1965.

Nevin, J., Hatching the wall-eyed pike, *Trans. Am. Fish. Soc.*, 16, 14, 1887.

Niemuth, W., Churchill, W., and Wirth, T., The walleye, its life history, ecology and management, *Publ. Wisc. Conserv. Dept.*, 227, 1, 1966.

Nickum, J. G., Intensive culture of walleyes: the state of the art, in *Selected Coolwater Fishes of North America*, Spec. Publ. No. 11, Kendall, R. L., Ed., American Fisheries Society, Washington, D.C., 1978, 187.

Olson, D. E., Improvement of Artificial Fertilization Methods at a Walleye Hatchery, Invest. Rep. Minn. St. Paul, DNR (310), 1971, 1.

Oseid, D. M. and Smith, L. L., Jr., Survival and hatching of walleye eggs at various dissolved oxygen levels, *Prog. Fish-Cult.*, 33, 81, 1971.

Paulus, R. D., Walleye Fry Food Habits in Lake Erie, Ohio Fish Monogr. 2, Ohio Department Natural Resources, Columbus, 1969, 1.

Petit, D., Effects of dissolved oxygen on survival and behavior of selected fishes of western Lake Erie, *Bull. Ohio Biol. Surv.*, 4, 1, 1973.

Pitlo, J., Walleye spawning habitat in pool 13 of the upper Mississippi River, *North Am. J. Fish. Manage.*, 9, 303, 1989.

Prentice, J. A. and Dean, W. J., Jr., Effect of temperature on walleye egg hatch rate, *Proc. Annu. Conf. S.E. Assoc. Fish Wildl. Agen.*, 31, 458, 1977.

Priegel, G. R., Reproduction and Early Life History of the Walleye in the Lake Winnebago Region, Tech. Bull. 45, Wisconsin Department of Natural Resources, Madison, 1970, 1.

Raisenen, G. A. and Applegate, R. L., Prey selection of walleye fry in an experimental system, *Prog. Fish-Cult.*, 45, 209, 1983.

Robins, R., *A List of Common and Scientific Names of Fishes from the U.S. and Canada*, American Fisheries Society, Bethesda, MD, 1991, 1.

Ryder, R. A., Effects of ambient light variations on behavior of yearling, subadult, and adult walleyes (*Stizostedion vitreum vitreum*), *J. Fish. Res. Bd. Can.*, 34, 1481, 1977.

Scherer, E., Effects of oxygen depletion and of carbon dioxide buildup on the photic behavior of the walleye (*Stizostedion vitreum vitreum*), *J. Fish. Res. Bd. Can.*, 28, 1303, 1971.

Scherer, E., Overhead-light intensity and vertical positioning of the walleye, *Stizostedion vitreum vitreum*, *J. Fish. Res. Bd. Can.*, 33, 289, 1976.

Scott, W. B. and Crossman, E. J., Freshwater fishes of Canada, *Fish. Res. Bd. Can. Bull.*, 184, 1, 1973.

Siefert, R. E. and Spoor, W.A., Effects of reduced oxygen on embryos and larvae of the white sucker, coho salmon, brook trout, and walleye, in, Proc. Int. Symp. Early Life History of Fish, Dunstaffnage Marine Research Laboratory, Oban, Scotland, May 17-23, 1973, 485.

Siegwarth, G. L. and Summerfelt, R. C., Growth comparison between fingerling walleyes and walleye × sauger hybrids reared in intensive culture, *Prog. Fish-Cult.*, 52, 100, 1990.

Smith, C. G., Egg production of walleyed pike and sauger, *Prog. Fish-Cult.*, 54, 32, 1941.

Smith, L. L., Jr. and Koenst, W. M., Temperature Effects on Eggs and Fry of Percoid Fishes, Environmental Protection Agency Report EPA-660/3-75-017, Washington, D.C., 1975, 1.

Smith, L. L., Jr. and Moyle, J. B., Factors influencing production of yellow pike-perch *Stizostedion vitreum vitreum* in Minnesota rearing ponds, *Trans. Am. Fish. Soc.*, 73, 243, 1945.

Smith, L. L., Jr. and Oseid, D. M., Effects of hydrogen sulfide on fish eggs and fry, *Water Res.*, 6, 71, 1972.

Walker, R. E. and Applegate, R. L., Growth, food, and possible ecological effects of young-of-the-year walleyes in a South Dakota prairie pothole, *Prog. Fish-Cult.*, 38, 217, 1976.

Waltemyer, D. L., The effect of tannin on the motility of walleye (*Stizostedion vitreum*) spermatozoa, *Trans. Am. Fish. Soc.*, 104, 808, 1975.

Waltemyer, D. L., Tannin as an agent to eliminate adhesiveness of walleye eggs during artificial propagation, *Trans. Am. Fish. Soc.*, 109, 731, 1977.

Webster, J., Trandahl, A., and Leonard, J., Historical Perspective of Propagation and Management of Coolwater Fishes in the U.S., in *Selected Coolwater Fishes of North America*, Spec. Publ. No. 11, Kendall, R. L., Ed., American Fisheries Society, Washington, D.C., 1978, 161.

Wolfert, D. R., Maturity and fecundity of walleyes from the eastern and western basins of Lake Erie, *J. Fish. Res. Bd. Can.*, 26, 1877, 1969.

Chapter 9

THE STRIPED BASS AND ITS HYBRIDS*

Jerome Howard Kerby

TABLE OF CONTENTS

---

*The U.S. Government is authorized to produce and distribute reprints for governmental purposes and to use any text or figures contained herein for other publications, notwithstanding any copyright that may appear hereon.

I. INTRODUCTION

The striped bass (*Morone saxatilis*) and its artificially produced hybrids are highly desirable sport and commercial fishes in the U.S. (Figure 1). The striped bass is an anadromous, euryhaline species that typically lives most of its life in marine and estuarine waters, south of the Roanoke River in North Carolina, and along the Gulf of Mexico coast. However, it is basically a riverine species that seldom moves into ocean salinities (Raney 1952; Barkuloo 1967; McIlwain 1968). The species was first established on the Pacific coast in 1879 and 1881, when 435 yearling fish were transported by rail from the Navesink and Shrewsbury Rivers in New Jersey (Raney et al. 1952). The Sacramento-San Joaquin population on the Pacific coast normally inhabits the river deltas or San Francisco Bay, but sometimes makes migrations along the coast that appear to be correlated with warmer ocean temperatures (Radovich 1963). Widespread landlocked populations also occur in freshwater reservoirs across the U.S. (Bailey 1975; Stevens 1984; Axon and Whitehurst 1985). The range of the species, which extends from the St. Lawrence River, Canada, to the St. John's River, FL on the Atlantic coast, from British Columbia to south of the Mexican border on the Pacific coast, and along the Gulf Coast, indicates that the species has a wide tolerance to a variety of environmental variables (Radovich 1963; Talbot 1966; Forrester et al. 1972).

Hybridization in fishes has been studied since the 1800s, but artificial hybridization has only recently been recognized and used as a tool to enhance certain desirable characteristics and for management purposes. Hybrids of female striped

FIGURE 1. Comparison of striped bass × white bass hybrid (palmetto bass; above) with striped bass. The hybrid was 395 mm in fork length (FL) and weighed 1058 g. The striped bass was 398 mm FL and weighed 824 g (From Kerby et al., *Trans. Am. Fish. Soc.*, 100, 787, 1971. With permission.)

bass × male white bass (*Morone chrysops*) were first produced by Robert E. Stevens in 1965 (Bishop 1968). Since then, almost every imaginable hybrid combination, including F_2 generations and various backcrosses and outcrosses, has been produced among the four *Morone* species in North America. Most *Morone* hybrids exhibit hybrid vigor with superior growth rate, greater disease and stress resistance, improved survival, and general hardiness (Bishop 1968; Logan 1968; Williams 1971; Ware 1975; Bonn et al. 1976; Kerby and Joseph 1979; Kerby et al. 1983a and b; Kerby 1986; Kerby et al. 1987b). Common names officially recognized by the American Fisheries Society (Robins 1991) for the hybrids are used herein (Table 1).

During the past 25 years, advances in culture and rearing techniques have opened new management opportunities in inland lakes and reservoirs and resulted in extensive new recreational fisheries for sport fishermen. In 1981, about 40 million striped bass and hybrid fingerlings were produced in 17 state and federal hatcheries for stocking in inland waters. By that same year, over 456 reservoirs (2.3 million ha) had been stocked with striped bass and/or hybrids, and successful fisheries were established in 279 reservoirs (1.6 million ha), which yielded annual recreational harvests in excess of 2.5 million kg of fish in 1981 (Stevens 1984;

TABLE 1
Common Names of Striped Bass Hybrids Recognized by the
American Fisheries Society

Female	Male	Common Name
Striped bass (*Morone saxatilis*)	White bass	Palmetto bass
White bass (*M. chrysops*)	Striped bass	Sunshine bass
Striped bass	White perch	Virginia bass
White perch (*M. americana*)	Striped bass	Maryland bass
Striped bass	Yellow bass	Paradise bass
	(*M. missippiensis*)	

From Robins, R., *A List of Common and Scientific Names of Fishes from the United States and Canada,* American Fisheries Society, Bethesda, MD, 1991, 1.

Axon and Whitehurst 1985). Stevens (1984) noted that hybrids are preferred to striped bass for smaller impoundments and that they appear to do better than striped bass in warmer waters.

In addition, significant enhancement efforts are underway in major estuaries on both the Atlantic and Pacific coasts of the U.S. as a part of major restoration efforts where precipitous declines of striped bass populations have occurred. Of these, current large-scale stocking programs are projected to extend through the 1990s in the Chesapeake Bay, Hudson River, and San Francisco Bay (Whitehurst and Stevens 1990). These efforts appear to be significantly impacting the populations in affected estuaries.

Research during the 1980s, which demonstrated that hybrids can be commercially cultured as a new high-value food fish, has resulted in the development of a new commercial industry (Kerby et al. 1983b; Woods et al. 1983; Smith et al. 1985; Kerby et al. 1987a; Kerby et al. 1987b; Huish et al. 1987; Harrell et al. 1988; Smith and Jenkins 1988a; Smith and Jenkins 1988b; Jenkins et al. 1989; Smith et al. 1989; Smith 1989). This chapter reviews the history of striped bass and hybrid culture and discusses some of the associated problems and potential.

II. EARLY HISTORY

The extent of the original striped bass resource in the U.S. was suggested by Captain John Smith who, in addition to remarking on its excellent flavor, noted that there were "such multitudes that I have seene stopped in the river close adjoining to my house with a sande at one tide as many as will loade a ship of 1000 tonnes" (Jordan and Evermann 1902). In 1639, the Massachusetts Bay Colony passed the first legislation in the New World to protect a fish species, and in 1670, a tax was levied on the fishery that partly funded the first public schools (Setzler et al. 1980). By the 1880s, however, "alarming" declines of "the most important and interesting of all the food fish of the Atlantic coast" were causing concern among fisheries workers (Worth 1884). Consequently, after the discovery and successful spawning of ripe striped bass in 1880 on the Roanoke River in NC (Worth 1882), the first striped bass hatchery was established in 1884 at Weldon, NC. In the first year of operation, about 298,000 larvae were hatched and 280,500 were released into the Roanoke River. Ripe adults were supplied to the hatchery by fishermen (Worth 1884). This hatchery has operated almost continuously throughout the intervening years (Tatum et al. 1966; Whitehurst and Stevens 1990). Other early hatcheries were also established, but because it was difficult or impossible to obtain ripe eggs, and because no benefits to commercial fisheries could be demonstrated, operations at all of the hatcheries except the one at Weldon were discontinued (Talbot 1966).

Although striped bass were first introduced into freshwater impoundments in New Jersey in the 1930s, few fish were taken from the lakes (Surber 1958). In 1941, however, Pinopolis dam was closed to form the 64,750-ha Santee-Cooper Reservoir in South Carolina, and by 1950, appreciable numbers of striped bass appeared in sport catches when it was evident that the population was undergoing an almost exponential increase. Contrary to early belief, spending time in salt water was obviously not a physiological requirement for reproduction. Because the successful population in Santee-Cooper Reservoir fed primarily on clupeid species such as gizzard shad, *Dorosoma cepedianum* (Scruggs and Fuller 1954; Stevens 1958; Fuller 1968), fishery managers decided to introduce striped bass as a biological control species in reservoirs where shad populations were causing problems. In 1954, South Carolina workers stocked adult and subadult fish into several of the state's larger reservoirs (Stevens 1958), and other states soon followed. Although the stocked fish usually thrived, they did not reproduce (Surber 1958; Sandoz and Johnston 1966; Stevens 1967; Bailey 1975). It was concluded that most reservoir systems did not provide a principal spawning requirement, i.e., sufficient upstream river length and currents strong enough to keep the semibuoyant eggs in suspension until they hatched (Stevens et al. 1965).

When it became apparent that striped bass could not reproduce successfully in the other large South Carolina reservoirs, a hatchery patterned after the one at Weldon, NC was constructed on the tailrace canal below Pinopolis Dam. More than 900 females were collected and examined in 1961, but none were found with free-flowing eggs (Stevens 1966). Of seven hormones tested from 1962 to 1964, human chorionic gonadotropin (HCG) was finally recommended because it was less expensive, acted more rapidly, and required only a single treatment (Stevens et al. 1965; Stevens 1966; Stevens 1967). Development of a satisfactory technique for hormone-induced spawning, and later refinement by Bayless (1972), represented the first major breakthrough in striped bass culture and allowed production of more than 100 million larvae per year at the South Carolina hatchery. This method made the capture of naturally ripe females unnecessary and led to the construction of striped bass hatcheries in many other states. Over 20 years later, HCG is still the hormone of choice in *Morone* culture because it is effective, economical, and readily available.

III. BROODSTOCK COLLECTION AND HANDLING

Broodstock for striped bass and hybrid production are normally collected on or near the spawning grounds, although some universities, governmental agencies, and particularly private aquaculturists are experimenting with creating and using domesticated *Morone* broodstock for commercial aquaculture.

Capture gear most often used includes electrofishing, hook-and-line, trap nets, and gill nets. In locations where it is legal, electrofishing (Figure 2) is preferred because it is highly efficient, and stress to broodfish is minimized if they are concentrated on or near the spawning grounds (Harrell and Moline 1988; Yeager et al. 1990). Reynolds (1983) and Berry et al. (1983) provided good discussions on electrofishing techniques and safety considerations.

Both stationary and drift gill nets are used to collect *Morone* broodstock, but they cause stress and physical damage, so are not a preferred capture method. Trammel nets can also be used effectively and may cause less stress to the fish (Yeager et al. 1990).

Several types of trap nets have also been used successfully. Pound nets are probably one of the most efficient, particularly in some coastal waters, but they are expensive and require experience to set. Hoop nets, which are traps designed for use in fast flowing water, are sometimes used during spawning runs and in upper tributary streams. Fyke nets represent a modified hoop net with one or two wings or leaders that guide the fish into the trap. They can be used in lakes and rivers with low to moderate currents (Hubert 1983; Yeager et al. 1990).

In many states, capture by hook-and-line is the only legal method for commercial aquaculturists. However, mortality rates of striped bass due to stress are often

FIGURE 2.   Electrofishing for striped bass. (Photograph courtesy of J. Howard Kerby, U.S. Fish and Wildlife Service.)

high, especially among gravid females (Yeager et al. 1990). White bass, white perch, and yellow bass are better able to handle this stress and are easier to capture than are adult female striped bass, which is a primary reason that sunshine bass are becoming more popular for commercial food fish production.

After capture and handling, broodfish must be handled and transported carefully to reduce stress. Because striped bass stress particularly easily, they must be treated with special care. Circular tanks are preferred over rectangular ones because the circular configuration allows unrestricted swimming and helps mitigate stress. Oxygen is preferred to compressed air, and recirculation provides a current that can help broodstock maintain balance and orientation (Yeager et al. 1990).

Fish are normally transported in 5 to 10 ppt salt (NaCl) or artificial seawater. An anesthetic approved by the U.S. Food and Drug Administration, such as tricaine methane sulfonate (MS-222), may be used to help mitigate stress. Temperature of hauling water should be close to that from which the fish were taken, and oxygen concentration should be kept above 7 mg/l. Antifoaming agents and prophylactic treatments for disease are often used during transport as well (Piper et al. 1982; Yeager et al. 1990). Throughout the process of capture, transport, and spawning, handling should be minimized to reduce stress. Good discussions of methods used for stress reduction are provided by Rees and Harrell (1990); Parker et al. (1990); and Yeager et al. (1990).

IV. PRODUCTION OF LARVAE

Two principal methods are used to produce larvae for culture. In the first, eligible females (see Rees and Harrell 1990, for eligibility criteria) are normally injected intramuscularly with 275 to 330 international units (IU) of HCG per kilogram of body weight (Stevens 1966; Bayless 1968; 1972; Bonn et al. 1976; Kerby et al. 1983a). Although 275 IU/kg appears to be the threshold dose required for ovulation, much higher doses apparently do not negatively impact ovulation success (Rees and Harrell 1990). Males may be injected with 110 to 165 IU/kg of HCG to increase semen volume. HCG injections can also be used to "bring back" males that are "drying up" (the semen becomes very viscous). The number of spermatozoa per unit volume is reduced, but fertilization percentages are similar to semen from treated and untreated fish (J. H. Kerby, unpublished data).

The latency period of females injected with HCG is temperature-dependent and may also be affected by the portion of the spawning season when the fish is injected. Between 20 and 28 h after injection, a small egg sample is taken with a 3-mm (outside diameter) glass or plastic catheter inserted through the urogenital opening into the ovary (Figure 3). Approximate time to ovulation is determined by microscopic examination of the eggs (Figures 4, 5, 6, and 7). Actual time of ovulation is verified when freely flowing eggs are produced after pressure is exerted on the abdomen. Accurate prediction of ovulation is extremely important because the eggs separate from the ovarian tissue and from the parental oxygen supply at ovulation. Anoxia results within about an hour if spawning does not

FIGURE 3. A 3-mm glass catheter is inserted into the ovary of a female striped bass to obtain an egg sample for use in predicting ovulation. (Photograph courtesy of Jack D. Bayless, South Carolina Wildlife and Marine Resources Department.)

occur and the eggs deteriorate (become "overripe") and cannot be fertilized (Stevens 1967; Bayless 1972; Kerby 1986). Techniques for accurately predicting ovulation were described by Rees and Harrell 1990).

Before spawning, female broodfish are usually either sacrificed or anesthetized. Quinaldine was formerly the anesthetic of choice, but is no longer recommended because of possible hazards to human health. Submersion in a 100 mg/l solution of MS-222 can be used to anesthetize fish before stripping, but care should be taken that the water is properly buffered and to prevent the anesthetic solution from dripping into the pan containing the eggs (Rees and Harrell 1990). Eggs then are manually stripped into a spawning pan (Figure 8). Semen is stripped into the pan, water is added, and the eggs, water, and semen are mixed for

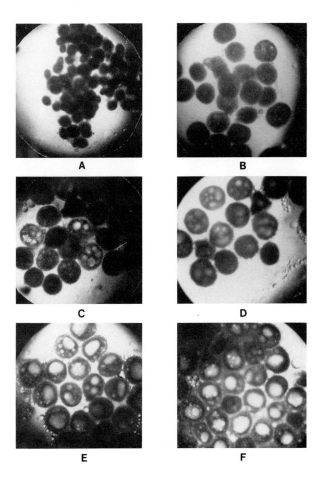

FIGURE 4. Development of striped bass eggs from immaturity to 11 h before ovulation. (A) Immature eggs; (B) 15 h before ovulation; (C) 14 h before ovulation; (D) 13 h before ovulation; (E) 12 h before ovulation; (F) 11 h before ovulation. (Photograph courtesy of Jack D. Bayless, South Carolina Wildlife and Marine Resources Department.)

2 to 3 min. Handling of the fish is facilitated somewhat by keeping a towel over their eyes. A modified method of fertilization described by Rees and Harrell (1990) involved stripping semen into slightly saline water and pouring the semen-water mixture over the eggs as they are expressed.

Following fertilization, the eggs are usually placed in modified McDonald hatching jars (Figure 9) at the rate of about 100,000 eggs per jar. A continuous flow of oxygenated water is introduced into the jars at a rate sufficient to keep them in constant motion (Figure 10). A "tube-within-a-tube" system prevents air bubbles from floating eggs out of the jar (Figure 9). Dead and unfertilized eggs turn opaque between 12 and 18 hours postfertilization and normally float out of the jars or "layer" on top of live eggs. In the latter case, they are siphoned off prior

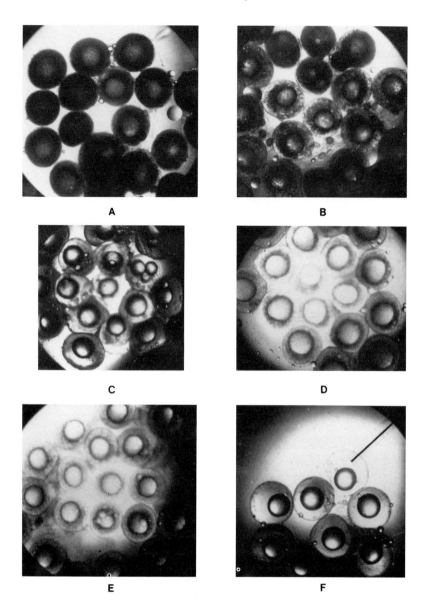

FIGURE 5. Development of striped bass eggs from 10 to 5 h before ovulation. (A) 10 h before ovulation, polarization complete; (B) 9 h before ovulation, yolk clearing; (C) 8 h before ovulation; (D) 7 h before ovulation; (E) 6 h before ovulation; (F) 5 h before ovulation. (Photograph courtesy of Jack D. Bayless, South Carolina Wildlife and Marine Resources Department.)

to hatch. After about 40 to 56 h (Figure 11), depending on temperature, the larvae hatch and swim over the lips of the jars into an aquarium. The equation $Y = 205 - 2.37X$, where Y = hours of incubation and X = water temperature (°F),

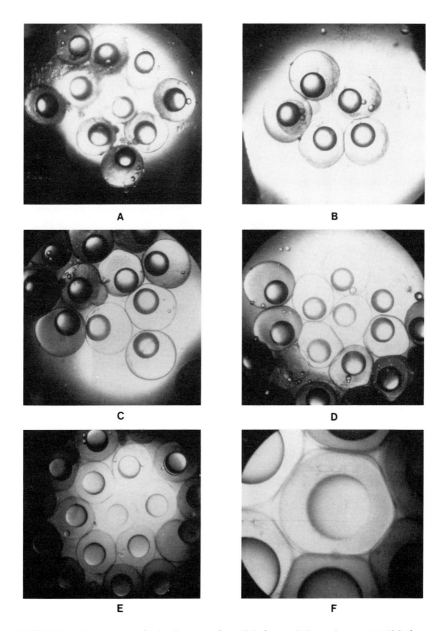

FIGURE 6. Development of striped bass eggs from 4 h before ovulation to ripeness. (A) 4 h before ovulation; (B) 3 h before ovulation; (C) 2 h before ovulation; (D) 1 h before ovulation; (E) Ripe eggs at ovulation; (F) Ripe eggs at ovulation. (Magnification × 50.) (Photograph courtesy of Jack D. Bayless, South Carolina Wildlife and Marine Resources Department.)

can be used to approximate time to hatch at any specific temperature (Bayless 1972; Rees and Harrell 1990). Larvae are normally held in aquaria for 2 to 5 d before they are shipped or stocked.

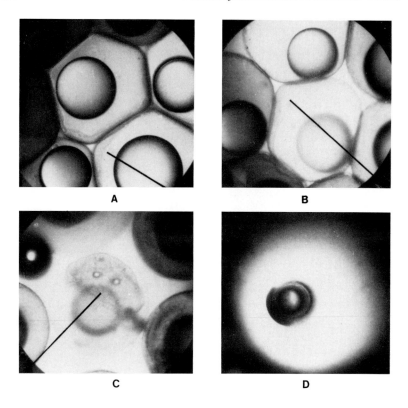

FIGURE 7. Development of striped bass eggs that have ovulated and become overripe. (A) Overripe eggs 1 h; note breakdown at inner surface of chorion (magnification × 50); (B) overripe eggs 1.5 h, breakdown at inner surface of chorion persists (magnification × 50); (C) overripe eggs 2 h; note deterioration confined to one-half of eff (magnification × 50); (D) overripe egg 16 h; dark areas appear white under microscope (magnification × 20). (Photograph courtesy of Jack D. Bayless, South Carolina Wildlife and Marine Resources Department.)

A variation of this method involves the use of circular tanks as incubators, particularly when the striped bass strain (e.g., Chesapeake Bay striped bass) produces eggs of low specific gravity that cannot be incubated in hatching jars because they float out of the jars with the slightest water flow (see Rees and Harrell 1990 for additional information).

The second larval production method, developed by Bishop (1975), induces the broodfish to spawn "naturally" in circular tanks (Figure 12). Both males and females are injected with HCG at the rates outlined above. Normally, 1 or 2 females and 2 to 4 males are placed in a circular tank 1.2 to 2.4 m in diameter and about 1.2 m deep. Water is supplied through 2 or more 13-mm (inside diameter) rigid tubes under slight pressure at a rate of 30 to 38 l/min, creating a circular velocity of 10 to 15 cm/s at the perimeter to keep the eggs in suspension. A 0.1-m diameter center standpipe, encircled by a fine-mesh screen (0.45 m in diameter), controls water depth and prevents the loss of eggs and larvae. Mesh size of the screen should be <500 μm (<200 μm for white bass, white perch, or yellow bass

FIGURE 8. Eggs from female striped bass are manually stripped into spawning pan. (Photograph courtesy of Jack D. Bayless, South Carolina Wildlife and Marine Resources Department.)

eggs) to retain the larvae. A bubble-curtain (usually constructed from perforated airline tubing) around the base of the screen keeps eggs and larvae from becoming entrapped on the screen and enhances egg suspension. The fish normally spawn in this system from 36 to 62 h after injection, depending on water temperature and the state of natural ripeness of the broodfish. The fish are not disturbed before or during spawning, and some culturists isolate the tanks with opaque partitions. After spawning, broodfish are removed from the tanks and released, normally in excellent health. The eggs are incubated in the tanks; larvae are subsequently collected with scoops or with a siphon attached to a large funnel constructed of fine mesh and placed so that the water in the tank flows through it. Details of these methods were described by Bayless (1972), Bishop (1975), Bonn et al. (1976), Rees and Harrell (1990), and Smith and Whitehurst (1990).

Both methods have advantages and disadvantages (Bishop 1975; Kerby 1986). The jar method is more labor intensive, requires greater expertise in predicting time of ovulation, and is more expensive. At least 3 persons are required on a 24-h/d basis for efficient operation. There is also higher eventual mortality of broodfish because of handling stress, whereas most are released unharmed after tank spawning and males can often be reused for subsequent spawning, if necessary (Smith and Whitehurst 1990). Because females spawn naturally when the tank spawning method is used, ovulation is usually complete

FIGURE 9.   Modified MacDonald Hatching jar showing "tube-within-a-tube" method of introducing water flow that prevents air bubbles from lifting eggs out of jars. (Photograph courtesy of Monte Stuckey, U.S. Fish and Wildlife Service.)

when eggs are released and fertilization rates are often better. On the other hand, because jar culture requires less space than tank culture, production per unit area is higher in the former. Larval production is also easier to predict when the jar method is employed, and the culturist has greater control over the developing eggs

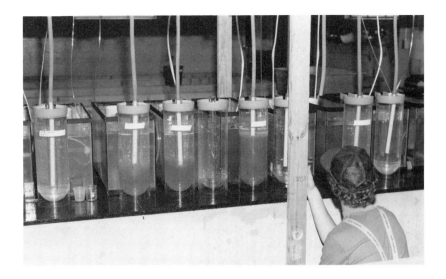

FIGURE 10.   Striped bass eggs incubating in modified MacDonald hatching jars.

and larvae than when they are in tanks. Because dead eggs often have a lower specific gravity than live eggs (depending on the race), they are either flushed out of the jars or can be removed by siphoning, which reduces fungus infections.

In the tank method, the removal of larvae from the tanks requires considerable time, but use of the screen-funnel-siphon arrangement reduces the amount of labor needed, and appropriate scheduling can alleviate this disadvantage. A major disadvantage of the tank method to date is that hybrids cannot be produced because female striped bass, even after they have ovulated, have not been induced to release their eggs in the presence of male white bass (Bishop 1975; Kerby 1986). Thus, hybrid production has heretofore required prediction of ovulation and manual-stripping procedures. This may be subject to change, however, at least with respect to producing palmetto bass, as striped bass females were recently induced to spawn naturally with male white bass through the use of gonadal releasing hormone implants, although fertilization rates were very low (Woods, 1992).

Comprehensive guidelines for spawning and larval production, complete with excellent photographs, are available in publications by Bayless (1972), Bonn et al. (1976), and Harrell et al. (1990a).

V. PRODUCTION OF FINGERLINGS

Efforts during the 1960s to establish significant striped bass populations in reservoirs by stocking larvae were largely unsuccessful. More than 25 reservoirs in 9 states were stocked with larvae, many in 2 or more successive years. In only two reservoirs were fisheries eventually established by larval stocking, and one of those was a hybrid population (McGill 1967; Bailey 1975). Thus, it was

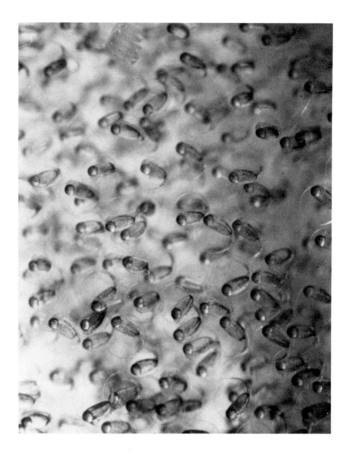

FIGURE 11.   Striped bass embryo immediately before hatch. (Photograph courtesy of Reginal Harrell, University of Maryland.)

concluded that the establishment of successful fisheries would require stocking of fingerlings (Sandoz and Johnston 1966). However, rearing the larvae proved to be more difficult than was originally envisioned because they are highly sensitive to a variety of factors. At certain stages, they "shock" easily and reasons for death are not readily discernible. Although much has been learned about their biology and requirements, there are still many unanswered questions.

Morone fingerlings are cultured in two principal system types. Ponds are presently most widely used, but more intensive tank and raceway culture is also being used successfully and shows promise. Both fresh water and brackish water have been utilized successfully.

A. POND CULTURE
1. Phase I Fingerling Production
The first notable success in rearing striped bass larvae to fingerlings occurred in 1964 at the Edenton National Fish Hatchery in North Carolina, where about

FIGURE 12. Circular tank used to spawn striped bass naturally. (Photograph courtesy of R. David Bishop, Tennessee Wildlife Resources Agency.)

30,000 fingerlings were produced (Anderson 1966). The hatchery manager recognized that striped bass culture was different from that of other warmwater species and required modified facilities and more intensive care. Subsequently, personnel at the Edenton hatchery explored and developed new culture methods and annually increased production. By 1970, yearly production approached a million fingerlings. Several reports describing results of the work provided the first detailed information on rearing striped bass in ponds (Regan et al. 1968; Ray and Wirtanen 1970; Wirtanen and Ray 1971). Over the years, many state and federal agencies have collected and synthesized data into two comprehensive guidelines on striped bass culture (Bonn et al. 1976; Harrell et al. 1990a).

Survival of larvae to the fingerling stage was normally less than 20% and often less than 10% during the early years of culture (McGill 1967; Sandoz and Johnston 1966; Reeves and Germann 1972; Harper and Jarman 1972; Barwick 1974; Hughes 1975). Increased experience and knowledge have resulted in increased harvests, but most hatchery managers still consider 30% survival satisfactory (Kerby 1986). During 1984 to 1985, average production at 12 southeastern hatcheries was about 118,600 (55 kg) striped bass per hectare and 135,900 (69.5 kg) hybrid striped bass per hectare (Brewer and Rees 1990), although rates as high as 197,000 (168 kg) phase I striped bass per hectare have been reported (Fitzmayer et al. 1986). Survival rates remain highly variable, and it is difficult to isolate causes of mortality.

Most culture now takes place in freshwater earthen ponds (Figure 13), although brackish water is used in some areas. The ponds may range in size from less than 0.04 to more than 2 ha, but most are in the 0.1 to 0.4 ha range and are usually 0.9 to 2.5 m deep.

FIGURE 13.   Ponds used for culture of striped bass and hybrid striped bass. (Photo courtesy of Monte Stuckey, U.S. Fish and Wildlife Service.)

Ponds are normally dried and disked during the winter to promote the break-down of organic matter. Rye grass or another cover crop is sometimes grown to reduce erosion and serve as a source of organic fertilizer, but should be used with caution (Braschler 1975; Bonn et al. 1976; Brewer and Rees 1990). Use of chemicals such as lime or herbicides may be advisable, depending on local circumstances. Pond filling is normally not recommended until shortly before stocking to reduce the effects of predacious insects and promote the availability of small zooplankton. Copepods and cladocerans are important to striped bass until they reach a length of 80 to 120 mm (Sandoz and Johnston 1966; Harper et al.

1969; Meshaw 1969; Harper and Jarman 1972; Humphries and Cumming 1972; Harrell et al. 1977). Sandoz and Johnston (1966) observed that early postlarvae held in aquaria selected cladoceran and copepod instars, and at first, avoided the larger female copepods. Harper and Jarman (1972) suggested that success in rearing striped bass larvae depended on the presence of an initial abundant supply of nauplii in the system.

Recent reviews (Boyd 1982; Geiger 1983a; Turner 1984; McCarty et al. 1986; Geiger and Turner 1990; Harrell and Bukowski 1990) provide excellent insights into pond fertilization practices and regimes for both fresh and brackish water ponds. Organic fertilizers have generally been preferred over inorganic types because the slower decay of organic materials seems to provide more sustained zooplankton production. Inorganic fertilizers, especially when used alone, often produce dense phytoplankton blooms that may result in high pH and can cause oxygen depletion, a dominant blue-green algae bloom, or both (Braschler 1975; Bonn et al. 1976).

Animal organics, such as manure and meat scraps, are sometimes used very successfully, but are not generally recommended because their inherent variability results in difficulties in standardizing application rates and because some produce a high biological oxygen demand that can cause oxygen depletion. The most commonly used organic fertilizers are various hays, plant pellets or meals (alfalfa, bermuda, milo, soybean, and cotton seed), rice brans, sugar refinery wastes (bagasse), wheat shorts, and yeast. Peanut, alfalfa, bermuda, and lespedeza hays are most commonly used. Application rates vary depending on physical structure and protein content. Hay is applied at higher rates than pellets and meals; soybean and cottonseed meals are applied at the lowest rates. Many culturists either soak or grind hay in a hammer mill prior to application to hasten decomposition. Subsequent applications may be required to help maintain the bloom (Stevens 1975; Bonn et al. 1976).

Geiger and Turner (1990) noted that the major differences between pond culture as described by Bonn et al. (1976) and pond culture as envisioned today for striped bass production revolve around concepts of modern zooplankton management. These include aggressive pond fertilization with combinations of organic and liquid inorganic fertilizers, inoculation or "seeding" of rearing ponds with preferred crustacean zooplankton, and monitoring zooplankton populations, their food, and basic water quality characteristics. The objectives are to maximize production of preferred zooplankton and to maintain the zooplankton populations as long as possible. To manipulate zooplankton communities, a basic understanding of zooplankton life history characteristics is needed. Geiger and Turner (1990) recommended that organic fertilizers used in zooplankton management be readily available; have low carbon to nitrogen ratios to permit rapid decomposition; be of small particle size to allow more rapid colonization by bacteria, protozoans, fungi, and algae (thereby producing more rapid decomposition and solubilization of nutrients); and be applied frequently (daily if possible) to reduce variation in zooplankton populations.

Liquid inorganic fertilizers, such as phosphoric acid and ammonium nitrate, are preferred over other forms because they are more soluble than granular fertilizers, thereby allowing more rapid uptake of key primary nutrients by phytoplankton. Geiger and Turner (1990) stipulated that liquid inorganic fertilizers should (1) contain nitrogen to enhance bacterial generation, which induces faster decomposition of organic fertilizers and helps prevent a dominance shift to blue-green algae); (2) contain sufficient amounts of soluble phosphorus to allow rapid uptake by unicellular green algae and minimize absorption into sediments; (3) be well-mixed with water and evenly dispersed over the pond (aeration devices should maximize distribution of fertilizer throughout the water column).

Timing and quantities of fertilizers and inoculation of plankton should be shifted to maximize crustacean zooplankton and their foods during the 2-week periods before and after stocking the fish (Geiger 1983a; Geiger 1983b; Geiger and Turner 1990).

For many years, striped bass larvae were normally stocked into ponds at 5 to 10 d of age, following swim-up and an initial period of feeding on brine shrimp (*Artemia* sp.) nauplii, in the belief that stocking older, more developed fry resulted in enhanced survival. Vessels for holding larvae until release into ponds include aquaria, raceways, and saran cages placed in ponds. Inslee (1977) suggested there may be enhancement of pond production subsequent to holding larvae in saran cages as compared to holding them in aquaria. Most culturists have now decided that holding larvae beyond 5-d posthatch is unnecessary and may be counterproductive, as some evidence suggests that holding fry may result in decreased swimbladder inflation. The recent trend is to stock 5-d-old fry directly into ponds without feeding them brine shrimp (Kerby 1986; Brewer and Rees 1990).

Larval stocking rates have also been extensively examined. Harper and Jarman (1972) and Parker (1979) reported that higher stocking produced survival rates comparable with those yielded by lower stocking rates; thus, the higher rates yielded the most fingerlings per hectare. This relationship holds true to a point, but since any system has a maximum carrying capacity in terms of biomass, larger numbers of fish may result in correspondingly smaller average sizes, as illustrated by data from Rees and Cook (1985b).

Stocking rates vary from 123,500 to 1.5 million larvae per hectare, with most agencies routinely stocking 247,000 to 618,000 per hectare (Bonn et al. 1976; Geiger and Parker 1985; Brewer and Rees 1990). However, each situation is different, and statistical manipulations of data by Rees and Cook (1985b) predicted an optimum stocking rate of about 1,875,000 larvae per hectare in their Georgia ponds to provide the largest number of 25.4 mm fingerlings. This does not mean that the rates provided the greatest percent survival. Those authors suggested that, under a given set of conditions, stocking rates can be manipulated to obtain a certain size of fingerling, specific average weights, maximum production, or a combination of all three.

Supplemental feeding of a salmon starter diet appears to enhance survival and growth of phase I fingerlings (25 to 60 mm long), especially if forage organisms have been depleted, and some culturists have incorporated feeding regimes into their standard phase I production practices (Brewer and Rees 1990). Experiments by Fitzmayer et al. (1986) indicated that continuous (hourly) feeding of formulated fry diets, compared to feeding twice daily, enhanced fingerling production. In the former treatment, survival and production averaged 52.8% (167.9 kg/ha) compared to 34.9% (110.6 kg/ha) in the latter treatment, and particles of prepared feed were found in 92% of the fish sampled from the treatment ponds.

Fingerlings are normally harvested 30 to 50 d after stocking at a size of 25 to 60 mm and weight of 0.2 to 2 g. They are normally stocked into reservoirs at that size, but may be restocked into culture ponds until fall or winter (phase II fingerlings), when they will be 75 to 250 mm long (3 to 200 g). Under good culture conditions, pond managers normally expect fish to weigh 100 to 150 g by the end of the first growing season.

2. Phase II Fingerling Production

Phase II culture ponds are generally similar to those used for phase I culture, but pond preparation is not nearly as involved. Fertilizers are not normally used for phase II production because artificial feeds will supply sufficient organic material to stimulate phytoplankton growth. Phytoplankton blooms are more important for shading out aquatic vegetation than for primary production of food because artificial diets replace zooplankton as the primary food (Smith et al. 1990).

If phase I fish are not relatively uniform in size, they should be carefully graded before putting them in holding tanks or raceways, as severe cannibalism can result if different sizes of fish are held together (Parker and Geiger 1984; Atstupenas and Wright 1987; Kerby and Harrell 1990). Grading during harvesting and loading can be easily accomplished with floating bar graders. Grading increases uniformity of fish, allows more even competition for food, and reduces cannibalism.

Fingerlings should be held at 10 ppt salinity continuously for 1 to 2 d to mitigate stress, if possible. Salinity should then be gradually reduced with hatchery water over an additional 2-d period until the original water has been completely replaced by ambient hatchery water. Any treatments with formalin, potassium permanganate, or other chemicals can be accomplished at this time (Kerby and Harrell 1990).

Phase I fingerlings fish should be trained to feed on a prepared diet before restocking for phase II production, if possible. Although nutritional requirements for striped bass and hybrids are not well defined, high quality, high protein salmon diets have worked well for phase I fingerlings through a #4 crumble size. When fish are large enough to take food larger than #4, they can usually be transferred to a high quality trout diet, which typically contains about 38% protein. Catfish or "pond" fish diets containing similar amounts of protein should not be used because they typically do not provide the quality that *Morone* spp. need. Klar and Parker (1989) found that various commercial trout and salmon diets yielded acceptable growth, but that striped bass fed a modified catfish diet and those fed a diet consisting of all fish by-products had significantly lower growth rates.

If fish are crowded in tanks or raceways under good environmental conditions, they should begin to accept feed within a day or two and be fully trained in approximately a week to 10 d (Kerby and Harrell 1990). They can then be returned to the ponds for grow-out to phase II and can be expected to reach a length of 170 to 250 mm by late winter (Humphries and Cumming 1973; Brewer and Rees 1990).

Stocking densities used for culture of phase II striped bass and hybrids range from 10,000 to 250,000 fish per ha (Kerby et al. 1983a; Kerby et al. 1983b; Kerby et al. 1987a; Parker and Geiger 1984; Atstupenas and Wright 1987). From 1986 to 1988, the average stocking density for hatcheries in the southeast was 54,361 fish per ha. If large numbers of fish averaging 25 to 41 g are desired, a higher density (61,775 to 98,800 per hectare) may be preferred. If fish weighing more than 0.11 kg are needed, a lower density (24,700 to 37,800 per hectare) may be more appropriate (Smith et al. 1990). Stocking at low densities (e.g., 12,350 fish per hectare) often results in greater size variation because the dominant larger fish consume most of the food (Atstupenas and Wright 1987). As a result, size distribution at harvest may be bimodal with a large percentage (80%) of small fish (20 g) and a small percentage of large fish (100 g). Survival can also be affected due to cannibalism. Careful grading of fish before stocking can largely prevent these effects (Kerby et al. 1983b; Kerby et al. 1987a). Stocking densities of 24,700 to 61,775 fish per hectare may result in more uniform size fish and better overall survival (Jenkins et al. 1989). Stocking at densities above 61,775 per hectare results in a large biomass of fish, which requires more feed, supplemental aeration, water exchange, and close monitoring of water quality and fish health.

B. INTENSIVE CULTURE

Culture of striped bass larvae to the fingerling stage in aquaria, troughs, tanks, or raceways has nearly as long a history as pond culture. Anderson (1966) and Tatum et al. (1966) were among the first investigators to attempt to culture striped bass in hatchery troughs or aquaria. Kerby's (1986) statement that this type of culture was largely experimental is still appropriate, and intensive tank culture for phase I fingerling production is used routinely by only a few organizations.

Intensive culture has some advantages, as well as problems. Such systems are space-efficient, can be precisely controlled, and allow for easy observation of the fish and for close monitoring of water quality. In addition, detection and treatment of diseases are facilitated. On the negative side, even closed recirculating systems require a constant supply of high-quality replacement water, and water use increases substantially if the system is semiclosed or open. Water quality can be difficult to maintain and must be monitored frequently so that corrective measures can be taken immediately, when necessary. Most closed systems are relatively complex, and the threat of mechanical or electrical failures requires extensive alarm and back-up systems. Cleaning of tanks and filters can be a major problem and may be highly labor-intensive. Because of the high densities of fish being maintained in such systems, disease and cannibalism can become significant problems, and provision of sufficient quantities of appropriate diets may be difficult prior to the phase I stage.

Early efforts to rear striped bass larvae to fingerlings in intensive systems were only marginally successful. Survival rates were often less than 1% and seldom more than 10% (Anderson 1966; Tatum et al. 1966; Rhodes and Merriner 1973; McIlwain 1976; Texas Instruments 1977; Lewis et al. 1981).

Modifications to a semiclosed system at the Mississippi Gulf Coast Research Laboratory resulted in the development of the first intensive system that provided acceptable survival percentages (Nicholson 1973; McIlwain 1976). The system now produces over 335,000 fingerlings annually (up to 46% survival) and is operated as a production facility for stocking coastal streams (Nicholson, 1982). A highly successful (by past standards) recirculating system was developed at Southern Illinois University (Lewis et al. 1981). During 7 years of development, survival increased from less than 10% in 1974 to 46% in 1980, a rate that compares favorably with pond production rates. However, the system is complex, consisting of a series of biofilters, pumps, ultraviolet sterilizers, and other devices, and requires constant monitoring of equipment and back-up systems. Maintenance of the biofilters and rearing procedures requires extensive commitments of labor. The investigators have provided comprehensive guidelines detailing the design and operation of their system and a good discussion of the principles underlying it (Lewis et al. 1981). Nicholson et al. (1990) noted that basic information still needed for phase I culture in intensive systems includes: (1) nutritional requirements, (2) methods to prevent cannibalism, (3) better understanding of the physiology of swim bladder inflation, (4) better methods for controlling disease, and (5) better handling techniques.

Intensive systems for production of phase II fingerlings have been more successful than for production of phase I fish, primarily because fingerlings 30 to 40 mm long can be readily trained to accept prepared feed, and they can be more easily handled and treated (Texas Instruments 1977; Powell 1973; Allen 1974; Collins et al. 1984; Kerby and Harrell 1990; Van Olst et al. 1981). Intensive culture of both striped bass and hybrid striped bass in flow-through tank and raceway systems is now routinely practiced by several commercial producers. Production is usually enhanced by addition of pure oxygen.

In another type of intensive culture for production of phase II fingerlings, floating cages have been placed in estuaries. Swingle (1972) and Powell (1973) reported that survival rates of striped bass cultured in 0.73 m^3 diameter vinyl-covered wire cages ranged from 43.2 to 97.4% in water with salinities ranging from 10.0 to 24.4 ppt.

VI. FOOD REQUIREMENTS

Stevens (1975) concluded that the production of fingerlings per unit area is directly proportional to the kinds and abundance of zooplankton available during the growth period, provided that larvae are healthy and water quality is maintained. The preferred foods of striped bass larvae and early juveniles in freshwater ponds are copepods, cladocerans, and insect larvae. Fish size has a significant effect on food selectivity. Early postlarvae select early instars of cladocerans and copepods; larger organisms are avoided (Sandoz and Johnston 1966; Harper et al. 1969; Humphries and Cumming 1972; Humphries and Cumming 1973). In general, copepods constitute the most important food of striped bass less than 30 mm long (Regan et al. 1968; Meshaw 1969; Harrell et al. 1977).

Cladocerans become increasingly important as the fish grow. Highest utilization occurs when the fish are 20 to 80 mm. Insects are important to fish of all sizes, but become the primary food for fish over 80 mm long (Regan et al. 1968; Harper et al. 1969; Meshaw 1969; Gomez 1970; Harper and Jarman 1972; Humphries and Cumming 1972). Results of selectivity studies are somewhat conflicting. Although *Cyclops* spp. were usually reported to be preferentially selected by smaller fish, rotifers (Brachionidae) were almost always not selected. However, Snow et al. (1982) judged rotifers to be an acceptable early diet for striped bass when other crustaceans were excluded from the pond. In wild populations, forage fish also become an important food by the time the striped bass reach 80 to 90 mm (Gomez 1970; Ager 1979). In general, palmetto bass food-selectivity appears to parallel that of striped bass (Woods et al. 1985b).

Food preferences of striped bass reared in brackish-water culture ponds have not been carefully described, although Heubach et al. (1963) and Markle and Grant (1970) described differences in the food of young-of-year fingerlings inhabiting waters of different salinities in estuaries. Heubach et al. (1963) found that as salinity increased, the dominant copepod species in the diet changed from *Cyclops* and *Diaptomus* to *Eurytemora*, *Acartia*, and *Pseudodiaptomus*. Amphipods, mysids, isopods, and polychaetes were also extensively consumed in saline waters (Heubach et al. 1963; Markle and Grant 1970). Minton and Harrell (1990) noted that if salinity is more than 5 ppt in brackish-water culture ponds, cladocerans and often freshwater cyclopoid copepods, such as *Cyclops*, will not be present. Instead, the population is initially dominated by rotifers, succeeded by calanoid and harpacticoid copepods.

Intensive culture presents problems since insufficient zooplankton of the preferred types can be produced in a secondary culture system with current culture techniques, and early postlarval striped bass will not readily accept or assimilate prepared diets (Kelley 1967; Bowman 1979; Falls 1983; Nicholson et al. 1990). Though expensive, brine shrimp nauplii can easily be hatched from dry cysts in 24 h and are readily accepted by striped bass, thus they are widely used in intensive culture. Some investigators have suggested that brine shrimp may not provide adequate nutrition for larval striped bass (Regan et al. 1968; Braschler 1975; Nicholson et al. 1990; Rees and Harrell 1990), and research is currently underway to develop methods to enrich nauplii with essential fatty acids and to slow naupliar development by maintaining them at low temperatures of 4 to 5°C (Lemarié 1991).

Sunshine bass larvae are too small to ingest brine shrimp nauplii, which necessitates that live zooplankton (particularly rotifers) be provided (Kerby and Harrell 1990).

A clearly defined "point of no return" for striped bass survival does not occur if food becomes available (Eldridge et al. 1977; Eldridge et al. 1981; Rogers and

Westin 1981). Rogers and Westin (1981) found that unfed larvae decreased in length and lost weight, but most survivors were able to capture and consume live food when they were near death. Most surviving larvae fed and grew at rates similar to those of groups fed earlier, but growth advantages of the earlier groups were maintained. Eldridge et al. (1981) reported survival of over 70% of the larvae when food was offered on or before 18 d after egg fertilization. However, the density of the nauplii is a key factor in promoting good survival and growth of the larvae (Lewis et al. 1981; Eldridge et al. 1981; Rogers and Westin 1981). Eldridge et al. (1981) and Daniel (1976) demonstrated a direct relation between survival and growth of larvae and brine shrimp concentrations.

Lewis et al. (1981) reported similar results in their rearing system, in that the number of striped bass that survived to a stockable size was directly related to quantity of nauplii fed. Initial daily densities of 50 to 60 nauplii per millimeter, followed by increased food density to 100 to 120 nauplii per millimeter when the fish reached 15 d of age were recommended. High densities reduced the need for larvae to actively search for food, and later the very high densities tended to reduce cannibalism.

Feeding should begin as soon as the digestive tract is developed, usually 4 or 5 d after hatching, depending on water temperature. Bonn et al. (1976) recommended feeding at least every 3 h, but preferred continuous feeding. McHugh and Heidinger (1977) found that larvae remain active and continue to feed at night and concluded that food should be provided continuously throughout the day and night. Nicholson et al. (1990) noted that larvae are offered nauplii at 15-min intervals, 24 h/d at the University of Maryland Crane Aquaculture Facility.

Continued survival beyond metamorphosis requires that fry be trained to accept prepared rations. Bonn et al. (1976) stated that prepared food with about the same particle size as brine shrimp nauplii should be included in the feeding regime when the fish are between 14- and 21-d old. Both prepared feed and brine shrimp should be fed until the 28th day of age, after which the brine shrimp can be phased out. Alternatively, Lewis et al. (1981) reported that larvae preferred smaller food particles and recommended particle sizes from 0.2 to 0.5 mm. They used a finely ground Tetra® commercial flake as the initial training food. When the larvae were 17-d old, a high-protein, pulverized salmon starter was added to the feed until the larvae were 31-d old. Thereafter, larvae were fed salmon starter exclusively. Frequent feeding (12 to 16 times daily) was recommended, during which feeding rates of 25 to 50% of body weight daily were employed. Feeding rates were gradually reduced to 5% of body weight daily when the fish were 40- to 45-d old.

A wide variety of commercial foods are available, including various brands of high-protein trout and salmon diets. Bonn et al. (1976) noted that a combination of dry trout food and whole processed fish make an excellent moist diet when mixed and reground. Various types of ground fish, shrimp, and liver have also been used, but cause water quality problems in intensive systems.

Nutritional requirements have not been well defined for striped bass, but salmonid diets containing 38 to 50% protein are normally used (Bonn et al. 1976; Valenti et al. 1976; Nicholson et al. 1990). Two rigorous studies of striped bass nutrition examined different levels of dietary protein and also interactive effects of dietary protein and lipid on growth, feed efficiency, and protein utilization. When formulated diets containing 34, 44, and 55% crude protein were fed to juvenile striped bass, the diet containing the highest protein percentage yielded the highest average weight gain and food conversion. Some protein sparing was observed when lipids were added at 7, 12, and 17% of the diet. Feed efficiency, protein efficiency ratio, and protein retained by the fish all increased as dietary protein increased (Millikin 1982; Millikin 1983). In an extension of Millikin's work, Berger and Halver (1987) reported that in one study maximum weight gain, feed efficiency, and protein efficiency were obtained with a diet containing 52% protein at 21 to 22°C. In a second study, when the diets contained approximately 44% protein, maximum growth was achieved with diets containing 33% dextrin and with diets containing 17% total lipid. Growth depression occurred if dietary lipid was increased beyond 17%.

VII. PREDATION

If care to exclude other fish species is exercised by filtering natural water prior to use, there are three primary sources of predation on cultured striped bass: predacious insects, cannibalism, and fish-eating birds. Aquatic insects, particularly back swimmers (Notonectidae) and phantom midge larvae (*Chaoborus* spp.), are the principal predators on larvae and small juveniles reared in ponds; whereas, cannibalism can become a source of high mortality in intensive culture systems. Additionally, predation on small striped bass larvae by a free-living copepod (*Cyclops bicuspidatus thomasi*) has been observed in brackish natural waters (Smith and Kernehan 1981) and could become a problem in culture. Birds, such as kingfishers, gulls, cormorants, herons, and ospreys, may become important predators in ponds and outdoor tanks and raceways after the fish become juveniles and subadults.

Humphries and Cumming (1973) pointed out that newly hatched striped bass larvae are especially vulnerable to predation because they cannot swim continuously. Tatum et al. (1966) observed phantom midge larvae killing larvae in laboratory experiments and suggested that they could pose a serious predation problem in ponds. Several methods for controlling predacious insects are available. Reeves and Germann (1972) drained ponds before they were stocked and treated them with potassium permanganate, while Huner and Dupree (1984) suggested sterilization with hydrated or slaked lime. Filling ponds immediately prior to stocking with larvae will allow the fish to grow too large to be available to predation when the insect populations become reestablished (Reeves and Germann 1972; Braschler 1975; Bonn et al. 1976; Brewer and Rees 1990). Applications of diesel fuel to the ponds leave a surface film that leads to suffocation of air-breathing insects. Various insecticides, such as ethyl and methyl parathion, trichorton and Baytex® have also been used under certain circumstances (Bonn et al. 1976; Brewer and Rees 1990), but most cannot be legally used for culture of food fish.

While cannibalism occurs in ponds, it has not been confirmed as a serious problem (Bayless 1972; Harrell et al. 1977; Parker 1980). However, it can clearly result in significant levels of mortality in intensive culture systems (Rhodes and Merriner 1973; McIlwain 1976; Bonn et al. 1976; Lewis et al. 1981; Braid and Shell 1981; Woods et al. 1981; Collins et al. 1984). High larval density, perhaps coupled with inadequate or insufficient food, seemingly promotes cannibalism. Further, growth rates of the cannibals are accelerated, thus progressively increasing their capacity to prey on their smaller siblings. Braid and Shell (1981) observed cannibalism when fry were only 6-d old; losses resulting from cannibalism ranged from 12 to 44% by the time the larvae were 16-d old. The authors observed that once fry had become cannibalistic, other fry became the favored food. Lewis et al. (1981) found that cannibalism can result in losses of 70 to 80% of the striped bass larvae in an intensive culture system. However, cannibalism was reduced when large quantities of brine shrimp nauplii were supplied.

Bonn et al. (1976) stated that cannibalism increased during transition from live to prepared food. Larger fish (termed "jumpers") remained deeper than the slower growing fish and persistently ate the smaller ones unless size differentiation was eliminated by grading. A similar phenomenon was observed among larvae in hatchery troughs and fingerlings up to 70 or 80 mm long in large circular tanks (Kerby, unpublished data). After differential growth had begun, the cannibals were seldom observed to feed on brine shrimp, suggesting that smaller fish were their preferred food. After the fingerlings were transferred to large circular tanks, behavioral patterns similar to those described by Bonn et al. (1976) were observed. The larger fish fed on the dry diet first, while the smaller fish remained on the edges of the feeding "frenzy." After a short time, the larger fish appeared satiated and dropped to lower levels in the water column while the smaller fish increased their feeding activity. During this period some of the larger fish moved rapidly up from the deeper levels and cannibalized smaller fish as they fed.

VIII. HYBRIDIZATION AND HYBRID PRODUCTION

The original objective of the *Morone* hybridization program initiated in the 1960s was to produce a fish that combined some of the more desirable characteristics of the parent species, such as size, longevity, food habits, and fighting ability of striped bass and the adaptability of white bass to exotic environments (Bayless 1972; Bonn et al. 1976; Kerby and Harrell 1990).

The palmetto bass has been artificially propagated and widely stocked in freshwater impoundments for control of shad (*Dorosoma* spp.) and as a food and sport fish since 1965. The sunshine bass is now being used similarly and is becoming increasingly popular with management biologists and commercial aquaculturists (Kerby and Harrell 1990).

Several additional crosses, such as hybrids between striped bass and white perch or yellow bass, and various backcrosses and outcrosses have also been produced, but only the sunshine bass is close to gaining the acceptance of the original cross (palmetto bass).

A. DESCRIPTIONS

Adults of *Morone* are easily distinguished morphologically. The hybrids, however, are generally intermediate in appearance, are more variable, and some of the crosses (especially backcrosses and second generation or F_2 crosses) are difficult to distinguish from the parents. Yet, F_1 hybrids can be readily distinguished from the parents by a number of morphometric and meristic traits (Bayless 1972; Williams 1976; Kerby 1972; Kerby 1979; Kerby 1980; Harrell 1984a; Waldman 1986; Harrell and Dean 1987; Harrell and Dean 1988). Accurate identification of backcrosses and F_2 hybrids may depend on biochemical (electrophoretic or isoenzyme) analyses (Avise and Van Den Avyle 1984; Todd 1986; Crawford et al. 1987; Forshage et al. 1988; Harvey and Fries 1988; Fries and Harvey 1989) or genetic analyses, such as mitochondrial or nuclear DNA (Chapman 1989; Mulligan and Chapman 1989).

In addition to the palmetto and sunshine basses, a few other crosses should be mentioned. The Virginia bass has been stocked in a few small impoundments and has been cultured and sold in commercial markets in experimental marketing trials (Huish et al. 1987). It is hardier and significantly more tolerant of stress than the palmetto bass, but growth rates are generally slower than those of palmetto bass and are highly variable (Kerby and Joseph 1979; Smith and Jenkins 1984; Huish et al. 1987; Kerby et al. 1987b).

The F_2 palmetto bass is less hardy than the F_1 or the parental species, survival is generally poor, and growth rates (and morphology) are highly variable (Smith et al. 1985; Kerby and Harrell 1990). It has been generally dismissed as an aquaculture candidate.

The striped bass × palmetto bass is the only backcross for which growth and survival characteristics under culture conditions are currently available. In preliminary experiments, Kerby et al. (1987b) found that it is hardy, with growth rates similar to those of palmetto bass.

The palmetto bass × striped bass is currently undergoing preliminary evaluations for growth and survival characteristics by the U. S. Fish and Wildlife Service. Larvae are relatively easy to produce and are vigorous. Assuming this backcross is hardy and has similar growth characteristics to the striped bass × palmetto backcross, it may be a candidate for commercial culture.

Trihybrids between palmetto bass and Virginia bass represent an interesting and potentially valuable cross if the hardiness of the white perch hybrids can be incorporated into the offspring without seriously affecting growth rates (Kerby and Harrell 1990).

B. HYBRID PRODUCTION

Like striped bass, most broodstock used for hybrid production are captured from the wild. However, it will become necessary for domesticated broodstock to play an increasing role in hybrid production in the future, particularly where hybrids are cultured commercially.

1. Larval Production

Methods for producing palmetto bass are similar to those described for striped bass, except that white bass males are used instead of male striped bass. HCG injection rates for male white bass are usually 220 to 240 IU per kg of body weight. These levels appear appropriate for white perch and yellow bass as well (Kerby and Harrell 1990). As noted, hybrids are presently produced by manually stripping ripe eggs and milt from the parental species. However, natural hybridization between white bass and white perch, white bass and yellow bass, natural backcrossing of hybrid striped bass with striped bass, and tank spawning of F_1 hybrids have been reported (Harrell 1984b; Jenkins and Smith 1987; Todd 1986; Forshage et al. 1988; Fries and Harvey 1989). Crawford et al. (1987) also reported suspected natural hybridization between striped bass and white bass in Arkansas. These reports suggest that it may be possible to induce tank spawning between species and hybrids if appropriate conditions are met. Recent experiments at the University of Maryland Crane Aquaculture facility have stimulated striped bass females to spawn in the presence of male white bass, suggesting that it may only be a matter of time before hybrids can be produced through natural spawning in tanks (Woods 1992).

"Wet" fertilization, where eggs and sperm are stripped simultaneously into a dish-pan containing 2 or 3 l of hatchery water, is preferred by many culturists for producing palmetto bass (Bayless 1972; Starling 1985), perhaps because copious amounts of urine that accompany semen while stripping white bass males may prematurely activate spermatozoa. Bayless (1972), Bonn et al. (1976), and Starling (1985) provided detailed descriptions of the procedures.

Methods for producing hybrids using females other than striped bass, F_2 crosses, and backcrosses using F_1 females are somewhat more complicated. Eggs from white bass, white perch, and yellow bass are much smaller than striped bass eggs and are highly adhesive, making typical jar culture difficult. Masses of eggs stick together and form a substrate that promotes fungus infections. Eggs from F_1 hybrids are intermediate in size between those of striped bass and the other *Morone* species. Most F_1 eggs are also adhesive, but are less so than those of nonstriped bass *Morone* species. Fertilization and spawning procedures should be handled similarly for those species and for hybrids (Kerby and Harrell 1990).

Ripe eggs from white bass, white perch, yellow bass, and F_1 hybrids are typically yellow to golden in color. In comparison, ripe striped bass eggs are various shades of green, ranging from very pale green to blue-green. When striped bass males are crossed with females of other species or F_1 hybrids, females are normally injected with doses of HCG ranging from 1100 to 2200 IU/kg (Bonn et al. 1976; Starling 1985; Kerby and Harrell 1990).

A small glass or polyethylene catheter (1.5 to 2.0 mm outside diameter) can be used to obtain eggs from white bass, white perch, and yellow bass to determine eligibility and predict ovulation. The plastic catheter described by Smith and Jenkins (1988a) probably causes less damage to oviduct tissue than glass. Egg development to maturation in these species is similar to that described for striped bass; Bayless (1972) described it in detail for white bass. However, in contrast to striped bass, development of white bass eggs is almost always mixed and when the majority of eggs ovulate a significant portion (which can be as high as 40%) are in various stages of development prior to ovulation (Bayless 1972). This condition probably occurs because females of these species are naturally intermittent spawners and may take several days to complete spawning.

Because the eggs are highly adhesive, they cannot be incubated normally in hatchery jars unless the adhesive material is neutralized. This material also forms a mucoid covering that renders the eggs almost opaque during incubation, which makes it very difficult to determine fertilization percentages and to observe embryonic development. Numerous methods have been used to reduce the adhesiveness and to clear or remove the material so that fertilization and development could be observed. Fuller's earth, silt, clay, starch, sodium sulfite, and tannic acid represent a few of the materials that have been tried with varying degrees of success (Kerby and Harrell 1990). The most successful technique was developed and modified over a period of several years, primarily by C. C. Starling and W. H. Revels, at the Richloam Fish Hatchery in Florida (Starling 1983; Starling 1985; Starling 1987; Rottmann et al. 1988) and was described in detail by Kerby and Harrell (1990). Briefly, it involves successive treatments of the fertilized eggs with two solutions. The first contains 20 g of NaCl, 15 g of urea, and a drop of antifoaming agent in 5 l of hatchery water. The second solution contains 0.75 g of tannic acid in 5 l of hatchery water. These treatments, combined with vigorous aeration, will help clear the eggs and will prevent them from adhering to each other, which allows incubation in standard hatching jars.

2. Phase I Fingerling Production

Techniques used for pond culture of hybrid larvae to phase I fingerlings are essentially the same as those described for striped bass. However, hybrid larvae produced from females of *Morone* species other than striped bass are about one-third smaller than striped bass or hybrids produced with striped bass eggs. These larvae are normally ready to feed when they are 4-d-old if water temperatures are ≥19°C and should be stocked by that time. Furthermore, they cannot be maintained or intensively cultured with brine shrimp nauplii because their mouths are too small to ingest the nauplii. Greater care should be exercised in pond filling and fertilization strategies. Because of their smaller size, time of stocking (in relation to pond filling and fertilization) is more critical to provide appropriate zooplankton. The fry are also more vulnerable to predation than larvae produced with striped bass eggs. In general, these smaller hybrids should be stocked in ponds that are in the early phases of a bloom, when rotifers and the smaller crustaceans (such as early copepod instars) are most abundant, rather than in more mature ponds where larger crustaceans such as adult copepods and cladocerans predominate (Kerby and Harrell 1990).

3. Phase II Fingerling Production

Techniques for phase II hybrid production are also similar to those described for striped bass. Optimum stocking rates for hybrid striped bass have not been well defined, although rates of 20,000 fingerlings per hectare and higher have been successful. Survival can normally be expected to exceed 80% barring oxygen depletion or other water quality, disease, or predation problems (Kerby et al. 1983b, 1987a; Jenkins et al. 1989). However, if fish are stocked at high densities, back-up aeration systems should be available. To maximize growth rates, feeding to satiation at least three times daily has been recommended (Kerby and Harrell 1990). Feeding rates normally should be estimated at $\geq 5\%$ of fish biomass per day if satiation feeding is not practiced (Tuncer et al. 1990).

4. Phase III Subadult and Adult Production

Culture of hybrids to subadult or early adult sizes (0.3 to 1.8 kg) is seldom practiced except when the objective is commercial food-fish production or research. Phase III culture can be accomplished in various ways. A cage study conducted in an Alabama estuary (Powell 1973) demonstrated survival over a 60-d period in excess of 98% and standing crops exceeding 89 kg/m^3. Williams et al. (1981) cultured both striped bass and palmetto bass to commercial size in 13.6 m^3 net pens in the Stono River estuary, South Carolina, with survival up to 88% and a standing crop of 15.8 kg/m^3. At North Carolina State University, hybrid striped bass were cultured to commercial size in three different culture systems (Kerby et al. 1983a; Kerby et al. 1987a; Woods et al. 1983; 1985a and b). Fish were reared in 0.1-ha freshwater earthen ponds, 38 m^3 flow-through circular pools, and 0.5 m^3 floating cages placed in a brackish-water estuary. Results in the three systems was variable, but generally good survival and growth were obtained so long as oxygen depletions and other problems did not occur. Mean survival in the cages was 95%, with standing crops up to 62 kg/m^3 at harvest. The pond studies yielded survival rates up to 87% and standing crops as high as 5765 kg/ha. More recently, hybrid production trials in South Carolina yielded survival rates from 93 to 96% and a standing crop of 9700 kg/ha (Smith 1989; Smith et al. 1989; 1990). Kerby and Harrell (1990) recommended stocking phase II fingerlings at approximately 10,000/ha, providing adequate aeration systems are available and sufficient volumes of water can be provided to the pond in the event of water quality problems. Assuming 80% survival, and an average weight of 0.9 kg at harvest, this would result in a standing crop of about 7200 kg/ha. High-quality commercial trout diets appear to provide adequate nutrition for rapid growth and high survival.

A number of recent publications provide detailed information regarding culture to phase III (Williams et al. 1981; Kerby et al. 1983a, 1983b, 1987a; Woods et al. 1983, 1985a; Kerby 1986; Jenkins et al. 1989; Smith et al. 1989; Kerby and Harrell 1990; Smith et al. 1990).

5. Broodstock Production

Striped bass males and hybrid striped bass (palmetto, sunshine, and Virginia bass) are relatively easy to culture to maturity in ponds. Under optimum growing conditions, male striped bass and hybrids of both sexes will mature and can be successfully spawned at 2 years of age, and males sometimes mature at 1 year of age. The same is true for white bass, although there has been much less experience in white bass culture. Maturation in *Morone* appears to be more a function of size than of age. Female hybrids begin to mature at approximately 0.45 to 0.7 kg and males at about 0.3 kg. In contrast, striped bass females normally require a minimum of 4 years and must weigh more than 2 kg before they begin to mature; it is probably not economical to spawn females that weigh less than 5.5 kg. Past experience in spawning female striped bass grown in ponds or tanks has produced inconsistent results (Kerby and Harrell 1990; Smith et al. 1990). Experience indicates that hybrids are easier to spawn and provide more consistent results than female striped bass that have been reared in ponds. Although F_2 hybrids are not good candidates for aquaculture, backcrosses of F_1 hybrids to striped bass may provide a suitable alternative to F_1 hybrids, based on a preliminary study (Kerby 1987b). Because female F_1 hybrids can mature in 2 years, are relatively easy to artificially spawn, and are better able to recover from the stress of artificial spawning than female striped bass, their utilization as "domesticated" broodstock would be an advantage if the progeny exhibit the vigor and growth rates of the F_1 parents.

White bass can also be reared for broodstock and wild fish can be acclimated to culture conditions, particularly if live forage is provided. Although techniques have not been well developed or described, it is unlikely that they will differ significantly from those used for hybrids produced with nonstriped bass females.

IX. FOOD FISH PRODUCTION

The striped bass is an excellent food fish and has historically been an important, high-value commercial product along the Atlantic coast. In the 1970s, when it became apparent that commercially exploited striped bass populations were declining and their commercial value was increasing, interest in culturing them as a commercial food fish began to develop (Kerby and Harrell 1990). Continued declines in natural populations and moratoria or catch quotas on commercial fisheries led to sustained high prices and served as an inducement to groups interested in commercial culture.

Researchers at North Carolina State University and at the South Carolina Marine Resources Research Institute began developing techniques for rearing striped bass and hybrid striped bass for commercial culture in the late 1970s (Williams et al. 1981; Kerby et al. 1983a and b, 1987a; Woods et al. 1983; Kerby 1986; Smith et al. 1985, 1989, 1990). As a result of these efforts, and as the hardiness and growth rates of the fish were recognized, interest became focused on hybrid striped bass. Other organizations, both government and private, are rapidly accelerating research designed to improve commercial culture production techniques for both striped bass and hybrids.

Wholesale prices have ranged as high as $8.80/kg for striped bass in the round (Kerby et al. 1983a; Carlberg et al. 1984), and live striped bass have been sold for $13.00/kg in New York and California (Swartz 1984; Carlberg and Van Olst 1987). Recent marketing trials (Huish et al. 1987; Smith 1989) and surveys of fresh-fish wholesalers (Carlberg and Van Olst 1987) have demonstrated that striped bass hybrids are also considered a high-value food fish that can be sold for $4.40 to 8.80/kg in the round to wholesale markets.

Smith et al. (1985) showed that hybrid striped bass could be cultured intensively under controlled conditions to weights over 0.23 kg in less than a year, with a standing crop at harvest of 43.2 kg/m^3. In later studies, Kerby et al. (1987a) and Smith et al. (1989) demonstrated that semi-intensive pond production of hybrids can range from 5600 to over 7500 kg/ha if adequate water quality is maintained.

Though hybrid striped bass can adapt to water with low hardness and alkalinity, they stress easily. Water with alkalinities ranging from 150 to 300 ppm (as $CaCO_3$) is preferable to water with lower alkalinities (Kerby and Harrell 1990).

Hybrids also have the advantage of being more temperature tolerant than striped bass, which allows effective culture in a wider geographical area. While the optimum culture temperature for striped bass appears to be about 24°C (Cox and Coutant 1981), hybrid striped bass appear to grow better at higher temperatures, with the optimum probably about 26.7 to 28°C (Kerby et al. 1987a; Woiwode and Adelman 1984; Woiwode 1989).

The information generated by researchers is now being put to use, and new fish farms are being constructed with the express purpose of farming hybrid striped bass. In 1987, Aquatic Systems, Inc., a tank culture operation on the west coast, marketed the first significant commercial crop, with production of 151,950 kg (Van Olst and Carlberg 1990). In 1988, the first privately produced commercial crops of pond-reared hybrid striped bass were harvested and marketed as a result of joint ventures with universities, the U.S. Department of Agriculture, and private producers. In 1989, private producers provided in excess of 8 million hybrid fingerlings to other growers for commercial production of food fish (Kerby and Harrell 1990). During the past 5 years, a number of commercial operations have been launched, and they are now supplying significant quantities of farm-raised fish to wholesalers annually. Hybrid striped bass are generally the fish of choice, and prices for farm-raised fish currently range from $4.95 to $8.00 per kilogram in the round (Smith 1992).

Information on various aspects of marketing and legal ramifications is becoming more available (Swartz 1984; Carlberg et al. 1987; Van Olst and Carlberg 1990; Dicks and Harvey 1989, 1990; Easley 1987; Jenkins 1987; Huish et al. 1987; Liao 1985; Liao and Smith 1987). McVey and Thomson (1990) reported that studies indicated a strong market demand for hybrid striped bass, with established markets for fish slightly over 0.4 kg and for fish in excess of 0.8 kg. Dicks and Harvey (1989, 1990) estimated annual production between 0.45 and 0.68 million kg in 1989 and 0.91 to 1.36 million kg after 1995. They forecast production as high as 9.1 million kg by the year 2000.

As with any commodity, market prices are expected to follow the law of supply and demand. As a product becomes more available, prices will undoubtedly fall.

However, with a little foresight, hybrid farmers can learn and profit from the lessons catfish farmers have already learned (i.e., that new markets must be continuously developed to ensure continued expansion). If these lessons are learned, this new aquaculture industry, developed primarily around striped bass hybrids, has the potential to rival the farm-raised catfish industry.

X. WATER QUALITY

Water quality requirements necessary for striped bass culture have not been clearly defined, though effects of specific variables have been examined and the generally favorable range in water quality, plus lethal limits for several variables, were reviewed with respect to larvae by Humphries and Cumming (1973). Mullis and Smith (1990) summarized water quality characteristics which should be minimally met. Striped bass appear to have a wide tolerance for several variables, and in general, the fish become tolerant to a wider range of water quality with increasing age.

A. TEMPERATURE

Temperature is one of the most important factors affecting striped bass survival and growth at all stages of development. It has a direct effect on fish metabolism and indirect effects on pH, dissolved oxygen, and nitrification (Kerby 1986; Nicholson et al. 1990). Shannon (1970) reported that the optimum spawning temperature range for striped bass in the Roanoke River (NC) is 16.7 to 19.4°C, though spawning occurred at temperatures as low as 12.8°C. Barkuloo (1967) reported that mature striped bass ovaries subjected to temperatures of 21 to 24°C changed from pale green to orange and then became opaque as the eggs died. In ripe males subjected to the same temperatures, live sperm were present, but motility was greatly reduced.

Striped bass eggs and larvae reportedly survive temperatures ranging from 12.8 to 23.9°C. At temperatures above or below this range, mortality increases dramatically (Albrecht 1964; Doroshev 1970; Shannon 1970; Otwell and Merriner 1975; Morgan et al. 1981). However, Bayless (1972) concluded that percentage hatch declined above 18.9°C, and Shannon and Smith (1968) found that larvae hatched above 21.1°C died within 70 h. Doroshev (1970) reported total larval mortality at temperatures below 10 and above 26°C.

Larvae and fingerlings exposed to temperature extremes survive better if they are acclimated. Davies (1973b) reported that, in general, larvae acclimated to lower temperatures more quickly than fingerlings, but acclimation to higher temperatures appeared to occur at about the same rate and was more rapid than acclimation to lower temperatures for both life stages. Fingerlings survived well at 4.4°C, but only 70% of the larvae survived at 10°C. All fingerlings died at temperatures above 35°C. Valenti et al. (1976) reported that fingerlings in saline water tolerated temperatures as low as 6°C without acute detrimental effects, but that chronic exposure to temperatures between 6 and 1°C limited growth and ultimately survival.

Recommended water temperatures for larvae until 9 d of age are between 14.4 and 21.1°C, and pond temperatures between 18 and 32°C are considered suitable for rearing, though growth is retarded at lower temperatures (Bonn et al. 1976). Larval temperature tolerance appears to increase with age, with 9- or 10-d-old fry able to tolerate temperatures up to 26.7°C (Brewer and Rees 1990). Bonn et al. (1976) stated that temperatures of 18 to 19°C are considered optimum for efficient food conversion. However, Cox and Coutant (1981) reported that growth of striped bass was most rapid and maintenance ration for 100 to 286 g juveniles was lowest when the temperature was near 24°C; no growth occurred at 33.5°C. Consumption of food increased as temperature increased to 25°C and decreased thereafter. Lewis et al. (1981) reported that temperatures as high as 32°C did not cause mortalities in their recirculating intensive culture system. Smith et al. (1990) noted that feeding activity decreases rapidly below 15.6 and above 32°C. Hybrid striped bass appear to be more thermally adaptable than striped bass, and optimum temperatures for growth seem to be higher. Woiwode and Adelman (1984) reported that the optimum growth temperature was 31°C for 8-g palmetto bass fingerlings and that growth continued (although at a reduced rate) at 35°C. Food conversion efficiency was best at 19°C, but growth was slower. In contrast, Woiwode (1989) reported that the optimum growth temperature for 130 g palmetto bass was about 26.7°C and that the optimum temperature for food conversion efficiency was 21.1°C (Nicholson et al. 1990).

B. DISSOLVED OXYGEN

Dissolved oxygen (DO) is the factor that is most often critical in striped bass culture. Turner and Farley (1971) found that mean survival of eggs decreased when DO fell to 5 mg/l at 18.4°C and was 50% of control survival when DO was 4 mg/l at 22.2°C. Incubation time increased at low oxygen concentrations and subsequent larval survival was reduced in direct proportion to the length of time eggs were exposed to low DO. A number of abnormalities have been reported in larvae from eggs incubated at 3 and 4 mg/l (Turner and Farley 1971; Harrell and Bayless 1983). Developmental abnormalities in embryos incubated at concentrations below 3 mg/l included truncation (club tail) and scoliosis in varying degrees, and total egg mortality occurred at concentrations near 1.5 mg/l. Time to hatch also increased with decreasing oxygen concentration (Harrell and Bayless 1983).

Hill et al. (1981a; 1981b) reported that juvenile striped bass avoided oxygen concentrations of 4.9 mg/l and lower. Chittenden (1971) described behavioral modifications of juveniles exposed to decreasing oxygen concentrations. Fish became restless, swam constantly in an increasingly violent manner, and then swam in spasmodic bursts at the surface as the oxygen concentration decreased. Subsequent behavioral stages included resting on the bottom, loss of equilibrium, and finally death. The first indications of restlessness occurred at DO concentrations between 1.3 and 3.7 mg/l. Death occurred from 0.5 to 1.04 mg/l.

Striped bass fingerlings in ponds have been reported to survive DO levels as low as 1.4 mg/l, but they were probably heavily stressed (Parker 1984). Klyashtorin and Yarzhombek (1975) concluded that oxygen concentrations below 4.0 to 4.5

mg/l, though perhaps not fatal, led to reduced food consumption, increased energy expenditure for respiration, and reduced growth rate.

Brewer and Rees (1990) recommended that DO values in ponds be maintained above 6 mg/l and urged caution in applying fertilizer to ponds with lower concentrations. Emergency aeration devices should be available in semi-intensive pond culture operations (Parker et al. 1990). Bonn et al. (1976) recommended that 5 and 4 mg/l be considered the minimum permissible levels for intensive and pond culture, respectively. Lewis et al. (1981) agreed with the 5 mg/l figure, but suggested that larvae require concentrations of 6 mg/l.

C. CARBON DIOXIDE

Carbon dioxide (CO_2) is a product of respiration. Because it also interferes with respiration, the minimum DO level required by fish increases with increasing CO_2 levels (Boyd 1979; Smith et al. 1990). Levels of CO_2 above 5 mg/l are considered dangerous, but can be reduced by aeration.

D. SALINITY

Although striped bass usually spawn in freshwater, Geiger and Parker (1985) reported brackish water ranging from 0.5 to 10 ppt positively influenced striped bass production. Albrecht (1964) observed that survival of both eggs and larvae was enhanced by low salinity (about 1.7 ppt), and moderate salinities (8.3 to 8.6 ppt) were not detrimental. Turner and Farley (1971) reported that egg survival after water-hardening in fresh water was higher when eggs were transferred to salinities up to 10 ppt compared with eggs water-hardened directly in higher salinity water. Bayless (1972) subjected 2-d-old larvae to various salinities. All larvae subjected to 28 ppt salinity died, but significant numbers of those held in 21 ppt survived and grew. Growth and survival were better in salinities of 3.5 to 14.0 ppt than in freshwater controls. Growth was best at 14.0 ppt, but survival was best at 10.5 ppt. When un-iodized NaCl was substituted for sea salts, survival decreased in 3.5 ppt and total mortality occurred at higher salinities.

Lal et al. (1977) also found low salinity to be beneficial to survival of eggs and larvae. Survival of eggs incubated in salinities of 3 to 17 ppt after a day in fresh water was better than that of freshwater controls, but progressively decreased for eggs hatched in salinities over 17 ppt. Larval growth was most rapid at 20 ppt, but survival was highest at 13 ppt. Tolerance to increased salinity improved with age, and striped bass fingerlings could be reared in "full strength" (33.8 ppt) immediately after metamorphosis without apparent stress (Lal et al. 1977). Van Olst et al. (1980) reported that larvae introduced into tanks of static brackish water (10 to 12 ppt) 12 h after they had hatched did not exhibit the frequent heavy mortalities that normally occurred between the second and fifth day of life. When 2 groups of larvae were stocked in freshwater ponds with about equal plankton densities after being held in either fresh or brackish water for 5 d, survival rates for the 2 groups after 22 d were 68% and nearly 100%, respectively. Colura et al. (1976) reported 55% survival of larval striped bass stocked in brackish-water ponds with salinities of 6 to 8 ppt.

Minton and Harrell (1990) noted that striped bass eggs can be incubated successfully in salinities up to 5 ppt, but chorion expansion is inhibited with increased salinity, and eggs usually do not water harden if salinity is over 10 ppt. In experiments designed to examine interactions between salinity and temperature, salinity (up to 10 ppt) did not affect percent hatch, but had a positive influence on survival (Morgan et al. 1981). However, Otwell and Merriner (1975) reported that temperature-salinity interactions were not statistically significant in survival of larvae older than 5 d. Other interactions between temperature and age, and salinity and age, were significant. Woods et al. (1983) reported an apparent salinity-temperature interaction with respect to growth rates of palmetto bass in brackish water. In general, higher temperatures were required to produce similar growth rates at increasing salinities.

Both striped bass and hybrids are tolerant of high salinities once the juvenile stage is reached. Virginia bass acclimated to 25 ppt were reared for 173 d without mortality (Kerby and Joseph 1979). In addition, Williams et al. (1981) cultured palmetto bass in salinities up to 28 ppt, and Wattendorf and Shafland (1982) demonstrated that juvenile palmetto bass can tolerate direct transfer to 36 ppt and subsequent return to fresh water.

Salinity can also affect other factors. Hazel et al. (1971) found that striped bass were more resistant to ammonia toxicity in brackish (about 11 ppt) water at 23°C than in either fresh or undiluted seawater. At 15°C, resistance was the same in fresh and brackish water, but was reduced in seawater.

Low salinities are recommended as measures to improve survival and reduce stress during harvest, handling, and transportation. Barwick (1974) found that, although plankton populations were reduced, mean survival of larvae to the fingerling stage was higher in ponds brought to 1 ppt salinity with salt (NaCl) than in untreated ponds. Hughes (1975) reported that fish handled better and were less stressed when salt was added to hatchery ponds before harvest. A 10 ppt reconstituted seawater solution was recommended by Bonn et al. (1976) for reduction of osmotic shock caused by handling stress during transportation of fingerlings. Lewis et al. (1981) recommended against the use of salt in recirculating systems because of corrosion problems and reduction in the efficiency of nitrifying bacteria in biofilters.

E. ALKALINITY, pH, AND TOTAL HARDNESS

Striped bass tolerances to pH have not been well-defined, although it is known that larvae are sensitive to pH changes and that tolerance is age-related. Doroshev (1970) reported 100% mortality after transfer of larvae to water with pH differences of only 0.8 to 1.0. However, Sager et al. (1986) found that 19- to 20-d-old larvae survived chronic exposure to pH levels below 6. Tatum et al. (1966) reported a 24-h LC_{50} value for striped bass fingerlings of pH 5.3, but Sager et al. (1986) observed that there was no significant mortality of fingerlings held in water with a pH $\leq$5.3 buffered with sodium bicarbonate. Hill et al. (1981a; 1981b) found that juvenile striped bass prefer slightly alkaline water and avoid water with pH below 6.6, which is consistent with findings that acid water results in a drop in oxygen-carrying capacity. Those authors found that aversion to acid conditions is so great that the fish will select areas with low DO levels to remain at a favorable pH (7.1 to 7.2). According to Humphries and Cumming (1973), the upper lethal limit has been placed at pH 10.0, but Regan et al. (1968) reported high survival of striped bass in ponds with that pH level. Lewis et al. (1981) reared striped bass in a recirculating system at pH values of 6.7 to 8.4 with no deleterious effects, but they preferred to maintain levels at 7.0 to 8.0 to ensure optimum conditions for biofiltration. Bonn et al. (1976) considered pH values of 7.5 to 8.5 to be optimum for striped bass culture and suggested that pH should not be outside the range of 7 to 9.5. Kraeuter and Woods (1987) considered a pH range of 6.7 to 8.5 optimum for intensive striped bass culture. Parker (1984) reported that, in a survey of 57 striped bass hatcheries, the water from the 11 most productive was slightly basic (mean pH 7.3) and well buffered (mean alkalinity 195 mg/l), whereas water from the other hatcheries was acidic (mean pH 6.4) and not as well buffered (mean alkalinity 81 mg/l). Parker also noted that development of dense plankton blooms in ponds with low alkalinity can cause drastic diel shifts because CO_2 is removed during the day to support photosynthesis. As a result, the pH may rise to 10 or 11. At night, if planktonic respiration produces CO_2 in excess of the buffering capacity of the pond, the pH can drop to 6 or less. Addition of agricultural lime ($CaCO_3$) in soft water or acid soil areas can help increase pond buffering capacity and reduce such drastic shifts by stabilizing pond pH (Parker 1984; Brewer and Rees 1990).

In intensive culture systems, alkalinity should be maintained at levels above 150 mg/l (as $CaCO_3$) to maintain the buffering capacity of the water and compensate for nitrous and nitric acids released by nitrifying bacteria (Lewis et al. 1981). Crushed oyster shell is often used as a means of providing carbonate ions, although crushed limestone in a filter is probably more effective than shell. Dolomite was recommended by Siddal (1974) since it contains magnesium as well as calcium.

Water high in alkalinity appears to benefit striped bass, probably because of decreased osmoregulatory stress and because well buffered water is less likely to undergo sudden pH changes. Hazel et al. (1971) could not use water with a hardness of 25 to 30 mg/l in bioassay experiments because survival of the controls was poor, but the fish thrived in the same water if salts were added. Osmotic shock following stress is thought to be a major cause of mortality. Water with total hardness ranging from 150 to 300 mg/l is generally considered excellent for striped bass culture (Bonn et al. 1976; Brewer and Rees 1990). Mauldin et al. (1988) recommended adding calcium to soft water ponds to improve survival. Nevertheless, Reeves and Germann (1972) found that increasing water hardness in rearing ponds from 20 to 150 mg/l with calcium sulfate resulted in lower survival percentages than were obtained in untreated ponds.

Responses of striped bass larvae to both temperature and pH can be altered by increasing the level of dissolved solids. Davies (1973a) reported that optimum conditions, calculated from a regression equation, appeared to be 17.6°C, pH of 7.5, and total dissolved solids at 185.7 mg/l (NaCl).

F. NITROGEN COMPOUNDS

High levels of ammonia, nitrite, and nitrate can result in significant changes in water quality. They are more common problems in intensive culture systems than in ponds. Ammonia is the primary nitrogenous metabolite produced by fish. The unionized form (NH_3) is much more toxic than the ionized form (NH_4^+), and the balance between the forms depends on temperature and pH. As pH increases, the ratio of NH_3 to NH_4^+ is increased; whereas, as the pH decreases, the ratio shifts in the opposite direction. As temperature increases, the ratio shifts in favor of NH_3. As an example, at a pH of 9.4 and 20°C, the ratio of NH_3 to NH_4^+ is 1:1, but at the same temperature and pH 7, almost all (99%) of the ammonia is in the NH_4^+ form, and at pH 12, 99% is in the form of NH_3 (Parker 1984). Lewis et al. (1981) concluded that ammonia is second only to low DO as a major cause of mortality in intensively cultured fish. Hazel et al. (1971) reported that striped bass 96-h LC_{50} values for total ammonia NH_4OH ranged from 1.5 to 2.8 mg/l. Reduced feeding and growth, gill damage, and decreased disease resistance are among the sublethal effects of ammonia toxicity. Bonn et al. (1976) stated that ammonia concentration should not exceed 0.6 mg/l in intensive culture systems, whereas Mullis and Smith (1990) recommended that the unionized form not exceed 0.02 mg/l.

Nitrite (NO_2^-) and nitrate (NO_3^-) are not normally problems in pond culture unless intensive stocking or heavily fertilization is practiced. However, diel shifts in pH can alter the relative toxicity of nitrogen compounds and may sometimes cause problems (Parker 1984). Little is known about the toxicity of nitrite and nitrate, though nitrite combines with hemoglobin and can cause methemoglobinemia or "brown blood disease," which reduces the capacity of hemoglobin to transport oxygen (Spotte 1979). Nitrate is not generally considered to be toxic in the ranges which commonly occur, even in intensive systems. Bonn et al. (1976) recommended that nitrite levels be kept below 0.2 mg/l. Striped bass reportedly tolerated nitrate concentrations over 800 mg/l, but larvae were stressed when levels reached 200 mg/l (Kerby et al. 1983b). Optimum levels for growth were below 38 mg/l (Bonn et al. 1976). According to Lewis et al. (1981), fingerlings survived exposure to nitrite concentrations of up to 1.4 mg/l as NO_2-N (4.6 mg/l as NO_2).

G. TURBIDITY AND LIGHT

Talbot (1966) observed that most streams where striped bass spawn can be characterized as turbid, but there is no evidence that turbidity is a requirement for successful spawning. Setzler et al. (1980), in a comprehensive literature review, found little to indicate that hatching success was affected by suspended sediment levels, but noted that developmental rate was slowed at levels over 1500 mg/l. Morgan et al. (1983) reported that up to 5250 mg/l of suspended sediment did not affect hatch of striped bass eggs, although development rates were slowed significantly at levels above 1500 mg/l. In contrast, Auld and Schubel (1978) reported that turbidities greater than 1000 and 500 mg/l were lethal to striped bass eggs and larvae, respectively.

Light intensity affects behavior and perhaps fry survival. Rees and Cook (1985a) found that direct sunlight reduced survival of palmetto bass larvae in aquaria and ponds and that the effects were linear. Partial shading of ponds improved survival.

Larvae and fingerlings often exhibit a hyperactive response to sudden increases in light intensity (McHugh and Heidinger 1978; Brewer and Rees 1990), and light shock and subsequent fingerling mortalities were reported by Hughes (1975). Bonn et al. (1976) noted that exposure to direct sunlight was deleterious to both larvae and fingerlings.

Photoperiod can influence growth rate and optimum temperature for growth of hybrid striped bass. When Woiwode (1989) altered photoperiod at the rate of 4.3 min/d starting from a 12-h light to 12-h dark cycle, 40-g hybrids grew significantly faster under increasing daylength than they grew under decreasing daylength. The optimum growth temperature shifted to a warmer temperature (about 27.8°C) as daylength increased, and to a cooler temperature (about 25.6°C) as daylength decreased. Kerby et al. (1987a) found a similar phenomenon in ponds. Growth of hybrid striped bass was almost static when mean monthly temperature was about 15°C during November when temperature and daylength were decreasing, but growth increased dramatically in March when the mean monthly temperature was also about 15°C, but daylength and temperature were increasing.

XI. OTHER FACTORS AND PROBLEMS

A. PROCUREMENT OF BROODSTOCK AND SPERM STORAGE

Adult striped bass broodstock are normally collected on or near their spawning grounds. However, hatcheries sometimes have difficulty in obtaining adequate supplies of ripe males (Bayless 1972; Bonn et al. 1976; Texas Instruments 1977). In some locations, males are generally abundant early in the season but scarce later on; in others, males are scarce throughout the season. Additionally, because white bass spawning typically peaks earlier than that of striped bass, collection of male white bass for hybrid production is sometimes difficult. Techniques for cryopreservation of striped bass sperm have been developed on an experimental basis (Kerby 1983, 1984; Kerby and Bodolus 1988). While fertilization rates are not normally as high as those obtained with fresh sperm, up to 87% fertilization has been obtained in small-scale experiments and up to 60% on a production basis from cryopreserved sperm. More than 6 million fry were produced at the Moncks Corner, SC, striped bass hatchery with cryopreserved sperm, and pond experiments demonstrated that survival and growth of the larvae compared favorably with those produced with fresh sperm (Kerby et al. 1985). Experimental results and specific procedures for using cryopreservation techniques for striped bass are available (Parker et al. 1990). However, the techniques still yield results that are too inconsistent to use with confidence for production.

B. GAMETE QUALITY

Poor egg and semen quality occur in various hatcheries (Kerby 1986; Harrell et al. 1990b). Egg quality is often lower early and late in the spawning season than during the peak, as evidenced by lower mean fertilization rates. Failure of the eggs to complete development within the females after hormone injection seems to be more frequent during the early and late portions of the spawning period as well, even though the females appear eligible for induced ovulation. Although the time of ovulation is occasionally miscalculated, resulting in overripe eggs, the expertise at established hatcheries is such that this seldom occurs. With present technology, egg quality cannot be predicted until cell division begins and fertilization percentages are determined. As a result, semen that may be available in limited quantities is frequently used on low quality eggs. Similarly, low quality semen may often be used on high quality eggs; motility alone is not a good criterion for assessment of sperm quality (Kerby 1983, 1984; Parker et al. 1990).

Various aberrations may also occur. Eggs are sometimes observed that, though they have a high fertilization rate, are characterized by fragile, brittle chorions that burst at the slightest pressure. Prolarvae that hatch from those eggs appear normal, but long term survival is questionable. Other abnormalities, such as failure of oil droplets to coalesce into a cohesive oil globule and detachment of oil globules or developing cells from the yolk, are sometimes seen. Because these phenomena are associated with eggs from specific females, they apparently do not result from conditions in the hatchery. Whether such conditions are inherited or are the results of disease, pollutants, or other factors is not known; further research is clearly needed.

C. STRESS, HANDLING, AND TRANSPORT

Striped bass are more sensitive to stresses associated with handling and transport than most warmwater species, and considerable care is required to successfully handle and transport them (Stevens 1958; Gray 1958; Regan et al. 1968; Braschler 1975; Hughes 1975; Bonn et al. 1976; Lewis et al. 1981; Parker et al. 1990). Also, some life stages are more sensitive to stress than others.

Eggs and larvae are transported in sealed plastic bags with water and pure oxygen. The bags are placed in styrofoam containers to prevent rapid temperature changes. Tatum et al. (1966) found that eggs could be transported more successfully when they were incubated for at least 6 h before being packed; 12 to 16 h of incubation provided the best results. At the 12- to 16-h stage, unfertilized eggs turn opaque and are more buoyant than live eggs and can be siphoned off. Bayless (1972) and Rees and Harrell (1990) recommended the holding of eggs for at least 24 h prior to shipment, because unfertilized eggs begin to break down and can cause deterioration of water quality. Bayless (1972) recommended shipping no more than 5000 eggs per liter of water. Hatching during transit should be avoided because water fouling can result from disintegration of chorions. Recommended temperatures during transit are 15.5 to 18.3°C (Bayless 1972).

Larvae 1 to 2 d of age may survive transport better than older fish (Bayless 1972; Bonn et al. 1976). Lewis et al. (1981) recommended shipping larvae at 1 or 5 d of age. Larvae should not be exposed to bright light because sudden exposure can result in mortality (Hughes 1975; Bonn et al. 1976; McHugh and Heidinger 1978). For this reason, many agencies stock larval rearing ponds at night. In any case, larvae should be tempered to allow them to adjust to changes in temperature and other aspects of water quality before being introduced into culture systems (Bonn et al. 1976; Lewis et al. 1981).

An apparently critical stage in development occurs during transformation from postlarva to juvenile. Humphries and Cumming (1973) observed that large numbers die during this period (about 21 to 30 d of age). Otwell and Merriner (1975) and Lewis et al. (1981) reported similar losses, although Lewis suggested that the cause may be inadequate nutrition and failure of some postlarvae to convert to prepared feed.

Careful handling of fingerlings during harvest is critical to survival. Striped bass are more excitable than other commonly cultured species and fatal shock may occur, especially if there are additional stresses such as diseases, poor water quality, or sudden exposure to bright light or noise (Regan et al. 1968; Humphries and Cumming 1973; Braschler 1975; Hughes 1975). Regan et al. (1968) reported an incident in which sea gulls harassed fish in the catch basin during harvest, causing frenzied motion and apparent physiological shock that resulted in an initial 25% mortality and complete subsequent mortality in the holding house. Some harvesting stress is apparently reduced if the operation occurs when pond temperatures are lowest and bright sunlight is not a factor. As a result, many agencies draw down their ponds during the night and harvest at night or at first light in the morning. Transfer of the fish to cooler, more highly oxygenated water than from which they were captured also seems helpful. Reeves (1974) noted that mortalities of phase II fingerlings were reduced when harvested from ponds at temperatures of 10°C or less.

Handling stress also results in reduced resistance to parasites and to later bacterial and fungal infections (Gray 1958; Humphries and Cumming 1973; Bonn et al. 1976). Treatment of ponds with potassium permanganate was recommended prior to harvest by Hughes (1975) to reduce parasitic infestations. Parker et al. (1990) recommended transporting striped bass in water with a total hardness of >100 mg/l (as $CaCO_3$) that included a salt concentration of 10 ppt; if it is necessary to transport in soft water (<20 mg/l $CaCO_3$), 50 to 1000 mg/l $CaCl_2$ should be added to the transport medium. Use of furacin and salt as prophylactic treatments following harvest and during hauling have also been recommended (Harper and Jarman 1972; Bonn et al. 1976; Lewis et al. 1981; Parker et al. 1990). Anesthetics, such as quinaldine and MS-222, have been used successfully (Regan et al. 1968; Bonn et al. 1976; Tomasso et al. 1980; Lewis et al. 1981; Parker and Davis 1982; Davis et al. 1982). Tomasso et al. (1980) found that a combination of MS-222 and salt reduced mortality associated with osmoregulatory dysfunction and resultant hypochloremia following handling and hauling stress in striped bass hybrids. Tomasso et al. (1981) found that the rate at which water quality deteriorates is directly related to the stress response; gradual deterioration allowed fish more time to adapt.

D. SWIMBLADDER INFLATION

Failure of the swim bladder to inflate has been recognized as a potential source of striped bass mortality (Doroshev 1970; Texas Instruments 1977; Doroshev and

Cornacchia 1979; Bulak and Heidinger 1980; Doroshev et al. 1981; Lewis et al. 1981; Bennett et al. 1987), but information is lacking on the etiology. Although not conclusively demonstrated, it appears that larvae subjected to intensive culture systems during the first 10 to 12 d of life may be more likely to have lower percentages of inflation than larvae stocked into ponds at 4 to 5 d of age (Harrell et al. 1990b). Doroshev and Cornacchia (1979) reported that the swim bladder normally inflates between the fifth and seventh day after hatching, simultaneously with initiation of feeding. Larvae appeared to gulp atmospheric gas for initial inflation, and the percentage of larvae with inflated swim bladders was higher when fish were reared in containers with well-aerated, turbulent water, but the experimental technique was inadequate to determine whether the larvae required access to the water surface for inflation. Normal inflation did not occur if delayed beyond the critical period. Chapman et al. (1988) also suggested that larval access to air is needed at the water surface for proper swim bladder inflation.

Lewis et al. (1981) observed that fish lacking an inflated swim bladder were smaller than normal and usually swam at or near the surface with their heads near the surface. Observations of larvae apparently attempting to break the surface film prompted the investigators to recommend frequent removal of surface film on culture tanks during the period when inflation occurs. Doroshev et al. (1981) reported that histological work suggests that the mode of swim bladder inflation may be secretory in nature, but the mechanism is still not understood. Fish lacking swim bladders in culture ponds (termed "stargazers") have been observed swimming at or near the surface with their heads near the water surface (Kerby 1986).

E. DISEASES

Most diseases that commonly afflict warmwater fish are also found in association with striped bass. Unnatural and stressful conditions that occur during culture encourage infections, and only hypoxia causes more mortality than disease in cultured fish (Bonn et al. 1976; Hughes et al. 1990). Most striped bass pathogens are normal inhabitants of water (Paperna and Zwerner 1976) and are unlikely to cause problems if the immune system of the fish is not compromised by a stressor. Bacterial infections and protozoan parasites cause the most problems, though metazoan parasites such as trematodes (both digenetic and monogenetic), copepods, and acanthocephalans can also be harmful. Fungal infections are common following handling and are usually secondary to other stressors.

Columnaris, caused by *Flexibacter columnaris*, is perhaps the most serious bacterial disease encountered by striped bass culturists. It occurs both in ponds and in intensive culture systems and is difficult to control (Kelley 1967; Regan et al. 1968; Ray and Wirtanen 1970; Bonn et al. 1976; Hawke 1976; McIlwain 1976; Mitchell 1984; Hughes et al. 1990). Vibriosis, caused by *Vibrio* spp., is ubiquitous in marine and brackish water and could represent a serious problem in those environments but for the fact that it can be prevented by vaccination (Williams et al. 1981; Van Olst et al. 1981). Additionally, terramycin and sulfamerazine are reported to be effective controls for vibriosis (Hughes et al. 1990).

A comprehensive review of diseases affecting striped bass and their identification and treatment is beyond the scope of this chapter. Bonn et al. (1976), Mitchell (1984), and Hughes et al. (1990) provided excellent descriptions of common diseases and recommended treatments. Paperna and Zwerner (1976) described numerous diseases occurring in striped bass in the Chesapeake Bay, and Setzler et al. (1980) summarized the more important parasitic diseases in tabular form. Amos (1985), Bullock (1971), and Post (1987) provided detailed explanations of methods for isolation and identification of fish pathogens.

Results of bioassays on a number of therapeutic chemicals to assess their toxicities were summarized by Wellborn (1969, 1971), Hughes (1975), and Bonn et al. (1976). Additionally, Schnick et al. (1989) and Griffin (1992) summarized the chemicals that have been approved for aquaculture and fishery management. Lewis et al. (1981) reviewed briefly the effects of several treatment chemicals on nitrifying bacteria in biofilters.

XII. THE FUTURE OF STRIPED BASS AND HYBRID CULTURE

Striped bass and hybrid striped bass have excellent potential for continued and expanded recreational fisheries across the U.S. and for commercial culture. There is currently such a significant market for fry and fingerlings from both private and governmental sources that producers are having difficulties meeting demand. The number of commercial growers increased to over 50 during the past few years and is continuing to increase (McVey and Thompson 1990). The new commercial food-fish industry based on hybrid striped bass is expected to continue to expand at a rapid rate.

Additionally, the technology developed over the last 25 years has made it possible to initiate significant enhancement programs for threatened estuarine populations of striped bass. Enhancement efforts began in San Francisco Bay in 1980 and in the Chesapeake Bay in 1985. By 1990, 4.4 million advanced fingerlings had been released in San Francisco Bay, and almost 2.8 million phase II fingerlings had been released in the Chesapeake Bay (Harrell et al. 1990b). It is too soon to evaluate the effects of these enhancement efforts on the populations, but it is conclusive that hatchery fish are contributing significantly to the stocks, particularly in years with poor natural year classes. However, Harrell et al. (1990b) cautioned that strategies should be developed to protect the genetic diversity of wild populations that are enhanced.

Much remains to be learned about the striped bass and its hybrids. For example, the genetics of various striped bass and white bass stocks must be carefully examined and evaluated so that a desirable strain of domestic broodstock can be developed. Desirable characteristics would include rapid growth, good survival, disease resistance, and increased tolerance to a wide range of environmental conditions. Research is also needed on genetic improvement. Since hybridization is possible, it may be feasible to breed additional hardiness and other desirable characteristics into domesticated stocks through selected backcrosses. Hybrid vigor has been demonstrated and should be researched in detail with respect to genetic manipulation of the parental species. Polyploidy may represent another tool for enhancement of growth and other characteristics (Kerby and Harrell 1990). Gene insertion techniques are as yet untested, but may ultimately be useful as tools for altering genetics to obtain more desirable characteristics.

Increased productivity should be better addressed. If fingerling production is to be increased in pond culture, an adequate supply of food must be provided and high water quality maintained. Strategies to control production of desired zooplankton and methods of increasing their production in ponds are needed. Another approach is to increase total production of zooplankton per unit area. This can be done by adding additional nutrients to the system in the form of organic or inorganic fertilizers. However, improved methods of alleviating increased biological and chemical oxygen demands that can result in severe deterioration of water quality are needed. Some progress has been made. For example, Parker (1980) reported that continuous aeration, combined with heavy fertilization and increased stocking densities, resulted in a 2.4-fold increase in fingerlings over that commonly obtained from conventional culture techniques. Average fingerling size was directly correlated with stocking densities and fertilization rate, but even with constant aeration, DO was a limiting factor at the highest fertilization and stocking rate. Mechanical aeration and addition of oxygenated water during critical periods can provide sufficient oxygenation to support standing crops at least as high as 7800 kg/ha of hybrid striped bass (Kerby et al. 1983a, 1987a; Smith et al. 1989).

As more "state-of-the-art" ponds containing plastic liners come into use, fertilization and management strategies must be developed and evaluated to best utilize them. Limited experience has demonstrated that these new ponds do not react like the more traditional earthen ponds, and compensations must be made.

Nutritional requirements of striped bass and hybrid striped bass need to be further addressed and optimum diets developed specifically for these fish. Trout and salmon diets, which result in large amounts of visceral fat, may be suboptimal or uneconomical for *Morone*. Both striped bass and hybrids, as well as their various life stages, should be addressed.

Though still problematical, progress has been made in intensive larval culture. However, only a few facilities have been successful in culturing significant numbers of phase I fingerlings in intensive facilities, and costs and frustration levels are high. Research must more accurately define and address larval physiological and environmental factors and design criteria that will provide the means for successful intensive larval culture.

In view of recent advances in intensive culture, it would be advantageous to explore methods by which the date of larval production could be controlled. Control of environmental and physiological factors such as light, temperature, and hormones is needed to allow producers to adjust timing of spawning to suit production needs. This could result in timing production with market demand. Progress has been made in advanced spawning of both striped bass and hybrids, but results to date have not yielded sufficient larval production to be economically feasible.

Stress and diseases continue to be significant problems in *Morone* culture. It is important to develop better methods for alleviating stress and for developing effective drugs and treatments for diseases. Recent rulings by the U.S. Food and Drug Adminstration have left very few available treatment options.

Gas bladder inflation, or more specifically the lack thereof, continues to be a frustrating problem in some hatcheries. Determination of the mechanism for inflation and development of methods for optimizing conditions inducive to inflation, are high priority research areas.

Energy requirements are a problem for any type of fish culture for the foreseeable future. It is desirable to reduce these costs without reducing production. Cheaper energy sources, such as wind and solar energy are being explored, but more extensive efforts are needed.

REFERENCES

Ager, L. M., Food Habits of Young-of-the-Year Striped Bass in Lake Sinclair, Project No. WC-4, GA Department of Natural Resources, Game and Fish Division, 1979, 1.

Albrecht, A. B., Some observations on factors associated with survival of striped bass eggs and larvae, *Calif. Fish Game*, 50, 100, 1964.

Allen, K. O., Notes on the culture of striped bass in tanks and small raceways, *Prog. Fish-Cult.*, 36, 60, 1974.

Amos, K. H., Ed., *Procedures for the Detection and Identification of Certain Fish Pathogens*, 3rd ed., Fish Health Section, American Fisheries Society, Corvalis, OR, 1985, 1.

Anderson, J. C., Production of striped bass fingerlings, *Prog. Fish-Cult.*, 28, 162, 1966.

Atstupenas, E. A. and Wright, L. D., Interim Rearing Guidelines for Phase II Striped Bass, U. S. Fish and Wildlife Service, Atlanta, GA, 1987, 1.

Avise, J. C. and Van Den Avyle, M. J., Genetic analysis of reproduction of hybrid white bass × striped bass in the Savannah River, *Trans. Am. Fish. Soc.*, 113, 563, 1984.

Axon, J. R. and Whitehurst, D. K., Striped bass management in lakes with emphasis on management problems, *Trans. Am. Fish. Soc.*, 114, 8, 1985.

Auld, A. H. and Schubel, J. R., Effects of suspended sediment on fish eggs and larvae: a laboratory assessment, *East Coast. Mar. Sci.*, 6, 153, 1978.

Bailey, W. M., An evaluation of striped bass introductions in the southeastern U.S., *Proc. S.E. Assoc. Game Fish Comm.*, 28, 54, 1975.

Barkaloo, J. M., Florida striped bass, *Fla. Game Freshwater Fish Comm. Fish. Bull.*, 4, 1, 1967.

Barwick, D. H., The effect of increased sodium chloride on striped bass fry survival in freshwater ponds, *Proc. S.E. Assoc. Game Fish Comm.*, 27, 415, 1974.

Bayless, J. D., Striped bass hatching and hybridization experiments, *Proc. S.E. Assoc. Game Fish Comm.*, 21, 233, 1968.

Bayless, J. D., Artificial Propagation and Hybridization of Striped Bass, *Morone saxatilis* (Walbaum), South Carolina Wildlife and Marine Resources Department, Columbia, 1972, 1.

Bennett, R. O., Kraeuter, J. N., Woods, L. C., III, Lipsky, M. M. and May, E. B., Histological evaluation of swim bladder non-inflation in striped bass larvae, *Morone saxatilis*, *Dis. Aquat. Org.*, 3, 91, 1987.

Berger, A. and Halver, J. E., Effect of dietary protein, lipid and carbohydrate content on the growth, feed efficiency and carcass composition of striped bass, *Morone saxatilis* (Walbaum), fingerlings, *Aquacult. Fish. Man.*, 18, 345, 1987.

Berry, C. R., Jr., Helm, W. T., and Neuhold, J. M., Safety in fishery field work, in *Fisheries Techniques*, Nielsen, L. A. and Johnson, D. L., Eds., American Fisheries Society, Bethesda, MD, 1983, 43.

Bishop, R. D., Evaluation of the striped bass (*Roccus saxatilis*) and white bass (*R. chrysops*) hybrids after two years, *Proc. S.E. Assoc. Game Fish Comm.*, 21, 245, 1968.

Bishop, R. D., The use of circular tanks for spawning striped bass (*Morone saxatilis*), *Proc. S.E. Assoc. Fish Game Comm.*, 28, 35, 1975.

Bonn, E. W., Bailey, W. M., Bayless, J. D., Erickson, K. E., and Stevens, R. E., Eds., *Guidelines for Striped Bass Culture*, Striped Bass Committee, Southern Division, American Fisheries Society, Bethesda, MD, 1976, 1.

Bowker, R. G., Baumgartner, D. J., Hutcheson, J. A., Ray, R. H., and Wellborn, T. L., Jr., Striped Bass *Morone saxatilis* (Walbaum): 1968 Report on the Development of Essential Requirements for Production, U. S. Fish and Wildlife Service, Atlanta, GA, 1968, 1.

Bowman, J. R., Survival and Growth of Striped Bass, *Morone saxatilis* (Walbaum), Fry Fed Artificial Diets in a Closed Recirculation System, Ph.D. dissertation, Auburn University, Auburn, AL, 1979, 1.

Boyd, C. E., *Water Quality in Warmwater Fish Ponds*, Auburn University Agricultural Experiment Station, Auburn, AL, 1979, 1.

Boyd, C. E., *Water Quality Management for Pond Fish Culture*, Elsevier, New York, 1982, 1.

Braid, M. R. and Shell, E. W., Incidence of cannibalism among striped bass fry in an intensive culture system, *Prog. Fish-Cult.*, 43, 210, 1981.

Braschler, E. W., Development of pond culture techniques for striped bass *Morone saxatilis* (Walbaum), *Proc. S.E. Assoc. Game Fish Comm.*, 28, 44, 1975.

Brewer, D. L. and Rees, R. A., Pond culture of phase I striped bass fingerlings, in *Culture and Propagation of Striped Bass and its Hybrids*, Harrell, R. M., Kerby, J. H., and Minton, R. V., Eds., Striped Bass Committee, Southern Division, American Fisheries Society, Bethesda, MD, 1990, 99.

Bulak, J. S. and Heidinger, R. C., Developmental anatomy and inflation of the gas bladder in striped bass, *Morone saxatilis*, *Fish. Bull.*, 77, 1000, 1980.

Bullock, G. L., The identification of fish pathogenic bacteria, in *Diseases of Fishes*, Snieszko, S. F. and Axelrod, H. R., Eds., T. F. H. Publications, Neptune City, NJ, book 2B, 1971, 1.

Bullock, G. L., Vibriosis in Fish, U. S. Fish and Wildlife Service Fish Disease Leaflet 77, Washington, D. C., 1977, 1.

Carlberg, J. M. and Van Olst, J. C., Processing and marketing, in *Hybrid Striped Bass Culture: Status and Perspective*, Hodson, R., Smith, T., McVey, J., Harrell, R., and Davis, N., Eds., UNC Sea Grant Publication UNC-SG-87-03, North Carolina State University, Raleigh, 1987, 73.

Carlberg, J. M., Van Olst, J. C., Massingill, M. J., and Hovanec, T. A., Intensive culture of striped bass: a review of recent technological developments, in *The Aquaculture of Striped Bass: A Proceedings*, McCraren, J. P., Ed., MD Sea Grant Publication UM-SG-MAP-84-01, University of MD, College Park, 1984, 89.

Chapman, D. C., Hubert, W. A., and Jackson, U. T., Influence of access to air and of salinity on gas bladder inflation in striped bass, *Prog. Fish-Cult.*, 50, 23, 1988.

Chapman, R. W., Spatial and temporal variation of mitochondrial DNA haplotype frequencies in the striped bass (*Morone saxatilis*) 1982 year class, *Copeia*, 1989, 344, 1989.

Chittenden, M. E., Jr., Effects of handling and salinity on oxygen requirements of the striped bass, *Morone saxatilis*, *J. Fish. Res. Bd. Can.*, 28, 1823, 1971.

Collins, C. M., Burton, G. L., and Schweinforth, R. L., High density culture of white bass × striped bass fingerlings in raceways using power plant effluents, in *The Aquaculture of Striped Bass: A Proceedings*, McCraren, J. P., Ed., MD Sea Grant Publication UM-SG-MAP-84-01, University of MD, College Park, 1983, 129.

Colura, R. L., Hysmith, B. T., and Stevens, R. E., Fingerling production of striped bass (*Morone saxatilis*), spotted seatrout (*Cynoscion nebulosis*), and red drum (*Sciaenops ocellatus*), in saltwater ponds, *Proc. World Maricult. Soc.*, 7, 79, 1976.

Cox, D. K. and Coutant, C. C., Growth dynamics of juvenile striped bass as functions of temperature and ration, *Trans. Am. Fish. Soc.,* 110, 226, 1981.

Crawford, T., Freeze, M., Fourt, R., Henderson, G., O'Bryan, G., and Phillip, D., Suspected natural hybridization of striped bass in two Arkansas reservoirs, *Proc. S.E. Assoc. Fish Wildl. Agen.,* 38, 455, 1987.

Daniel, D. A., A Laboratory Study to Define the Relationship Between Survival of Young Striped Bass (*Morone saxatilis*) and Their Food Supply, Administrative Report No. 76-1, California Department of Fish and Game, 1976, 1.

Davies, W. D., The effects of total dissolved solids, temperature, and pH on the survival of immature striped bass: a response surface experiment, *Prog. Fish-Cult.,* 35, 157, 1973a.

Davies, W. D., Rates of temperature acclimation for hatchery reared striped bass fry and fingerlings, *Prog. Fish-Cult.,* 35, 214, 1973b.

Davis, K. B., Parker, N. C., and Suttle, M. A., Plasma corticosteroids and chlorides in striped bass exposed to tricaine methanesulfonate, quinaldine, etomidate, and salt, *Prog. Fish-Cult.,* 44, 205, 1982.

Dicks, M. and Harvey, D., Aquaculture situation and outlook report, Aqua-1, Economic Research Service, U. S. Department of Agriculture, Washington, D. C., 1989, 11.

Dicks, M. and Harvey, D., Aquaculture situation and outlook report, Aqua-4, Economic Research Service, U. S. Department of Agriculture, Washington, D. C., 1990, 13.

Doroshev, S. I., Biological features of the eggs, larvae and young of the striped bass [*Roccus saxatilis* (Walbaum)] in connection with the problem of acclimatization in the U.S.S.R., *J. Ichthyol.,* 10, 235, 1970.

Doroshev, S. I. and Cornacchia, J. W., Initial swim bladder inflation in the larvae of *Tilipia mossambica* (Peters) and *Morone saxatilis* (Walbaum), *Aquaculture,* 16, 57, 1979.

Doroshev, S. I., Cornacchia, J. W, and Hogan, K., Initial swim bladder inflation in the bladder of physoclistus fishes and its importance in larval culture, *Rapp. P. V. Réun. Cons. Int. Explor. Mer,* 178, 495, 1981.

Easley, J. E., Economic research: the striped bass-white bass hybrid, in *Hybrid Striped Bass Culture: Status and Perspective,* Hodson, R., Smith, T., McVey, J., Harrell, R., and Davis, N., Eds., UNC Sea Grant Publication UNC-SG-87-03, North Carolina State University, Raleigh, 1987, 83.

Eldridge, M. B., King, D. J., Eng, D., and Bowers, M. J., Role of the oil globule in survival and growth of striped bass (*Morone saxatilis*) larvae, *Proc. West. Assoc. Game Fish Comm.,* 57, 303, 1977.

Eldridge, M. B., Whipple, J. A., Eng, D., Bowers, M. J., and Jarvis, B. M., Effects of food and feeding factors on laboratory-reared striped bass larvae, *Trans. Am. Fish. Soc.,* 110, 111, 1981.

Falls, W. W., Food Habits and Feeding Selectivity of Larvae of the Striped Bass *Morone saxatilis* (Walbaum) (Osteichthys; Percichthidae) Under Intensive Culture Conditions, Ph.D. dissertation, University of Southern Mississippi, Hattiesburg, 1983, 1.

Fitzmayer, K. M., Geiger, J. G., and Van Den Avyle, M. J., Acute toxicity effects of simazine on *Daphnia pulex* and larval striped bass, *Proc. S.E. Assoc. Fish Wildl. Agen.,* 36, 146, 1985.

Fitzmayer, K. M., Broach, I., and Estes, R. D., Effects of supplemental feeding on growth, production, and feeding habits of striped bass in ponds, *Prog. Fish-Cult.,* 48, 18, 1986.

Forrester, C. R., Peden, A. E., and Wilson, R. M., First records of the striped bass, *Morone saxatilis,* in British Columbia waters, *J. Fish. Res. Bd. Can.,* 29, 337, 1972.

Forshage, A. A., Harvey, W. D., Kulzer, K. E., and Fries, L. T., Natural reproduction of white bass × striped bass hybrids in a Texas reservoir, *Proc. S.E. Assoc. Fish Wildl. Agen.,* 40, 9, 1988.

Fries, L. T. and Harvey, W. D., Natural hybridization of white bass with yellow bass in Texas, *Trans. Am. Fish. Soc.,* 118, 98, 1989.

Fuller, J. C., Jr., South Carolina's Striped Bass Story, South Carolina Wildlife Resources Department, 1968, 1.

Geiger, J. G., A review of pond zooplankton production and fertilization for the culture of larval and fingerling striped bass, *Aquaculture,* 35, 353, 1983a.

Geiger, J. G., Zooplankton production and manipulation in striped bass rearing ponds, *Aquaculture*, 35, 331, 1983b.

Geiger, J. G. and Parker, N. C., Survey of striped bass hatchery management in the southeastern U.S., *Prog. Fish-Cult.*, 47, 1, 1985.

Geiger, J. G. and Turner, J. C., Pond fertilization and zooplankton management techniques for production of fingerling striped bass and hybrid striped bass, in *Culture and Propagation of Striped Bass and its Hybrids*, Harrell, R. M., Kerby, J. H., and Minton, R. V., Eds., Striped Bass Committee, Southern Division, American Fisheries Society, Bethesda, MD, 1990, 79.

Gomez, R., Food habits of young-of-the-year striped bass, *Roccus saxatilis* (Walbaum), in Canton reservoir, *Proc. Okla. Acad. Sci.*, 50, 79, 1970.

Gray, D. L., Striped bass for Arkansas?, *Proc. S.E. Assoc. Game Fish Comm.*, 11, 287, 1958.

Griffin, B. R., Status of chemicals for use in warmwater fish production, *Aquaculture Mag.*, 18(1), 76, 1992.

Harper, J. L. and Jarman, R., Investigation of striped bass, *Morone saxatilis* (Walbaum), culture in Oklahoma, *Proc. S.E. Assoc. Game Fish Comm.*, 25, 501, 1972.

Harper, J. L., Jarman, R., and Yacovino, J. T., Food habits of young striped bass, *Roccus saxatilis* (Walbaum), in culture ponds, *Proc. S.E. Assoc. Game Fish Comm.*, 22, 373, 1969.

Harrell, R. M., Identification of hybrids of the *Morone* complex (Percichthyidae) by means of osteological patterns, meristics, and morphometrics, Ph.D. dissertation, University of South Carolina, Columbia, 1984a, 1.

Harrell, R. M., Tank spawning of first generation striped bass × white bass hybrids, *Prog. Fish-Cult.*, 46, 75, 1984b.

Harrell, R. M. and Bayless, J. D., Effects of suboptimal oxygen concentrations on developing striped bass embryos, *Proc. S.E. Assoc. Fish Wildl. Agen.*, 35, 508, 1983.

Harrell, R. M. and Bukowski, The culture, zooplankton dynamics and predator-prey interactions of Chesapeake Bay striped bass, *Morone saxatilis* (Walbaum), in estuarine ponds, *Aquacult. Fish. Man.*, 21, 25, 1990.

Harrell, R. M. and Dean, J. M., Pterygiophore interdigitation patterns and morphometry of larval hybrids of *Morone* species, *Trans. Am. Fish. Soc.*, 116, 719, 1987.

Harrell, R. M. and Dean, J. M., Identification of juvenile hybrids of *Morone* based on meristics and morphometrics, *Trans. Am. Fish. Soc.*, 117, 529, 1988.

Harrell, R. M. and Moline, M. A., Plasma corticosteroid and chloride dynamics of striped bass as indicators of stress when comparing two capture techniques, *J. World Aquacult. Soc.*, 19, 35A, 1988.

Harrell, R. M., Loyacano, H. A., Jr., and Bayless, J. D., Zooplankton availability and selectivity of fingerling striped bass, *Ga. J. Sci.*, 35, 129, 1977.

Harrell, R. M., Meritt, D. W., Hochheimer, J. N., Webster, D. W. and Miller, W. D., Overwintering success of striped bass and hybrid striped bass held in cages in MD, *Prog. Fish-Cult.*, 50, 120, 1988.

Harrell, R. M., Kerby, J. H., and Minton, R. V., Eds., *Culture and Propagation of Striped Bass and Its Hybrids*, Striped Bass Committee, Southern Division, American Fisheries Society, Bethesda, MD, 1990a, 1.

Harrell, R. M., Kerby, J. H., Smith, T. I. J., and Stevens, R. E., Striped bass and striped bass hybrid culture: the next twenty-five years, in *Culture and Propagation of Striped Bass and its Hybrids*, Harrell, R. M., Kerby, J. H., and Minton, R. V., Eds., Striped Bass Committee, Southern Division, American Fisheries Society, Bethesda, MD, 1990b, 253.

Harvey, W. D. and Fries, L. T., Identification of *Morone* species and congeneric hybrids using isoelectric focusing, *Proc. S.E. Assoc. Fish Wildl. Agen.*, 41, 251, 1988.

Hawke, J. P., A survey of the diseases of striped bass, *Morone saxatilis* and pompano, *Trachinotus carolinus* cultured in earthen ponds, *Proc. World Maricult. Soc.*, 7, 495, 1976.

Hazel, C. R., Thomsen, W., and Meith, S. J., Sensitivity of striped bass and stickleback to ammonia in relation to temperature and salinity, *Calif. Fish Game*, 57, 154, 1971.

Heubach, W., Toth, R. J., and McCready, A.M., Food of young-of-the-year striped bass (*Roccus saxatilis*) in the Sacramento-San Jaoquin River System, *Calif. Fish Game*, 49, 224, 1963.

Hill, L. G., Schnell, G. D., and Matthews, W. J., Locomotor responses of the striped bass, *Morone saxatilis*, to environmental variables, *Am. Midl. Nat.*, 105, 139, 1981a.

Hill, L. G., Schnell, G. D., and Matthews, W. J., Locomotor Responses of the Striped Bass, *Morone saxatilis* (Walbaum), Culture Investigations in Louisiana with Notes on Sensitivity of Fry and Fingerlings to Various Chemicals, Fish. Bull. 113, Louisiana Wildlife and Fisheries Commission, Baton Rouge, 1981b, 1.

Hubert, W. A., Passive capture techniques, in *Fisheries Techniques*, Nielsen, L. A. and Johnson, D. L., Eds., American Fisheries Society, Bethesda, MD, 1983, 95.

Hughes, J. S., Striped bass, *Morone saxatilis* (Walbaum), Culture Investigations in Louisiana Notes of Sensitivity of Fry and Fingerlings to Various Chemicals, Fish. Bull. 113, Louisiana Wildlife and Fisheries Commission, Baton Rouge, 1975, 1.

Hughes, J. S., Wellborn, T. L., and Mitchell, A. J., Parasites and diseases of striped bass and hybrids, in *Culture and Propagation of Striped Bass and its Hybrids*, Harrell, R. M., Kerby, J. H., and Minton, R. V., Eds., Striped Bass Committee, Southern Division, American Fisheries Society, Bethesda, MD, 1990, 217.

Huish, M. T., Kerby, J. H., and Hinshaw, J. M., Commercial production and sale of striped bass hybrids in a North Carolina farm pond, *J. World Aquacult. Soc.*, 18, 8A, 1987.

Humphries, E. T. and Cumming, K. B., Food habits and feeding selectivity of striped bass fingerlings in culture ponds, *Proc. S.E. Assoc. Game Fish Comm.*, 25, 522, 1972.

Humphries, E. T. and Cumming, K. B., An evaluation of striped bass fingerling culture, *Trans. Am. Fish. Soc.*, 102, 13, 1973.

Huner, J. V. and Dupree, H. K, Pond management, in *Third Report to the Fish Farmers*, Hunter, J. V., and Dupree, H. K, Eds., U.S. Fish and Wildlife Service, Washington, D.C, 1984, 17.

Inslee, T. D., Holding striped bass larvae in cages until swim-up, *Proc. S.E. Assoc. Fish Wildl. Agen.*, 31, 422, 1977.

Jenkins, W. E., Laws and regulations, in *Hybrid Striped Bass Culture: Status and Perspective*, Hodson, R., Smith, T., McVey, J., Harrell, R., and Davis, N., Eds., UNC Sea Grant Publication UNC-SG-87-03, North Carolina State University, Raleigh, 1987, 93.

Jenkins, W. E. and Smith, T. I. J., Natural and induced production of striped bass hybrids in tanks, *Proc. S.E. Assoc. Fish Wildl. Agen.*, 39, 255, 1987.

Jenkins, W. E., Smith, T. I. J., Stokes, A. D., and Smiley, R. A., Effect of stocking density on production of advanced juvenile hybrid striped bass, *Proc. S.E. Assoc. Fish Wildl. Agen.*, 42, 56, 1989.

Jordan, D. S. and Evermann, B. W., *American Food and Game Fishes*, Doubleday, Page and Company, New York, 1902, 1.

Kelley, J. R., Jr., Preliminary report on methods for rearing striped bass, *Roccus saxatilis* (Walbaum), fingerlings, *Proc. S.E. Assoc. Game Fish Comm.*, 20, 341, 1967.

Kerby, J. H., Feasibility of Artificial Propagation and Introduction of Hybrids of the *Morone* Complex into Estuarine Environments, with a Meristic and Morphometric Description of the Hybrids, Ph.D. dissertation, University of Virginia, Charlottesville, 1972, 1.

Kerby, J. H., Meristic characteris of two *Morone* hybrids, *Copeia*, 1979, 513, 1979.

Kerby, J. H., Morphometric characteris of two *Morone* hybrids, *Proc. S.E. Assoc. Fish Wildl. Agen.*, 33, 344, 1980.

Kerby, J. H., unpublished data.

Kerby, J. H., Cryogenic preservation of sperm from striped bass, *Trans. Am. Fish. Soc.*, 112, 86, 1983.

Kerby, J. H., Cryopreservation of striped bass spermatozoa: problems and progress, in *The Aquaculture of Striped Bass: a Proceedings*, McCraren, J. P., Ed., MD Sea Grant Publication UM-SG-MAP-84-01, University of MD, College Park, 1984, 205.

Kerby, J. H., Striped bass and striped bass hybrids, in *Culture of Nonsalmonid Freshwater Fishes*, CRC Press, Boca Raton, FL, 1986, 127.

Kerby, J. H. and Bodolus, D. A., Cryogenic Preservation of Sperm from Striped Bass, Final Report, Contract 14-16-0009-79-132, U.S. Fish and Wildlife Service, 1988, 1.

Kerby, J. H. and Harrell, R. M., Hybridization, genetic manipulation, and gene pool conservation of striped bass, in *Culture and Propagation of Striped Bass and its Hybrids*, Harrell, R. M., Kerby, J. H., and Minton, R. V., Eds., Striped Bass Committee, Southern Division, American Fisheries Society, Bethesda, MD, 1990, 159.

Kerby, J. H. and Joseph, E. B., Growth and survival of striped bass and striped bass × white bass hybrids, *Proc. S.E. Assoc. Game Fish Comm.*, 32, 715, 1979.

Kerby, J. H., Burrell, V. G., Jr., and C. E. Richards, Occurrence and growth of striped bass × white bass hybrids in the Rappahannock Rivber, Virginia, *Trans. Am. Fish. Soc.*, 100, 787, 1971.

Kerby, J. H., Bayless, J. D., and Harrell, R. M., Growth, survival, and harvest of striped bass produced with cryopreserved spermatozoa, *Trans. Am. Fish. Soc.*, 114, 761, 1985.

Kerby, J. H., Woods, L. C., III, and Huish, M. T., Culture of the striped bass and its hybrids: a review of methods, advances, and problems, in *Proc. Warmwater Fish Culture Workshop*, Stickney, R. R., and Meyers, S. P., Eds., Spec. Publ. No. 3, World Mariculture Society, 1983a, 23.

Kerby, J. H., Woods, L. C., III, and Huish, M. T., Pond culture of striped bass × white bass hybrids, *J. World Maricult. Soc.*, 14, 613, 1983b.

Kerby, J. H., Hinshaw, J. M., and Huish, M. T., Increased growth and production of striped bass × white bass hybrids in earthen ponds, *J. World Aquacult. Soc.*, 18, 35, 1987a.

Kerby, J. H., Huish, M. T., Klar, G. T., and Parker, N. C., Comparative growth and survival of two striped bass hybrids, a backcross, and striped bass in earthen ponds, *J. World Aquacult. Soc.*, 18, 10A, 1987b.

Klar, G. T. and Parker, N. C., Evaluation of five commercially prepared diets for striped bass, *Prog. Fish-Cult.*, 51, 115, 1989.

Klyashtorin, L. B. and Yarzhombek, A. A., Some aspects of the physiology of the striped bass, *Morone saxatilis*, *J. Ichthyol.*, 15, 985, 1975.

Kraeuter, J. N. and Woods, C., III, Culture of striped bass and its hybrids: first feeding to six months, in Hybrid Striped Bass Culture: Status and Perspective, Hodson, R., Smith, T., McVey, J., Harrell, R., and Davis, N., Eds., UNC Sea Grant Publication UNC-SG-87-03, North Carolina State University, Raleigh, 1987, 23.

Lal, K., Lasker, R., and Klujis, A., Acclimation and rearing of striped bass larvae in sea water, *Calif. Fish Game*, 63, 210, 1977.

Lemarié, D. P., U.S. Fish and Wildlife Service, personal communication, 1991.

Lewis, W. M., Heidinger, R. C., and Tetzlaff, B. L., Tank Culture of Striped Bass, Fisheries Research Laboratory, Southern Illinois University, Carbondale, 1981, 1.

Liao, D. S., The economic and market potential for hybrid bass aquaculture in estuarine waters: a preliminary evaluation, *J. World Maricult. Soc.*, 16, 151, 1985.

Liao, D. S. and Smith, T. I. J., Preliminary market analysis for cultured hybrid striped bass, Symp. *Markets for Seafood and Aquacultural Products,* Charleston, SC, August 19 to 21, 1987, 1.

Logan, H. J., Comparison of growth and survival rates of striped bass and striped bass × white bass hybrids under controlled environments, *Proc. S.E. Assoc. Game Fish Comm.*, 21, 260, 1968.

Markle, D. F. and Grant, G. C., The summer food habits of young-of-the-year striped bass in three Virginia rivers, *Chesapeake Sci.*, 11, 50, 1970.

Mauldin, A. C., II, Grizzle, J. M., Young, D. E., and Henderson, H. E., Use of additional calcium in soft-water ponds for improved striped bass survival, *Proc. S.E. Assoc. Fish Wildl. Agen.*, 40, 163, 1988.

McCarty, C. E., Geiger, J. G., Sturmer, L. N., Gregg, B. A., and Rutledge, W. P., Marine finfish culture in Texas: a model for the future, in *Fish Culture in Fisheries Management,* Stroud, R. H., Ed., Fish Culture and Fisheries Management Sections, American Fisheries Society, Bethesda, MD, 1986, 249.

McGill, E. M., Jr., Pond water for rearing striped bass fry, *Roccus saxatilis* (Walbaum), in aquaria, *Proc. S.E. Assoc. Game Fish Comm.*, 20, 331, 1967.

McHugh, J. J. and Heidinger, R. C., Effects of light on feeding and egestion time of striped bass fry, *Prog. Fish-Cult.,* 39, 33, 1977.

McHugh, J. J. and Heidinger, R. C., Effects of light shock and handling shock on striped bass fry, *Prog. Fish-Cult.,* 40, 82, 1978.

McIlwain, T. D., Distribution of the striped bass, *Roccus saxatilis* (Walbaum), in Mississippi waters, *Proc. S.E. Assoc. Game Fish Comm.,* 21, 254, 1968.

McIlwain, T. D., Closed recirculating system for striped bass production, *Proc. World Maricult. Soc.,* 7, 523, 1976.

McVey, E. M. and Thompson, N., Culture of Striped and Hybrid Striped Bass, U.S. Department of Agriculture, Washington, D.C., 1990, 1.

Meshaw, J. C., Jr., A Study of Feeding Selectivity of Striped Bass Fry and Fingerlings in Relation to Plankton Availability, Master's thesis, North Carolina State University, Raleigh, 1969, 1.

Millikin, M. R., Effects of dietary protein concentration on growth, feed efficiency, and body composition of age-0 striped bass, *Trans. Am. Fish. Soc.,* 111, 373, 1982.

Millikin, M. R., Interactive effects of dietary protein and lipid on growth and protein utilization of age-0 striped bass, *Trans. Am. Fish. Soc.,* 112, 185, 1983.

Minton, R. V. and Harrell, R. M., The culture of striped bass and hybrids in brackish water, in *Culture and Propagation of Striped Bass and its Hybrids,* Harrell, R. M., Kerby, J. H., and Minton, R. V., Eds., Striped Bass Committee, Southern Division, American Fisheries Society, Bethesda, MD, 1990, 243.

Mitchell, A. J., Parasites and diseases of striped bass, in The Aquaculture of Striped Bass: A Proceedings, McCraren, J. P., Ed., MD Sea Grant Publication UM-SG-MAP-84-01, University of MD, College Park, 1984, 177.

Morgan, R. P., II, Rasin, V. J., Jr., and Capp, R. L., Temperature and salinity effects on development of striped bass eggs and larvae, *Trans. Am. Fish. Soc.,* 110, 95, 1981.

Morgan, R. P., II, Rasin, V. J., Jr., and Noe, L. A., Sediment effects on eggs and larvae of striped bass and white perch, *Trans. Am. Fish. Soc.,* 112, 220, 1983.

Mulligan, T. J. and Chapman, R. W., Mitochondrial DNA analysis of Cheasapeake Bay white perch, *Morone americana, Copeia,* 1989, 679, 1989.

Mullis, A. W. and Smith, J. M., Design considerations for striped bass and striped bass hybrid hatching facilities. The culture of striped bass and hybrids in brackish water, in *Culture and Propagation of Striped Bass and Its Hybrids,* Harrell, R. M., Kerby, J. H., and Minton, R. V., Eds., Striped Bass Committee, Southern Division, American Fisheries Society, Bethesda, MD, 1990, 7.

Nicholson, L. C., Culture of striped bass (*Morone saxatilis*) in raceways under controlled conditions, presented at *West. Assoc. State Game Fish Comm.,* 1973.

Nicholson, L. C., Woods, L. C., III, and Woidwode, J. G., Intensive culture techniques for the striped bass and its hybrids, in *Culture and Propagation of Striped Bass and its Hybrids,* Harrell, R. M., Kerby, J. H., and Minton, R. V., Eds., Striped Bass Committee, Southern Division, American Fisheries Society, Bethesda, MD, 1990, 141.

Nicholson, L. C., personal communication, 1982.

Otwell, W. S. and Merriner, J. V., Survival and growth of juvenile striped bass, *Morone saxatilis,* in a factorial experiment with temperature, salinity and age, *Trans. Am. Fish. Soc.,* 104, 560, 1975.

Paperna, I. and Zwerner, D. E., Parasites and diseases of striped bass, *Morone saxatilis* (Walbaum), from the lower Chesapeake Bay, *J. Fish Biol.,* 9, 267, 1976.

Parker, N. C., Channel catfish production in continuously aerated ponds, in Proc. Fish Farming Conf. Annu. Conv., Fish Farmers of Texas, Texas A&M University, College Station, 1979, 39.

Parker, N. C., Striped bass culture in continuously aerated ponds, *Proc. S.E. Assoc, Fish Wildl. Agen.,* 33, 353, 1980.

Parker, N. C., Culture requirements for striped bass, in The Aquaculture of Striped Bass: A Proceedings, McCraren, J. P., Ed., University of MD Sea Grant Publ. UM-SG-MAP-84-01, College Park, 1984, 29.

Parker, N. C. and Davis, K. B., Requirements of warmwater fish, in *Proc. Bioeng. Symp. Fish Culture,* Fish Culture Section Special Publication No. 1, American Fisheries Society, Bethesda, MD, 1981, 21.

Parker, N. C. and Geiger, J. G., Production methods for striped bass, in Third Report to the Fish Farmers: The Status of Warmwater Fish Farming and Progress in Fish Farming Research, Dupree, H. K. and Huner, J. V., Eds., U.S. Fish and Wildlife Service, Washington, D.C., 1984, 106.

Parker, N. C., Klar, G. T., Smith, T. I. J., and Kerby, J. H., Special considerations in the culture of striped bass and striped bass hybrids, in *Culture and Propagation of Striped Bass and its Hybrids,* Harrell, R. M., Kerby, J. H., and Minton, R. V., Eds., Striped Bass Committee, Southern Division, American Fisheries Society, Bethesda, MD, 1990, 191.

Piper, R. G., McElwain, I. B., Orme, L. E., McCraren, J. P., Fowler, L. G., and Leonard, J. R., *Fish Hatchery Management,* U. S. Fish and Wildlife Service, Washington, D.C., 1982, 1.

Post, G. W., *Textbook of Fish Health,* 2nd Ed., T.F.H. Publications, Neptune City, NJ, 1987, 1.

Powell, M. R., Cage and raceway culture of striped bas in brackish water in AL, *Proc. S.E. Assoc. Game Fish Comm.,* 26, 345, 1973.

Radovich, J., Effect of ocean temperature on the seaward movement of striped bass, *Roccus saxatilis,* on the Pacific coast, *Calif. Fish Game,* 49, 191, 1963.

Raney, E. C., The life history of the striped bass, *Roccus saxatilis* (Walbaum), in *The Striped Bass (Roccus saxatilis),* Raney, E. C., Tresselt, E. F., Hollis, E. H., Vladykov, V. D., and Wallace, D. H., Eds., *Bull. Bingham Oceanogr. Coll.,* 14, 5, 1952.

Raney, E. C., Tresselt, E. F., Hollis, E. H., Vladykov, V. D., and Wallace, D. H., Eds., The Striped Bass *(Roccus saxatilis), Bull. Bingham Oceanogr. Coll.,* 14, 5, 1952.

Ray, R. H. and Wirtanen, L. J., Striped Bass, *Morone saxatilis* (Walbaum), 1969 Report on the Development of Essential Requirements for Production, U.S. Fish and Wildlife Service, Atlanta, GA, 1970, 1.

Rees, R. A. and Cook, S. F., Effects of sunlight intensity on survival of striped bass × white bass fry, *Proc. S.E. Assoc. Fish Wildl. Agen.,* 36, 83, 1985a.

Rees, R. A. and Cook, S. F., Evaluation of optimum stocking rate of striped bass × white bass fry in hatchery rearing ponds, *Proc. S.E. Assoc. Fish Wildl. Agen.,* 37, 257, 1985b.

Rees, R. A. and Harrell, R. M., Artificial spawning and fry production of striped bass and hybrids, in *Culture and Propagation of Striped Bass and its Hybrids,* Harrell, R. M., Kerby, J. H., and Minton, R. V., Eds., Striped Bass Committee, Southern Division, American Fisheries Society, Bethesda, MD, 1990, 43.

Reeves, W. C., Effects of feeding regimes and sources of fish on production of advance fingerling striped bass, *Proc. S.E. Assoc. Game Fish Comm.* 27, 540, 547, 1974.

Reeves, W. C. and Germann, J. F., Effects of increased water hardness, source of fry and age at stocking on survival of striped bass fry in earthen ponds, *Proc. S.E. Assoc. Game Fish Comm.,* 25, 542, 1972.

Regan, D. M., Wellborn, T. L., and Bowker, R. G., Striped Bass, *Roccus saxatilis* (Walbaum), Development of Essential Requirements for Production, U.S. Fish and Wildlife Service, Atlanta, GA, 1968, 1.

Reynolds, J. B., Electrofishing, in *Fisheries Techniques,* Nielsen, L. A. and Johnson, D. L., Eds., American Fisheries Society, Bethesda, MD, 1983, 147.

Rhodes, W. and Merriner, J. V., A preliminary report on closed system rearing of striped bass sac fry to fingerling size, *Prog. Fish-Cult.,* 35, 199, 1973.

Robins, R., *A List of Common and Scientific Names of Fishes from the U.S. and Canada,* American Fisheries Society, Bethesda, MD. 1991, 1.

Rogers, B. A. and Westin, D. T., Laboratory studies on effects of temperature and delayed initial feeding on development of striped bass larvae, *Trans. Am. Fish. Soc.,* 110, 100, 1981.

Rottmann, R. W., Shireman, J. V., Starling, C. C., and Revels, W. H., Eliminating adhesiveness of white bass eggs for the hatchery production of hybrid striped bass, *Prog. Fish-Cult.,* 50, 55, 1988.

Sager, D. R., Woods, L. C., III, and Kraeuter, J. N., Survival of *Morone saxatilis* in low pH oligohaline waters, *J. Wash. Acad. Sci.,* 76, 237, 1986.

Sandoz, O. and Johnston, K. H., Culture of striped bass, *Roccus saxatilis* (Walbaum), *Proc. S.E. Assoc. Game Fish Comm.,* 19, 390, 1966.

Schnick, R. A., Meyer, F. P., and Gray, D. L., A guide to approved chemicals in fish production and fishery resource management, University of Arkansas Cooperative Extension Service and U.S. Fish and Wildlife Service Publication MP 241, Little Rock, 1989, 1.

Scruggs, G. D., Jr. and Fuller, J. C., Jr., Indications of a freshwater population of striped bass, *Roccus saxatilis* (Walbaum), in Santee-Cooper Reservoir, *Proc. S.E. Assoc. Game Fish Comm.,* 8, 64, 1954.

Setzler, E. M., Boynton, W. R., Wood, K. V., Zion, H. H., Lubbers, L., Mountford, N. K., Frere, P., Tucker, L., and Mihursky, J. A., Synopsis of biological data on striped bass, *Morone saxatilis* (Walbaum), NOAA Technical Report, NMFS Circular 433, National Marine Fisheries Service, Washington, D.C., 1980, 1.

Shannon, E. H., Effect of temperature changes upon developing striped bass eggs and fry, *Proc. S.E. Assoc. Game Fish Comm.,* 23, 265, 1970.

Shannon, E. H. and Smith, W. B., Preliminary observations on the effect of temperature on striped bass eggs and sac fry, *Proc. S.E. Assoc. Game Fish Comm.,* 21, 257, 1968.

Siddall, S. E., Studies of closed marine culture systems, *Prog. Fish-Cult.,* 36, 8, 1974.

Smith, J. M. and Whitehurst, D. K., Tank spawning methodology for the production of striped bass, in *Culture and Propagation of Striped Bass and its Hybrids,* Harrell, R. M., Kerby, J. H., and Minton, R. V., Eds., Striped Bass Committee, Southern Division, American Fisheries Society, Bethesda, MD, 1990, 73.

Smith, R. E. and Kernehan, R. J., Predation by the free-living copepod *Cyclops bicuspidatus thomasi,* on larvae of the striped bass and white perch, *Estuaries,* 4, 81, 1981.

Smith, T. I. J., The culture potential of striped bass and its hybrids, *World Aquacult.,* 20(1), 32, 1989.

Smith, T. I. J. and Jenkins, W. E., Controlled spawning of F_1 hybrid striped bass (*Morone saxatilis* × *Morone chrysops*) and rearing F_2 progeny, *J. World Maricult. Soc.,* 15, 147, 1984.

Smith, T. I. J., South Carolina Resources Research Institute, personal communication, 1989.

Smith, T. I. J. and Jenkins, W. E., Culture and controlled spawning of striped bass (*Morone saxatilis*) to produce stiped bass and striped bass × white bass (*Morone chrysops*) hybrids, *Proc. S.E. Assoc. Fish Wildl. Agen.,* 40, 152, 1988a.

Smith, T. I. J. and Jenkins, W. E., Broodstock development and spawning of striped bass, *Morone saxatilis,* and white bass, *M. chrysops, J. World Aquacult. Soc.,* 19, 65A, 1988b.

Smith, T. I. J., Jenkins, W. E., and Minton, R. V., Production of advanced fingerling and subadult striped bass hybrids in earthen ponds, in *Culture and Propagation of Striped Bass and its Hybrids,* Harrell, R. M., Kerby, J. H., and Minton, R. V., Eds., Striped Bass Committee, Southern Division, American Fisheries Society, Bethesda, MD, 1990, 121.

Smith, T. I. J., Jenkins, W. E., and Snevel, J. F., Production characteristics of striped bass (*Morone saxatilis*) and F_1, F_2 hybrids (*M. saxatilis* and *M. chrysops*) reared in intensive tank systems, *J. World Aquacult. Soc.,* 16, 57, 1985.

Smith, T. I. J., Jenkins, W. E., Stokes, A. D., and Smiley, R. A., Semi-intensive pond production of market-size striped bass (*Morone saxatilis*) × white bass (*M. chrysops*) hybrids, *World Aquacult.,* 20(1), 81, 1989.

Snow, J. R., Al-Ahmad, T. A., and Parsons, J. E., Rotifers as a production diet for striped bass fingerlings, *Proc. S.E. Assoc. Fish Wildl. Agen.,* 34, 280, 1982.

Spotte, S., *Fish and Invertebrate Culture; Water Management in Closed Systems,* 2nd ed., John Wiley & Sons, New York, 1979, 1.

Starling, C. C., Fish Hatcheries Review and Annual Progress Report for 1983-84, Florida Game and Fresh Water Fish Commission, Tallahassee, 1983, 1.

Starling, C. C., Striped bass × white bass *Morone* spp. hybrids (sunshine bass) production at Florida state fish hatcheries, in Fish Hatcheries Review and Annual Report for 1984-1985, Florida Game and Fresh Water Fish Commission, Tallahassee, 1985, 65.

Starling, C. C., Fish culture research and development, in Fish Hatcheries Review and Annual Report for 1986-1987, Florida Game and Fresh Water Fish Commission, Tallahassee, 1987, 27.

Stevens, R. E., The striped bass of the Santee-Cooper reservoir, *Proc. S.E. Assoc. Game Fish Comm.*, 11, 253, 1958.

Stevens, R. E., Hormone-induced spawning of striped bass for reservoir stocking, *Prog. Fish-Cult.*, 28, 19, 1966.

Stevens, R. E., A final report on the use of hormones to ovulate striped bass, *Roccus saxatilis* (Walbaum), *Proc. S.E. Assoc. Game Fish Comm.*, 18, 525, 1967.

Stevens, R. E., Current and future considerations concerning striped bass culture and management, *Proc. S.E. Assoc. Game Fish Comm.*, 28, 69, 1975.

Stevens, R. E., Striped bass culture in the U.S., *Commer. Fish Farmer*, 5(3), 10, 1979.

Stevens, R. E., Historical overview of striped bass culture and management, in *The Aquaculture of Striped Bass: A Proceedings*, McCraren, J. P., Ed., University of MD Sea Grant Publ. UM-SG-MAP-84-01, College Park, 1984, 1.

Stevens, R. E., May, O. D., Jr., and Logan, H. J., An interim report on the use of hormones to ovulate striped bass (*Roccus saxatilis*), *Proc. S.E. Assoc. Game Fish Comm.*, 17, 226, 1965.

Surber, E. W., Results of striped bass (*Roccus saxatilis*) introductions into freshwater impoundments, *Proc. S.E. Assoc. Game Fish Comm.*, 11, 273, 1958.

Swartz, D., Marketing striped bass, in *The Aquaculture of Striped Bass: A Proceedings*, McCraren, J. P., Ed., University of MD Sea Grant Publ. UM-SG-MAP-84-01, College Park, 1984, 233.

Swingle, W. E., AL's marine cage culture studies, *J. World Maricult. Soc.*, 3, 75, 1972.

Talbot, G. B., Estuarine environmental requirements and limiting factors for striped bass, in *A Symposium on Estuarine Fisheries*, Smith, R. F., Swartz, A. H., and Massman, W. H., Eds., American Fisheries Society Spec. Publ. No. 3, Bethesda, MD, 1966, 37.

Tatum, B. L., Bayless, J. D., McCoy, E. G., and Smith, W. B., Preliminary experiments in the artificial propagation of striped bass, *Roccus saxatilis*, *Proc. S.E. Assoc. Game Fish Comm.*, 19, 374, 1966.

Texas Instruments Incorporated, Feasibility of Culturing and Stocking Hudson River Striped Bass, an Overview 1973 to 1975, Texas Instruments Incorporated, Ecological Services, Dallas, TX, 1977, 1.

Todd, T. N., Occurrence of white bass-white perch hybrids in Lake Erie, *Copeia*, 1986, 196, 1986.

Tomasso, J. R., Davis, K. B., and Parker, N. C., Plasma corticosteroid and electrolyte dynamics of hybrid striped bass (white bass × striped bass) during netting and hauling, *Proc. World Maricult. Soc.*, 11, 303, 1980.

Tomasso, J. R., Davis, K. B., and Parker, N. C., Plasma corticosteroid dynamics in channel catfish, *Ictalurus punctatus* (Rafinesque), during and after oxygen depletion, *J. Fish Biol.*, 18, 519, 1981.

Tuncer, H., Harrell, R. M., and Houde, E. D., Comparative energetics of striped bass (*Morone saxatilis*) and hybrid (*M. saxatilis × M chrysops*) juveniles, *Aquaculture*, 86, 387, 1990.

Turner, C. J., Striped bass culture at Marion Fish Hatcher, in *The Aquaculture of Striped Bass: A Proceedings*, McCraren, J. P., Ed., Maryland Sea Grant Publication UM-SG-MAP-84-01, University of Maryland, College Park, 1984, 59.

Turner, J. L. and Farley, T. C., Effects of temperature, salinity, and dissolved oxygen on the survival of striped bass eggs and larvae, *Calif. Fish Game*, 57, 268, 1971.

Valenti, R. J., Aldred, J., and Liebell, J., Experimental marine cage culture of striped bass in northern waters, *Proc. World Maricult. Soc.*, 7, 99, 1976.

Van Olst, J. C. and Carlberg, J. M., Commercial culture of hybrid striped bass: status and potential, *Aquacult. Mag.*, 16(1), 49, 1990.

Van Olst, J. C., Carlberg, J. M., Massingill, M. J., Hovanec, T. A., Cochran, M. D., and Doroshev, S. I., Methods for Intensive Culture of Striped Bass Larvae at Central Valleys Hatchery, Progress Report, California Fish and Game, University of California at Davis and Aquaculture Systems International, Elk Grove, CA, 1981, 1.

Waldman, J. R., Diagnostic value of *Morone* dentition, *Trans. Am. Fish. Soc.*, 115, 900, 1986.

Ware, F. J., Progress with *Morone* hybrids in fresh water, *Proc. S.E. Assoc. Game Fish Comm.*, 28, 48, 1975.

Wattendorf, R. J. and Shafland, P. L., Observations on salinity tolerance of striped bass × white bass hybrids in aquaria, *Prog. Fish-Cult.*, 44, 148, 1982.

Wawronowicz, L. J. and Lewis, W. M., Evaluation of the striped bass as a pond-reared food fish, *Prog. Fish-Cult.*, 41, 138, 1979.

Wellborn, T. L., Jr., The toxicity of nine therapeutic and herbicidal compounds to striped bass, *Prog. Fish-Cult.*, 31, 27, 1969.

Wellborn, T. L., Jr., Toxicity of some compounds to striped bass fingerlings, *Prog. Fish-Cult.*, 33, 32, 1971.

Whitehurst, D. K. and Stevens, R. E., History and overview of striped bass culture and management, in *Culture and Propagation of Striped Bass and its Hybrids,* Harrell, R. M., Kerby, J. H., and Minton, R. V., Eds., Striped Bass Committee, Southern Division, American Fisheries Society, Bethesda, MD, 1990, 1.

Williams, H. M., Preliminary studies of certain aspects of the life history of the hybrid (striped bass × white bass) in two South Carolina reservoirs, *Proc. S.E. Assoc. Game Fish Comm.*, 24, 424, 1971.

Williams, H. M., Characteristicsd for distinguishing white bass, striped bass and their hybrid (striped bass × white bass), *Proc. S.E. Assoc. Game Fish Comm.*, 29, 168, 1976.

Williams, J. E., Sandifer, P. A., and Lindbergh, J. M., Net-pen culture of striped basss × white bass hybrids in estuarine waters of South Carolina: a pilot study, *J. World Maricult. Soc.*, 12, 98, 1981.

Wirtanen, L. J. and Ray, R. H., Striped Bass, *Morone saxatilis* (Walbaum): 1970 Report on the Development of Essential Requirements for Production, U.S. Fish and Wildlife Service, Atlanta, GA, 1971, 1.

Woiwode, J. G., The effects of temperature, photoperiod and ration size on the growth and thermal resistance of the hybrid striped × white bass, Ph.D. dissertation, University of Minnesota, St. Paul, 1989, 1.

Woiwode, J. G. and Adelman, I. R., Growth, food conversion efficiency, and survival of the hybrid white × striped bass as a function of temperature, in *The Aquaculture of Striped Bass: A Proceedings,* McCraren, J. P., Ed., University of MD Sea Grant Publ. UM-SG-MAP-84-01, College Park, 1984, 143.

Woods, L. C., University of North Carolina, personal communication, 1992.

Woods, L. C., III, Kerby, J. H., and Huish, M. T., Estuarine cage culture of hybrid striped bass, *J. World Maricult. Soc.*, 14, 595, 1983.

Woods, L. C., III, Kerby, J. H., and Huish, M. T., Culture of hybrid striped bass to marketable size in circular tanks, *Prog. Fish-Cult.*, 47, 147, 1985a.

Woods, L. C., III, Lockwood, J. C., Kerby, J. H., and Huish, M. T., Feeding ecology of hybrid striped bass (*Morone saxatilis* × *M. chrysops*) in culture ponds, *J. World Maricult. Soc.*, 16, 71, 1985b.

Woods, L. C., III, Kerby J. H., Huish, M. T., Gafford, G. M., and Rickards, W. L., Circular tank for intensive culture of hybrid striped bass, *Prog. Fish-Cult.*, 43, 199, 1981.

Worth, S. G., The artificial propagation of the striped bass (*Roccus lineatus*) on Albemarle Sound, *Bull. U.S. Fish Comm.*, 1, 174, 1882.

Worth, S. G., Report upon the propagation of striped bass at Weldon, N.C., in the spring of 1884, *Bull. U.S. Fish Comm.*, 4, 225, 1884.

Yeager, D. M., Van Tassel, J. E., and Wooley, C. M., Collection, transportation, and handling of striped bass brood stock, in *Culture and Propagation of Striped Bass and its Hybrids,* Harrell, R. M., Kerby, J. H., and Minton, R. V., Eds., Striped Bass Committee, Southern Division, American Fisheries Society, Bethesda, MD, 1990, 29.

Chapter 10

BAITFISH

James T. Davis

TABLE OF CONTENTS

8633-9/93/$0.00 + $.50
© 1993 by CRC Press, Inc.

I. INTRODUCTION

At least 20 species of fishes have been raised commercially as bait (Martin 1968). In the U.S., only three warmwater species are of commercial importance. They are the golden shiner (*Notemigonus crysoleucas*), fathead minnow (*Pimephales promelas*), and goldfish (*Carassius auratus*). Other species that are of local importance as bait include killifish (*Fundulus* spp.), chub suckers (*Erimyzon* spp.), stone rollers (*Campostoma* spp.), tilapia (*Tilapia* spp.), top minnows (*Poecilia* spp.), and shiners (*Notropis* spp.). The information presented here is confined to the first three species due to their commercial importance and the availability of research and production data.

Baitfish production in the U.S. increased from approximately 10,000 ha, as reported by Martin in 1968, to over 22,500 ha, reported by a Soil Conservation Service survey conducted in 1978. Most of this acreage is in the state of Arkansas with over 9300 ha reported in 1986 by Collins. The golden shiner is captured in the wild but is primarily produced on fish farms in the midwest and south. Value of golden shiners in Arkansas during 1987 was reported by Farwick to be $18.4 million. The fathead minnow is reared in the midwestern U.S. in shallow lakes, but because of increasing demand, the species is being intensively cultured in Arkansas, Louisiana, and Texas.

Goldfish, particularly the common variety, have been employed as bait for many years. "Culls" from the fancy varieties, which are produced for the aquarium and water garden trade, have been used for bait. This has been accompanied by an increased demand for "cull" goldfish as food items for predatory fish in the aquarium trade.

Certain aspects of the baitfish industry make it a highly competitive business. Bait of a desirable size is a perishable commodity which does not lend itself to stockpiling or the maintenance of large inventories. For that reason, competition is extremely keen in the marketing of bait to urban areas. Many producers and distributors indicate that the trade is too crowded and profits are too low. Most agree, however, that profits to bait producers can be quite high under favorable conditions, even with intense competition.

II. GENERAL CULTURE CONSIDERATIONS

The life histories of the golden shiner and fathead minnow have been well documented (Giudice et al. 1981; Flickinger 1971), as has that of the goldfish, both in the U.S. and Japan (Martin 1982; Yoshitaka 1983). Culture ponds for baitfish are of the type generally utilized for the culture of other fishes (Giudice 1968; Smith 1968; refer also to Chapter 1). Special considerations in pond management which differ among the three species of interest are discussed below.

Water supply for a baitfish culture operation is even more critical than for other warmwater species. Under most conditions, a first concern for site selection should be to ascertain the water supply. Most operators prefer a large spring with a water flow in excess of approximately 185 l/min for each hectare of pond surface. Since there are few areas where spring water is available in the volumes necessary to support a large minnow farm, most bait producers construct their facilities in areas with plentiful supplies of cool well water. Artesian wells are preferred because of the reduced costs involved in pumping, but as with flowing springs, artesian water supplies are limited.

All of the species under discussion reach a peak in spawning intensity when the water temperature is above 20°C, so water at or above that temperature is sought by minnow and goldfish culturists. In addition, water of about 20°C is preferred for use in holding and hauling tanks. Therefore, one water temperature is desirable for all uses in most areas.

One problem with the use of well water involves the occasional presence of contaminants. Carbon dioxide and hydrogen sulfide can usually be removed by aeration prior to introduction of the water into culture chambers. Iron can be flocculated in the same manner. Iron-rich water introduced directly into holding tanks after aeration should be sand filtered to remove the iron precipitate. This will reduce the chance of gill clogging in fish exposed to the suspended iron. If the water is introduced directly into ponds, sand filtration is often bypassed and the iron is allowed to settle *in situ*.

Surface waters such as streams or reservoirs are also used for baitfish culture. Water supplies of those types are generally considered to be much less satisfactory than springs or wells. Contamination from pesticides; runoff from feedlots, stables, and municipalities; and domestic sewage effluents may be present. In addition, most surface water supplies contain fish and other biological contaminants.

If surface water is to be used, it should be pumped through a sand filter or a fine-mesh screen material such as saran cloth. The latter is relatively durable and permits more water flow than sand filters, but it will allow the passage of many parasite and disease organisms. Also, frequent and regular monitoring of screens is required so clogging or ruptures can be quickly discovered and affected screens replaced.

If undesirable fish are inadvertently admitted to culture ponds, they should be removed by draining the pond or by treating with a 5% rotenone solution applied at the rate of 2 mg/l. Following rotenone treatment, a minimum of 2 weeks should be allowed for inactivation of the chemical prior to baitfish stocking. The water should be tested by placing a few fish in a container that is placed in the pond for 24 h before stocking. The container should be perforated to allow exposure of the fish to the pond water. If the fish survive for 24 h under those conditions it is usually safe to stock the pond with the desired crop.

Surface water is often unsatisfactory because of high silt loads, which are difficult and expensive to remove. The usual treatment procedure is to allow the silt to settle out of the water by holding it in a reservoir for 12 to 48 h before placing it in the culture ponds. Water from reservoirs can be a problem because of temperature. During the summer, reservoir surface water may be too warm, while deep water may be too cool and devoid of oxygen. Finally, surface waters are usually too cold during the winter to prevent ice formation on culture ponds, particularly in northern parts of the baitfish production region. Seining cannot be conducted when ponds are ice-covered.

Water from municipalities has been used in some instances, but has not proven to be economically feasible for long periods because of the expense involved in chlorine removal. Chlorine can be effectively removed by charcoal filtration or the addition of sodium thiosulfate, but careful monitoring for residual chlorine is required when either method is employed, and both methods can add significantly to production costs.

Water pH has a decided effect on the efficacy of baitfish production as has been discussed by several authors (Giudice et al. 1981; Flickinger 1971; Martin 1982; Yoshitaka 1983). Generally, baitfish can tolerate pH in the range of 6.0 to 9.0, but improved production has been obtained when pH is between 6.5 and 8.0. Control of pH is possible through the use of acid or alkaline-based fertilizers. Additional expense may be involved and experienced management is essential.

In certain areas, water of low pH and, concomitantly, low alkalinity occur. The pH and alkalinity of such waters can usually be increased through the application of agricultural limestone or gypsum, with the amounts used dependent upon soil and water analyses. For best production, both alkalinity and hardness should be between 50 and 300 mg/l.

Dissolved oxygen (DO) requirements for baitfish are the same as for other species as discussed in Chapter 1; that is, the DO level should be maintained in excess of 3.0 mg/l at all times. While all three species under discussion can survive lower levels, stress under reduced oxygen concentrations will result in reduced fish growth.

Salinity is of great importance in the culture of golden shiners. In waters of only 2 ppt salinity, production is severely curtailed. While goldfish tolerate 2 ppt, total production will be reduced (Murai and Andrews 1977). Few data have been collected on the effect of salinity on fathead minnow production, but fish culturists report reduced production in water having salinities over 1.5 ppt.

III. REPRODUCTION AND GENETICS

The culture method used will determine the technology which can be applied to breeding. Golden shiners, fathead minnows, and goldfish can all be spawned in extensive or intensive culture systems. Generally, when extensive systems are employed, the young fish are allowed to remain in the ponds with the adults. Harvest is usually by lift nets or traps, and fish which are too small for sale are returned to the ponds for further growth. Pond sizes for extensive spawning and rearing range from 2 to 20 ha.

In intensive rearing systems, the eggs or fry are removed from spawning ponds and transferred to nursery ponds and thence to growout ponds. Spawning ponds of small size, often less than 0.1 hectare, are utilized. They are heavily stocked with broodfish which have been selected for maximum production. Since very few detailed genetic studies have been completed, the selection is usually based on the manager's knowledge that the strain of fish withstands handling well and tends to be prolific.

Broodstock selection for fathead minnow production is usually done with a mechanical grader. The males are normally larger than the females and care is taken to avoid selection of only the largest individuals for use as broodstock. At the same time, larger females usually lay more and larger eggs. In extensive culture, stocking rates of 1250 to 5000 adults per hectare (3 to 5 females per male) are normal. Higher stocking rates will usually result in the occurrence of stunted fish due to overpopulation in the ponds.

In intensive culture, broodfish stocking densities from 35,000 to 60,000 per hectare are commonly used, though research has demonstrated that 50,000 adults at the ratio of 5 females to 1 male results in best production (Flickinger 1971). Females can be expected to produce from 1000 to 10,000 eggs, depending on size. Under the proper conditions, up to 7.5 million minnows per hectare can be expected, if young fish are removed several times during the summer.

Little organized effort has been devoted to selection of improved strains of fathead minnows. In part, this is due to the short lifespan of the species, normally less than 2 years, but it is also an indication that most culturists are satisfied with the results being obtained.

Golden shiners are also grown in both extensive and intensive systems. Stocking rates for ponds used in extensive culture are generally about 5000 fish per hectare and the fish selected are usually of mixed sizes (designated "pond run") to avoid the possibility of skewed sex ratios, which could occur with mechanical grading since golden shiner females grow more rapidly than the males.

When intensive culture is employed with egg transfer, broodfish stocking densities of 400 to 500 kg/ha are common. The sex ratio for those broodfish is also "pond run". Most growers believe that larger females lay more eggs, but that the smaller females are more active and less likely to be plagued by an ovarian parasite which inhibits egg production. For planning purposes, up to 10,000 eggs per female is considered normal, but most growers report fewer than 3000 young per female stocked.

In recent years, some efforts to select desirable strains of broodfish have been undertaken by the larger growers. No data have been released on the success of those endeavors, and apparently no research agencies have undertaken in-depth, long-term genetic studies.

Methods employed for spawning goldfish are much the same as those for golden shiners. The number of broodfish stocked depends on their size, which directly affects the number of eggs produced per female. For extensive systems, from 250 to 750 broodfish are stocked per hectare, while 800 to 1000 kg/ha are stocked in intensive culture.

The genetics of goldfish have been studied in detail, and there have been many varieties developed, primarily for the aquarium trade. Variety development has been the result of selective breeding and some crossbreeding. Some of the more common goldfish breeds are the lionhead, comet, shubunkin, calico, and common. Procedures for breeding the varieties are usually passed from culturist to culturist, though some of the methodology has been published (Yoshitaka 1983)

IV. POND MANAGEMENT AND SPAWNING PROCEDURES

A. FATHEAD MINNOWS

Fathead minnows have the unique habit of spawning on the underside of vegetation or other objects found in the water. Therefore, aquatic plants are commonly allowed to remain in culture ponds when the extensive technique is utilized. There is a danger in leaving too many plants, however, since oxygen depletions may occur from decaying vegetable material. Utilization of plants as natural spawning substrates is widely employed, but has the disadvantage that fish overpopulation and stunting sometimes occur. Because most of the broodfish die after spawning, there is little interference between adults and their offspring. In at least some culture installations, it is believed that the older fingerlings compete with fry for food. That situation may explain some of the variable production levels which have been reported (Saylor 1973).

In intensive culture, the fry transfer method is commonly employed. To increase spawning, area rock, pieces of bricks, and boards are used to supplement natural spawning sites. Old lumber is often stapled to wires stretched parallel across the pond. Boards should be provided at the rate of 41 cm of 2.5 × 10 cm lumber per 100 male broodfish. Some culturists furnish cubicles for the males to use for territories during nest protection, but that technique has not led to increased fry production (Flickinger 1972).

Spawning may begin when water temperatures are as low as 12.8°C (Andrews and Flickinger 1973) and continue as long as the water temperature remains below 29°C or returns to below that temperature (Flickinger 1971). In some areas of the western U.S., fall spawning of fathead minnows is practiced, but survival may be limited.

Egg incubation requires 5 to 7 d depending on water temperature. After absorption of their yolk sacs, the fry feed on phytoplankton for about 2 weeks after which they will readily consume prepared diets. During the first 2-week period of feeding, the fry are usually captured in lift nets or seines and transferred to rearing ponds. Some data are available which indicate that more fry can be produced from the same area of spawning ponds if the ponds are drained and all fry removed at least twice during each growing season (Saylor 1973). Other methods are utilized in the production of fathead minnows for forage in hatchery situations (Nagel 1976; Guest 1977; Benoit and Carlson 1977; Norman-Boudreau and Daggett 1989).

When rearing ponds are employed, the fry are directly stocked from spawning ponds. The rearing ponds are fertilized in advance of stocking and receive fry at the rate of 125,000 to 750,000 per hectare. The accepted practice is to count the number of fish in a kilogram (determined by weight added to a previously tared container) and then calculate the total weight required to reach the desired stocking density for each pond. The number of fry in a kilogram may be rather high, so 0.1-kg sample counts are sometimes employed.

Fish should be weighed in a small amount of water and handled carefully to avoid stress. As a general rule, stocking numbers make allowance for 25 to 50%

mortality during the growing season. To reach salable size within a single growing season, about 250,000 fish per hectare is the maximum stocking rate. If all, or a portion, of the second growing season is to be used to produce marketable fish, two or three times as many minnows can be reared within the growout ponds (Flickinger 1971). Overwinter survival of fish weighing more than 0.08 g (10,000 fish per kilogram) is usually quite high, but the postspawning mortality of broodfish often exceeds 80%.

B. GOLDEN SHINERS AND GOLDFISH

Since the methods used for production of golden shiners and goldfish are virtually the same, the two species are discussed together. In the extensive culture or free-spawning method, pond water level is lowered during the late winter or early spring to stimulate the growth of grass along the shoreline. Many growers plant rye grass in that area.

When the ponds are refilled, the plants become spawning sites. Aquatic plants will also serve that purpose, as will rice. Care must be exercised when using the latter, however, as the vegetation may become so dense as to preclude effective management of the pond. If plant growth is inadequate, on the other hand, hay or straw may be placed near the shoreline as spawning substrate. When that method is used, the material should be staked down to prevent it from piling up at one end of the pond in response to wind-generated movement.

Spawning activity may decline for a variety of reasons, but is normally dependent upon water quality conditions. The usual method of promoting active spawning is to add cool water and rapidly raise the water level in the pond. The technique has also been employed to extend the spawning season in small ponds. Other methods of inducing spawning include the use of chemicals to "shock" the fish. While such methods are commonly employed, there is little literature on the subject (Giudice et al. 1981).

If possible, the broodfish should be removed from the pond as they begin to compete with young-of-the-year fish for food and space. Many culturists sell the spent brooders as large bait minnows, but a few retain them for reuse as broodfish.

Intensive culture of golden shiners and goldfish is usually practiced using the egg transfer method. When that technique is employed, it is essential to remove all natural vegetation from spawning ponds so that egg deposition can be controlled. Research indicates that the presence of aquatic vegetation, as well as certain temperatures and photoperiods, can cause spontaneous ovulation (Stacey et al. 1979). Soil sterilants are often applied around pond margins to preclude sprouting of aquatic or terrestrial vegetation.

When the fish are ready to spawn, mats are placed in the pond. Their locations are prepared in advance using gravel to level the area where the mats are to be placed. Spawning mats are approximately 30 × 60 cm and are constructed by sandwiching Spanish moss between 10 × 10 cm mesh pieces of welded wire. The tops and bottoms of each frame are wired or hog-ringed together.

It is important to place spawning mats 2.5 cm below the water surface and to put them on level substrates. In most cases several mats are placed end-to-end parallel with the shoreline. The mats are ready for transfer when they are uniformly covered with eggs. Allowing too many eggs on a given mat encourages fungus growth which will reduce the hatching rate, and transferring mats with too few eggs per mat is wasteful.

The number of mats which are placed in a spawning pond and length of time between placement and removal will depend on the spawning activity of the broodfish. Goldfish usually spawn just after dawn and continue spawning until the sunlight strikes the pond, while golden shiners may spawn throughout the day.

Mat densities in spawning ponds should not exceed the number that can be covered with eggs within a few hours. If the fish spawn more rapidly than expected, filled mats can be removed and replaced with fresh ones or new mats can be stacked on top of filled ones to prevent egg predation. Such predation is a constant problem in goldfish production ponds. Too many mats in a pond will result in underutilization (Giudice et al. 1981). At the same time, too few mats may cause the fish to use alternative sites for spawning.

When the mats are filled, they are moved to rearing or nursery ponds. Estimation of egg numbers on mats is usually a function of experience. If the mats are well-filled, from 125 to 200 mats are placed in each hectare of rearing pond for golden shiners. Morrison and Burtle (1989) reported fry survival exceeding 1,250,000 from 10 well-filled mats and suggested that 3 to 7 mats per hectare could provide the needed stock for normal shiner production operations. As many as 400 mats are often used per hectare in goldfish ponds. It is common practice to leave the mats in the rearing ponds for 10 d, after which they are removed and thoroughly washed. They can then be reintroduced into spawning ponds as needed.

Fry transfer is also practiced with golden shiners and goldfish because fry numbers can readily be determined and rigid control over stocking rates is possible. Determination of fry numbers is accomplished utilizing the weight-count method outlined above or by counting the number of fry in a 30-ml sample and calculating the volume displacement of fish required to stock a given pond.

The number of fry stocked per hectare depends upon the time available between stocking and marketing, size of fish desired, level of management employed, and length of the growing season. For golden shiners, stocking rates of 125,000 to 500,000 per hectare may be used, while goldfish are usually stocked at from 50,000 to 2,500,000 per hectare. The higher stocking rates are used when fish are being stockpiled in ponds over winter for growout the following season. The fish are restocked at reduced densities in the spring and can be expected to grow rapidly to market size.

V. FEEDING AND NUTRITION

The food available for baitfish fry depends, to a large extent, on the abundance of plankton available in spawning and rearing ponds. Preparation for fish usually involves allowing ponds to dry for several days after which they are rapidly filled with water. Liquid fertilizer is added as the water enters the ponds. In some instances, plankton from an adjacent pond is seeded into newly filled ponds. Within 4 or 5 d, the spawning mats are placed into the nursery ponds. When the eggs hatch about 5 d later, there is a ready supply of food available to the fry. For fathead minnows, where spawning mats are not employed, pond preparation begins about 10 d prior to stocking.

In most ponds, the water is fertilized at a rate sufficient to produce a Secchi disc reading of about 20 cm. That level of fertility is maintained by intermittent addition of fertilizer until the fish reach about 2.5 cm, after which the bloom is allowed to decline.

As soon as the fry begin to come to the water surface they should be offered prepared feed. Feed should be made available a minimum of twice daily, and many growers feed four to eight times daily. The normal feeding method is to provide finely ground feed that will float. The feed is distributed along the upwind side of the pond, from which it will spread over the pond surface. Sufficient feed should be applied to ensure that it is available to all parts of the pond.

During the initial stages of feeding, total consumption may not occur. Nonetheless, feeding should be continued, even if the fish are being overfed, since it is important that the fish receive ample feed whenever they are willing to eat. Most producers feed fry from four to six times a day, but after the fish are eating on a regular basis, twice daily feeding is sufficient.

As the fish increase in size, the feed should be gradually changed from a finely ground product to one which is more coarsely ground. Subsequently, crumble-sized particles may be fed, and finally, the fish can be offered pelleted feed. Gatlin and Phillips (1988) reported increased feed efficiency when using crumbles in aquaria, but were unable to demonstrate significant differences in weight gained in ponds among meal, crumbled pellets, floating pellets, or sinking pellets. At the same time, there was a significant increase in production over that in ponds where no feed was provided.

When a change from one feed particle size to another is made, that change should be gradual. One part of the new particle size should be mixed with four parts of the previous feed size until the fish are observed consuming the new particle size. Thereafter, the ratio of new particle size to old can be gradually increased daily until only the new size is fed. Fish appear able to detect changes in feed ingredients and may ignore new formulations for a period of time.

The amount of feed to be offered baitfish in ponds depends to a large extent on what the fish will readily consume and the time available between onset of feeding and marketing. For most rapid growth, the fish should be fed all they will consume in a 2-h period twice daily. The amount can be reduced if less than maximum growth is desired. For the beginning grower, it is recommended that when feeding is initiated fry should be fed at 10% of their body weight and that the amount should be gradually reduced to 3% of fish body weight by the time the fingerlings are 7 cm in length.

Reports from commercial growers indicate that higher stocking rates, and consequently higher feeding rates, do not guarantee increased harvests. In addition, higher feeding rates generally lead to lower food-conversion rates. This is an area which should be studied extensively if increased efficiency is a production goal.

The nutritional requirements of baitfish have not been extensively studied (Giudice et al. 1981). It is generally assumed that no complete feeds are available and that it will be necessary to balance any prepared feed with natural food. Ludwig (1989) reported that in ponds where fish were offered commercial feed with 29.5% protein and 1.5% fat at 3% of body weight daily, they continued to feed on zooplankton. Therefore, it is essential to encourage plankton blooms in production ponds. For most growers, this is best accomplished with a combination of organic and inorganic fertilizers. Depending upon pond condition, from 1000 to 2500 kg/ha of manure and 10 to 30 l of inorganic liquid fertilizer (11-37-0) are applied with the incoming water. Additional liquid fertilizer is added as needed to maintain a Secchi disc reading between 15 and 40 cm.

Fry feeds are generally higher in protein (38%) than grower feeds (26 to 32%). Formulas are available from several sources which comply with the protein and particle size standards that are in use in the industry (Giudice et al. 1981, Giudice 1968).

Because goldfish are usually more valuable than other baitfish species and it is essential to get them on prepared feeds quickly to promote rapid growth, many growers add hard-boiled egg yolks (filtered through cotton cloth) or commercial-grade egg yolk. These items may be substituted for other ingredients in starter rations or they may be fed alone for the first 2 to 3 d after the fry begin to feed (Giudice 1968; Yoshitaka 1983).

One of the main recommendations on feeding minnows and goldfish is that the fish, except as newly feeding fry, should not be overfed. After the first several days of feeding, the rate should be reduced if the ration is not completely consumed. When the weather is very hot the fish should be fed during the coolest part of the day. If the air temperature is above 30°C or the sky is overcast, the amount of feed offered should be reduced. Most producers limit the amount of feed made available to the fish to 30 kg/ha/d during the summer.

VI. VEGETATION CONTROL

The best method of vegetation control is prevention. Most production ponds have few problems with aquatic vegetation, and control is effected by closely mowing the grass near pond edges. If problems develop, various methods of vegetation control are available (Dupree and Huner 1984). For intensive culture, complete vegetation removal from spawning ponds is required. Chemical control is often supplemented by mechanical removal or biological control.

VII. HARVESTING AND HANDLING

Harvesting of minnows, especially shiners, during the summer months requires a great deal of patience and attention to detail. Any time the air temperature is above 30°C or the surface water temperature is above 22°C, the movement of fish should be confined to the coolest parts of the day. This usually means that if seining is to be conducted, all the work should be accomplished prior to 1000 h. If the fish are being captured with lift nets, the work can continue until later in the day if the water in the hauling tank has been cooled to or below 18°C and it is well-oxygenated.

When seining, care should be taken to avoid capturing more fish than are required. Seining an entire pond to capture a small quantity of fish will often result in excessive stress and subsequent epizootics may develop. As the seine is landed, it is usually shortened until it can be bagged and then moved out from shore to deeper water where the fish will not be so crowded. Then the seine is staked out with steel rods.

During crowding, the fish undergo a period of excitement and severe stress. The fish tend to brush against one another and against the net, causing scale loss. Damage to the fish provides not only an avenue for entrance of disease organisms, but may also cause such extensive mutilation that the fish are no longer salable.

Under circumstances of extreme crowding the fish may become so excited that suffocation results. It is generally recommended that no more than 250 kg of fish be seined at one time to avoid overcrowding.

Fish are removed from the seine with a long-handled dip net with a mesh size of about 5 mm. Nets which will hold not more than 1.5 kg are recommended. The fish are carefully placed in buckets partially filled with cool water from the hauling tank. The buckets are moved directly to the hauling tank and the fish gently introduced therein.

Seines for use in harvesting minnows and goldfish have been discussed by other authors, and the technology is well documented (Martin 1968; Giudice et al. 1981; Giudice 1968; Flickinger 1971). Perhaps the most important element is that the seine should be of soft material so damage to the fish can be reduced. The same condition holds for other nets utilized in the capture of baitfish.

Lift nets are often used to harvest minnows and goldfish. These nets are usually fastened by lines to a long pole from which the nets are lowered into the water. Feed is used to attract fish over the net. When the operator perceives that sufficient numbers of fish are present, the net is rapidly lifted and the fish are captured. The procedure of removing the fish from the bag of the lift net is the same as that described for seines.

There are many designs for hauling tanks utilized to carry fish from ponds to holding facilities (Martin 1968; Giudice et al. 1981; Giudice 1968; Flickinger 1971). Actual design is dependent on the quantity of fish to be hauled and the distance of the trip. Holding-tank water temperature should be maintained below 22°C, and the tanks should not be overloaded. The major criterion for short hauls is that no more than 1 kg of fish should be placed in each 10-l of water during hot weather. If the trip from the pond bank to the holding facility is of more than 20 min duration, the weight of fish in the hauling tank should be reduced by 10% for each additional 20 min.

Following harvest, most fish are held without handling for at least 24 h. During that period the fish are kept in cool water and recover from the stress to which they were exposed during harvesting. The recovery period enables them to better withstand subsequent handling. An additional advantage of the delay before secondary handling is that the fish have an opportunity to clear their intestinal tracts. This helps prevent water quality deterioration during the grading process.

In general, 10-l of water per kilogram of fish is sufficient for up to a week in holding tanks if the water is well-aerated. Normally, one agitator is recommended for each 50 kg of fish in a holding tank. The agitator should be of at least 1/25 hp and should operate continuously. In addition, 10% of the water in the tank should be replaced at least once daily with new water of the same temperature as that in the holding tank. Most culturists flush their tanks with a complete change of water after the first 6 h and daily thereafter.

Most minnows and feeder goldfish are graded prior to sale. Mechanical graders are used for this procedure and are available in a variety of shapes and sizes. Generally, grader bars are made with a desired spacing between them. Fish below a given size pass through each grader, while larger fish are retained. After the fish are graded, they are counted to determine the number per thousand as that is the measure under which the fish are marketed. Details of the grading procedure have been documented in the literature (Martin 1968, Giudice et al. 1981).

Holding facilities are where most mortalities of minnows and goldfish can be documented, therefore special care should be taken to minimize mortalities during holding. Details of recommended practices aimed at reducing holding mortalities have been discussed by Martin (1968) and Johnson (1981).

Movement from holding tanks to retail establishments requires the same type of hauling equipment utilized to transport the fish from ponds to holding facilities. For long distance hauling, larger trucks are often utilized since they can carry more fish at a smaller cost per unit weight than small trucks. Alternatively, fish can be shipped in plastic bags charged with oxygen as discussed by Harry (1968). One of the primary considerations is that the fish should not be exposed to rapid changes in temperature during holding, handling, or hauling.

VIII. DISEASES AND PREDATORS

Bacterial and fungal infections may kill large numbers of baitfish, but are usually associated with stress and/or injury. Those problems may also occur following environmental degradation.

Most of the losses in baitfish ponds are attributable to parasites, the major ones of which are protozoans. Many parasites are common to both minnows and goldfish and can cause extensive losses. The major external parasites are *Trichodina, Ichthyobodo, Chilodonella, Cryptobia*, and *Ichthyophthirius*. Most are relatively easy to recognize with the aid of a microscope. Control and management recommendations have been outlined by Giudice et al. (1981) and Bishop (1968).

Sporozoan parasites are among the most difficult to control. The major ones are *Mitraspora cyprini* in goldfish, *Myxobolus notemigoni* (milk scale disease) in golden shiners, and *M. argenteus* in fathead minnows (Giudice et al. 1981, Bishop

1968). *Pleistophoro ovariae* affects the ovaries of golden shiners and may adversely affect production in that species. Control is best achieved by utilizing 1-year-old broodfish.

Of the internal parasites, the flukes are the most damaging to fish populations. Control is dependent to a large extent on good management practices as reviewed by Giudice (1968). Tapeworms and round worms are a problem in certain areas, but are not generally found in sufficient numbers to cause high mortalities among baitfish.

Crustacean parasites are a recurring problem. In the goldfish industry, the fish louse *Argulus* is a major cause of losses in some hatcheries. The anchor parasite is a major source of losses in the other two bait species.

Predators are a major problem for all baitfish producers. Predation occurs at all life stages of the fish. Fish, mink, raccoons, birds, frogs, snakes, turtles, insects, and alligators all adversely impact baitfish populations. Control of each of these requires careful management, and there is a further danger with some predators because they may be vectors of diseases. Control is usually achieved by physical removal of the offending species. This can be costly, but in many instances can make the difference between profit and loss. More information is available in publications from various state fish and game agencies (Giudice 1968).

IX. MARKETING

One of the continuing problems perplexing the baitfish industry is the variety of sizes requested by fishermen. Most food-fish producers have a fixed group of sizes that are needed to supply the market. Baitfish producers normally sell fish ranging from 1 kg up to 15 kg per 1000 fish. In addition, there is a ready market for larger shiners from 20 to 40 fish per kilogram. The smaller bait sizes are called "crappie" minnows and the larger ones are used for "whopper" bass or for surf fishing.

Marketing represents a discipline in which little information is available for baitfish producers. It is an acknowledged fact that there are many outlets for baitfish, but knowledge on how to sell to wholesalers and retailers is usually discovered by the individual grower working alone. Pricing is usually dependent upon demand and the extent to which competition exists within a particular area. In regions where there are numerous suppliers competing for the same market, heavy reliance is placed on the services supplied to the buyer. For example, a minnow supplier may be required to visit a large retailer on a tightly fixed schedule and may also be expected to deliver other types of bait (e.g., worms and crickets) as well as fishing tackle to the retailer (Goodman 1984). In other areas, service is secondary to price and/or supply. Most wholesaling is done from trucks while retailing is often done using plastic bags or by placing the baitfish directly into minnow buckets furnished to the buyer.

The future of the baitfish industry seems bright in that each year there is an increase in the number of fishing licenses sold, and many fishermen are convinced that the best way to assure a good catch is by using live bait. There seems to be ample marketing opportunity for the large operator who can haul fish for long distances and dependably supply holding facilities. At the same time, there is room in the industry for small producers who develop local markets and furnish high quality baitfish to the public.

REFERENCES

Andrews, A. K. and Flickinger, S. A., Spawning requirements and characteristics of the fathead minnow, *Proc. S.E. Assoc. Fish Game Comm.*, 27, 759, 1973.

Benoit, D. A. and Carlson, R. W., Spawning success of fathead minnows on selected artificial substrates, *Prog. Fish-Cult.* 39, 67, 1977.

Bishop, H., Parasites and diseases of common bait fishes, in *Proc. Commercial Bait Fish Conf,* Texas A&M University, College Station, 1968, 31.

Collins, C.M., Present status of fish farmers in Arkansas, *Aquaculture Mag.*, 12(3), 50, 1986.

Dupree, H.K. and Huner, J. W., Third Report to the Fish Farmers, U.S. Fish and Wildlife Service, Washington, D.C., 1984, 1.

Farwick, J., Arkansas commercial fishery industry survey, Federal Aid to Commercial Fisheries PL 88-309, National Marine Fisheries Survey Project 2-432-R, Arkansas Game and Fish Commission, Little Rock, 1988, 1.

Flickinger, S. A., Investigation of pond spawning methods for fathead minnows, *Proc. S.E. Assoc. Fish Game Comm.*, 26, 376, 1972.

Flickinger, S.A., Pond Culture of Bait Fishes, Colorado Cooperative Extension Service, Fort Collins, 1971, 1.

Gatlin, D. M., III and Phillips, H. F., Effects of diet form on golden shiner (*Notemigonus crysoleucas*) performance, *J. World Aquacult. Soc.*, 19, 47, 1988.

Giudice, J. J., Gray, D. L., and Martin, J. M., Manual for Bait Fish Culture in the South, University of Arkansas, Cooperative Extension Service and U.S. Fish and Wildlife Service, Washington, D.C., 1981, 1.

Giudice, J.J., The culture of bait fishes. in *Proc. Commercial Bait Fish Conf.*, Texas A&M University, College Station, 1968, 13.

Goodman, R., Personal communication, 1984.

Guest, W.C., Technique for collecting and incubating eggs of the fathead minnow, *Prog. Fish-Cult.*, 39, 188, 1977.

Harry, G., Handling and transporting golden shiner minnows, in *Proc. Commercial Bait Fish Conf.,* Texas A&M University, College Station, 1968, 37.

Johnson, S. K., Maintaining Minnows — A Guide for Retailers, Texas Agricultural Extension Service Bull. B-1365, College Station, 1981, 1.

Ludwig, G. M., Effect of golden shiners on plankton and water quality in ponds managed for intensive production, *J. World Aquacult. Soc.*, 20, 46, 1989.

Martin, J.M., The minnow farming industry, in *Proc. Commercial Bait Fish Conf.*, Texas A&M University, College Station, 1968, 5.

Martin, J.M., Goldfish Farming, U.S. Fish and Wildlife Service, Washington, D.C., 1982, 1.

Morrison, J. R. and Burtle, G. J., Hatching of golden shiner eggs in hatchery tanks and subsequent fry survival in rearing ponds, *Prog. Fish-Cult.*, 51, 229, 1989.

Murai, T. and Andrews, J. W., Effects of salinity on the eggs and fry of the golden shiner and goldfish, *Prog. Fish-Cult.*, 39, 121, 1977.

Norman-Boudreau, K. and Daggett, G. R., Improved design for fathead breeding chambers, *Prog. Fish-Cult.,* 51, 111, 1989.

Nagel, T., Technique for collecting newly hatched fathead minnow fry, *Prog. Fish-Cult.,* 38, 137, 1976.

Saylor, M.L., Effect of harvesting methods on production of fingerling fathead minnows, *Prog. Fish-Cult.,* 35, 110, 1973.

Smith, E.R., Minnow pond construction and water quality, in *Proc. Commercial Bait Fish Conf.,* Texas A&M University, College Station, 1968, 7.

Stacey, N.E., Cook, A. F., and Peter, R. E., Spontaneous and gonadotropin-induced ovulation in goldfish *Carassius auratus* L.: effects of external factors, *J. Fish Biol.,* 15, 349, 1979.

Yoshitaka, A., Goldfish Culture, in *Modern Methods of Aquaculture in Japan*, Elsevier, New York, 1983, 79.

INDEX

A

Acidity, see pH
Aerators, 27–28
    catfish culture, 44
    largemouth bass, 181–182
Aeromonas hydrophila
    pike and muskellunge, 207
    tilapia, 104–105
Aeromonas punctata, 132–133
Albinism, 57
Algae
    catfish, 54–55, 70
    pH, 288
    striped bass, 269–270
    tilapia, 90–91
Alkalinity, see pH
Alum, 47
Aluminum sulfate, 47
American Fisheries Society, 3
Amino acids, see Proteins
Ammonia (NH_4^+)
    aquaculture, 24–25
    bacteria, 14, 16
    catfish, 45
    largemouth bass, 188–189
    striped bass, 289
    tilapia, 87–88
    walleye, 233–234
Anaphylactic reactions, 105
Anchor worms, 133–134
Androgen, see Testosterone
Anesthetic
    largemouth bass, 180
    pike and muskellunge, 201
    striped bass, 257, 258, 292
Antibiotics
    carp, 132
    catfish, 66
    largemouth bass, 153
    striped bass, 293
Antifreeze genes, 99
Aquaculture
    background, 2–3
    philosophy, 3–5
Aquaria, 95
Arkansas, catfish, 35, 37
Ascorbic acid, 63, 103

B

Bacteria
    bass, 174–175, 293
    carp, 132–133
    catfish, 46, 66–67
    closed recirculating water systems, 14,
        16, 17
    pike and muskellunge, 204, 207
    tilapia, 104–105
    walleye, 244
Bacterial hemorrhagic septicemia, 132–133
Baitfish
    culture systems, 308–310, 312–314
    disease and predators, 318–319
    harvesting and handling, 316–318
    marketing, 319–320
    nutrition, 314–316
    reproduction and genetics, 310–311
Bass
    Guadalupe, 187, 189
    largemouth
        broodstock, 147–158
        egg and fry development, 162–176
        growout procedures, 176–179
        handling and transport, 179–182
        spawning, 158–162
        tilapia and, 99
        water quality, 188–189
    palmetto, 277–282
    smallmouth, 182–184, 189
    spotted, 187
    striped
        broodstock, 255–257, 290
        characteristics and distribution,
          252–254
        disease, 293–294
        early history, 254–255
        fingerling production, 265–273
        food fish production, 282–284
        future, 294–296
        hybridization and hybrid production,
          277–282
        nutrition, 273–276
        predators of, 276–277
        reproduction, 257–265
        stress, handling, and transport,
          291–292

O

Off-flavors in catfish, 70
Oil in catfish cultivation, 54
Oils, dietary, 63
Oreochromis, see Tilapia
Osmolarity, 180–181
Oxygen, dissolved (DO)
  aquaculture, 26–28
  baitfish, 310
  bass
    largemouth, 178, 181, 188, 189
    striped, 285–286, 295
  catfish, 43–45
  pike and muskellunge, 211
  recirculating water system, 16–17
  sunfishes, 189
  tilapia, 87
  walleye, 233, 235
  yellow perch, 224
Ozonation, 18

P

Paddlewheel aerators, 27–28
Paddlewheels in egg hatching, 52
Pantothenic acid, 65–66, 103
Parasites
  baitfish, 318–319
  bass, 153–154, 292
  carp, 133–134
Pathogenic organisms, see Disease
Pellets, pressure vs. extruded, 58
Perca flavescens, see Perch, yellow
Perch, yellow
  culture, 224–225
  distribution and characteristics, 216–217
  growth, 223–224
  larval development, 221–223
  reproduction, 217–221
Permits, 5, 8
pH
  ammonia, 25
  baitfish, 310
  bass, 189, 283, 287–288
  biofilters, 14
  catfish, 47
  monitoring, 23–24
  nitrite toxicity, 26
  sunfishes, 189
  tilapia, 88
Phosphorus, 64, 103
Photoperiod, 223, 290

Photosynthesis, 26–27
Pigment, 103–104
Pike, northern
  fingerling production, 204–211
  reproduction, 200–204
Pimephales promelas (fathead minnow),
    see Baitfish
Pituitary
  carp, 126
  catfish, 52–53
  pike and muskellunge, 201
  walleye, 236
Plankton, see Algae; Zooplankton
Plant production, 90
Plants, carp, 131
Poland, carp culture, 121, 122
Polyculture
  bass, 147
  buffalo, 135
  carp, 124–125
  catfish, 40–41
  tilapia, 90
Polyploidy, 57
Pomoxis sp., see Crappie
Ponds
  baitfish, 308–309, 312–314
  bass
    largemouth, 158–159, 163–164,
      167–171, 176–177
    smallmouth, 183–184
    striped, 266–272, 280–282, 295
  buffalo, 134–135
  carp, 122–123, 125–126
  catfish, 37–38
  construction and characteristics, 5–13
  pike and muskellunge, 204–207
  tilapia, 88–89, 94–96
  walleye, 237–239
  yellow perch, 225
Poultry manure, 91–92
Povidone-iodine, 207
Power plant lakes, 85
Predators
  of baitfish, 319
  of bass, 154, 276–277
  pike and muskellunge as, 206
  of tilapia, 95–96
Pressure-pelleted feeds, 58
Processing
  catfish, 69–70
  polyculture, 41
Progesterone, 201–202
Protease, 236